Xi Li

Radio Access Network Dimensioning for 3G UMTS

VIEWEG+TEUBNER RESEARCH

Advanced Studies Mobile Research Center Bremen

Herausgeber | Editors:

Prof. Dr. Otthein Herzog
Prof. Dr. Carmelita Görg
Prof. Dr.-Ing. Bernd Scholz-Reiter

Das Mobile Research Center Bremen (MRC) erforscht, entwickelt und erprobt in enger Zusammenarbeit mit der Wirtschaft mobile Informatik-, Informations- und Kommunikationstechnologien. Als Forschungs- und Transferinstitut des Landes Bremen vernetzt und koordiniert das MRC hochschulübergreifend eine Vielzahl von Arbeitsgruppen, die sich mit der Entwicklung und Anwendung mobiler Lösungen beschäftigen. Die Reihe „Advanced Studies" präsentiert ausgewählte hervorragende Arbeitsergebnisse aus der Forschungstätigkeit der Mitglieder des MRC.

In close collaboration with the industry, the Mobile Research Center Bremen (MRC) investigates, develops and tests mobile computing, information and communication technologies. This research association from the state of Bremen links together and coordinates a multiplicity of research teams from different universities and institutions, which are concerned with the development and application of mobile solutions. The series "Advanced Studies" presents a selection of outstanding results of MRC's research projects.

Xi Li

Radio Access Network Dimensioning for 3G UMTS

VIEWEG+TEUBNER RESEARCH

Bibliographic information published by the Deutsche Nationalbibliothek
The Deutsche Nationalbibliothek lists this publication in the Deutsche Nationalbibliografie; detailed bibliographic data are available in the Internet at http://dnb.d-nb.de.

Dissertation Universität Bremen, 2009

mrc
Mobile Research Center

Gedruckt mit freundlicher Unterstützung des
MRC Mobile Research Center der Universität Bremen

Printed with friendly support of
MRC Mobile Research Center, Universität Bremen

1st Edition 2011

Editorial Office: Ute Wrasmann | Anita Wilke

Vieweg+Teubner Verlag is a brand of Springer Fachmedien.
Springer Fachmedien is part of Springer Science+Business Media.
www.viewegteubner.de

Cover design: KünkelLopka Medienentwicklung, Heidelberg
Printed on acid-free paper
Printed in Germany

ISBN 978-3-8348-1398-5

Preface

This thesis was written during my research as a doctorate candidate at the Communication Networks Group (ComNets) at the University of Bremen within TZI (Center for Computer Science and Information Technology). Throughout the years at ComNets I have always enjoyed the very inspiring, encouraging, deeply rewarding, and exceptionally collegial research atmosphere.

To the greatest extent I would like to thank Prof. Dr. rer. nat. habil. Carmelita Görg for her supervising this thesis, for her guidance, and her encouragements. She has given me a lot of great advice and inspiring ideas, not only to my thesis but also to other scientific work. She has provided me an open and challenging research environment in the ComNets group, where it is a pleasure to conduct research and to take teaching responsibilities. Furthermore, she has given me a lot of opportunities to participate in national and international workshops and conferences, which helped me keep an open mind and view on the ongoing research fields that have inspired me in many ways for my research work.

Besides, I am also deeply grateful for the comments and suggestions by Prof. Dr.-Ing. Ralf Lehnert, who has been my Professor during my master study at Dresden University of Technology and who also is the second examiner of this thesis. During the years of my master study, he gave me a lot of great advice and guidance on performing the scientific work as well as on many other technical aspects. I very much appreciate him for taking his valuable time to have discussions with me on my thesis. The discussions with him were of great help to me.

In addition, I would like to thank Dr. Andreas Timm-Giel, who was always very open for discussions and has given me a great amount of valuable suggestions over the past years. I without a doubt greatly benefited from this, both in my research work and personally. His encouragements and advice have been a great support for me.

Further thanks go to my colleagues at the institute whom I have not only considered my co-workers but also my friends: especially Wenmin Chen, Chunlei An, Liang Zhao, Yi Chen, Koojana Kuladinithi, Asanga Udugama, Bernd-Ludwig Wenning, and Markus Becker, for sharing many technical and sometimes not-so-technical discussions. Additionally, I also very much thank Thushara Lanka Weerawardane, Yasir Naseer Zaki and Umar Toseef for having numerous technical discussions and exchange of research ideas.

Finally, I would like to thank my family, especially my mum for encouraging and supporting me all the time. Special thanks to Claas for his continual support, his patience and understanding throughout the past many years.

Xi Li

Abstract

This thesis studies the dimensioning for UMTS Radio Access Networks. In this thesis, dimensioning is investigated with specific focus on the transport network of the Iub interface, which connects the Node B with the RNC. This interface is considered as one of the most important economic factors for the UMTS network dimensioning. In order to cover large urban and rural areas, a large number of Node Bs are required and thus the transport resources for the Iub interface become considerably costly. The ultimate goal of this thesis is to investigate important aspects related to the UMTS radio access network dimensioning, and to propose novel analytical methods to provide suitable estimations on the required transport capacity for the UMTS radio access network in order to achieve maximum utilization of the transport resources.

In order to provide a comprehensive investigation on the dimensioning of the Iub interface, various traffic types, different evolutions of UMTS radio access networks as well as different transport solutions, QoS mechanisms and network topologies are studied in this thesis. In the framework of this thesis, UMTS Rel99 is considered as the basic UMTS network. Furthermore, evolved UMTS networks such as HSDPA and HSUPA, and the evolved transport from ATM to IP are investigated. For each evolution of the UMTS radio access network, its specific protocol stacks, important features, its specific traffic- and resource control functions and their impacts on the Iub dimensioning are studied.

For the dimensioning process, two basic types of traffic are distinguished, elastic and circuit switched traffic. They are associated with non real time data applications and delay-sensitive real time services, which are identified as the two main traffic classes served in the current UMTS networks. The fundamental property of elastic traffic is its rate adaptability, which is caused by the feedback mechanism of TCP. In this thesis, the theory of processor sharing is applied to the dimensioning of the Iub interface for elastic traffic for satisfying its desired end-to-end application QoS. To consider the specific UMTS functions and network structures, the basic processor sharing model has been significantly extended in this thesis to dimension the radio access networks under various scenarios of traffic, QoS framework, resource control functions, different transport technologies and network structures. In case of circuit-switched traffic, which needs to meet a guaranteed blocking probability, the classical Erlang models are applied. In addition, a number of queuing models are proposed in this thesis to dimension the Iub link for guaranteeing a required transport network QoS.

For validating the developed analytical models as well as for performance evaluation, several simulation models were developed in this thesis to model different UMTS radio access networks. By performing extensive simulations and analysis of the simulation results, important dimensioning rules are derived and the proposed analytical dimensioning models are demonstrated. Through validating with simulation results, it is demonstrated that the proposed analytical models in this thesis are able to capture relevant characteristics and provide accurate dimensioning results, and thus can be applied for UMTS radio access network dimensioning. At the end, a dimensioning tool is developped in this thesis, containing all developed analytical models. This tool can perform dimensioning for various traffic scenarios, transport solutions, QoS mechanisms and network topologies for different UMTS radio access networks. Overall, the investigations and the analytical dimensioning models presented in this thesis can help network service providers to optimize their network infrastructure to reduce costs while still being able to provide the desired quality of service.

Contents

Abstract VII

List of Figures XV

List of Tables XXI

List of Abbreviations XXIII

1 Introduction 1

1.1 Motivation 1
1.2 Scope of UMTS Network Dimensioning 2
1.3 Important Challenges for UMTS Network Dimensioning 4
1.4 Contributions of this Thesis 7
1.5 Thesis Overview 11

2 UMTS Evolution 15

2.1 Background of UMTS 15
2.2 UMTS Network Architecture 16
2.3 Evolution of UMTS 18
2.3.1 UMTS Release 99 19
2.3.2 High Speed Downlink Packet Access (HSDPA) 20
2.3.3 High Speed Uplink Packet Access (HSUPA) 21
2.3.4 Introduction of "All-IP" UMTS Networks 22
2.3.4.1 IP-Based UTRAN 23
2.3.4.2 Pseudo-Wire Emulation (PWE) Technology 25
2.3.5 LTE and SAE 26

3 UMTS Terrestrial Radio Access Network (UTRAN) 29

3.1 UTRAN Architecture and Interfaces 29
3.2 Generic Protocol Model for UTRAN Interfaces 31
3.3 UMTS Release 99 32
3.3.1 UTRAN Protocol Stack 33
3.3.2 Channels in the UTRAN 36
3.3.3 Iub Interface 38
3.3.3.1 Protocol Stack for the Iub User Plane 38
3.3.3.2 ATM 40
3.3.3.3 AAL2 Protocol 41
3.3.3.4 Transport Requirements on the Iub Interface 43
3.3.4 Resource Management 44
3.3.4.1 Radio Access Bearer (RAB) 44

3.3.4.2 Radio Resource Management (RRM) ... 46
3.3.4.3 Transport Resource Management (TRM) ... 49
3.4 IP-Based UTRAN ... 51
3.4.1 IP-Based UTRAN User Plane Protocol Architecture ... 51
3.4.2 QoS Support in the IP-Based UTRAN ... 52
3.4.2.1 Introduction of the DiffServ QoS Scheme ... 52
3.4.2.2 Mapping of UMTS QoS to DiffServ PHBs ... 56
3.4.2.3 DiffServ-Based QoS Architecture in IP-Based UTRAN ... 56

4 Framework of UMTS Network Dimensioning 59
4.1 Objectives of UMTS Network Dimensioning ... 59
4.1.1 Network Costs ... 59
4.1.2 Quality of Service ... 59
4.1.2.1 User-Relevant QoS ... 60
4.1.2.2 Network-Relevant QoS (Network Performance) ... 61
4.2 Important Issues for UMTS Network Dimensioning ... 61
4.2.1 Traffic Analysis ... 63
4.2.2 Network Topology ... 65
4.2.3 Traffic Control ... 66
4.2.4 Resource Control ... 67
4.2.5 Routing and Traffic Engineering ... 67
4.2.6 Summary ... 68
4.3 Framework for UMTS Network Dimensioning ... 68
4.4 Dimensioning Approaches ... 72
4.5 Conclusion ... 75

5 Introduction of Simulation Models 77
5.1 Requirements of Simulation Models ... 77
5.2 Simulation Model Framework ... 78
5.3 Simulation Environment ... 80
5.4 Simulation Model of UMTS Rel99 ... 80
5.5 Simulation Model of IP-Based UTRAN ... 88
5.5.1 Exact IP-Based UTRAN Simulation Model ... 88
5.5.2 Simplified IP-Based UTRAN Simulation Model ... 89
5.5.3 Summary ... 90
5.6 Simulation Model of HSPA ... 90
5.7 Simulation Model of Combined UMTS Rel99 and HSPA ... 91
5.8 Traffic Models ... 92
5.8.1 Voice Traffic Model ... 92
5.8.2 Web Browsing Traffic Model ... 94
5.8.3 File Transfer Traffic Model ... 96

6 Dimensioning for Single ATM-Based Iub 99
6.1 Overview and Objectives ... 99
6.2 Dimensioning for Circuit-Switched Traffic ... 100
6.2.1 Erlang-B Formula ... 100
6.2.2 Multi-Dimensional Erlang-B Formula ... 105

6.2.3 Iub Dimensioning with the Erlang-B Formula 109
6.2.4 Validation of Erlang Model 111
6.2.5 Dimensioning Results 114
6.2.6 Summary 116
6.3 Dimensioning for Elastic Traffic 116
6.3.1 Transmission Control Protocol (TCP) 117
6.3.2 Related Work 120
6.3.3 Processor Sharing Model for Elastic Traffic 121
6.3.4 Application of M/G/R-PS Model for Dimensioning Iub Interface 122
6.3.5 Basic M/G/R-PS Model 124
6.3.6 Extended M/G/R-PS Model 127
6.3.7 Suggested Extensions on M/G/R-PS Model for Iub Dimensioning 129
6.3.7.1 Extension for case with Single RAB and not Applying CAC 130
6.3.7.2 Extension for case with Single RAB and Applying CAC 132
6.3.7.3 Extension for case with Multiple RABs and not Applying CAC 133
6.3.7.4 Extension for case with BRA and not Applying CAC 136
6.3.7.5 Extension for case of Mixing with Circuit-Switched Traffic 137
6.3.8 Validation of Extensions of Processor Sharing Models 139
6.3.8.1 Simulation Environment 139
6.3.8.2 Single RAB without Applying CAC 140
6.3.8.3 Single RAB with CAC 143
6.3.8.4 Multiple RABs without Applying CAC 145
6.3.8.5 BRA without Applying CAC 148
6.3.8.6 Elastic Traffic Mixed with Circuit-Switched Traffic 149
6.3.9 Dimensioning Procedures 150
6.3.10 Dimensioning Results 153
6.3.11 Summary 157
6.4 Dimensioning for Transport Network Performance 157
6.4.1 System Models for the Iub Interface 158
6.4.1.1 Delay Analysis on the Iub 158
6.4.1.2 System Models 160
6.4.2 UMTS Iub Traffic Characteristics 162
6.4.2.1 Squared Coefficient of Variation 162
6.4.2.2 Self-Similarity of Elastic IP Traffic 163
6.4.3 Analytical Dimensioning Models with MMPP 164
6.4.3.1 Markov Modulated Poisson Process (MMPP) 165
6.4.3.2 MMPP(2)/D/1 166
6.4.3.3 MMPP(2)/D/1 – Non-Preemptive Priority 169
6.4.3.4 Iub Dimensioning with MMPP 171
6.4.3.5 Validations 173
6.4.4 Analytical Dimensioning Models with BMAP 177
6.4.4.1 Batch Markovian Arrival Process (BMAP) 177
6.4.4.2 BMAP/D/1 182
6.5 Dimensioning for HSPA 183

7 Dimensioning for Single IP-Based Iub 187

7.1 Overview and Objectives 187
7.2 Dimensioning for Circuit-Switched Traffic 188

7.3 Dimensioning for Elastic Traffic ... 188
7.3.1 Analytical Models for the IP-Based Iub Dimensioning ... 189
7.3.2 Validation of Analytical Dimensioning Models ... 193
7.3.3 Dimensioning Procedure ... 196
7.3.4 Dimensioning Results ... 197
7.4 Comparing the ATM and IP-Based Iub ... 198
7.5 Summary ... 202

8 Dimensioning for Multi-Iub RAN Scenario 203
8.1 Overview and Objectives ... 203
8.1.1 Dimensioning of Individual Links ... 204
8.1.2 Overbooking of the Iub Backbone Link ... 204
8.2 Dimensioning for Circuit-Switched Traffic ... 206
8.2.1 Network Dimensioning Approach ... 206
8.2.2 Applicability of the Dimensioning Approaches ... 209
8.2.3 Summary ... 211
8.3 Dimensioning for Elastic Traffic in ATM-Based UTRAN ... 211
8.3.1 Estimation of End-to-End Application Performance ... 212
8.3.2 Dimensioning Approaches ... 217
8.3.3 Dimensioning Results ... 220
8.3.4 Summary ... 223
8.4 Dimensioning for Elastic Traffic in IP-Based UTRAN ... 223
8.4.1 Dimensioning Approaches ... 224
8.4.2 Dimensioning Results ... 227
8.4.3 Summary ... 232
8.5 Overbooking of the Iub Backbone Link ... 233
8.5.1 Overbooking of ATM-Based Iub Backbone Link ... 233
8.5.2 Overbooking of IP-Based Iub Backbone Link ... 237
8.5.3 Comparing ATM and IP-Based Multi-Iub RAN ... 242
8.5.4 Summary ... 244

9 Conclusion and Outlook 247

Appendix 251
A.1 ATM Cell Format ... 251
A.2 AAL2 Packet Formats ... 251
A.3 ATM Service Categories and QoS Parameters ... 253
A.4 Ethernet ... 255
A.5 Random Early Detection and Weighted RED ... 255
A.6 AMR ... 258
A.7 Self-Similarity ... 260
A.8 ATM Workstation and Server ... 261
A.9 Introduction for HSPA (HSUPA/HSDPA) ... 262
A.10 Simulation Model of HSPA ... 267
A.11 Multiplexing Schemes for the IP-Based UTRAN ... 269
A.12 Exact IP-Based UTRAN Simulation Model ... 271
A.13 Impact of TCP Slow Start ... 273
A.14 Processor Sharing among Multiple RABs ... 273

A.15 Dimensioning for HSPA Traffic 274
A.16 Use of Traffic Separation for Transport of HSPA and Rel99 282
A.17 Dimensioning for Mixed Traffic Cases (ATM-Based Iub) 295
A.18 Validating Dimensioning Approaches for Multi-Iub IP RAN 296
A.19 Investigating Overbooking in the ATM-Based UTRAN 299
A.20 Statistical Evaluation: Confidence Interval 304
A.21 Statistical Evaluation: Limited Relative Error 307
A.22 Comparing ATM and IP-Based UTRAN 308

Bibliography **313**

List of Figures

1.1 Generic view of a UMTS network ... 3
1.2 Investigated network structures ... 10
2.1 UMTS Network Structure ... 17
2.2 IP-based UNTRAN Protocol Stack (CIP, LIPE, AAL2, PPPmux) ... 24
2.3 Carrier Ethernet-based UTRAN using PWE3 ... 26
2.4 LTE radio access network ... 28
3.1 UMTS terrestrial radio access network: nodes and interfaces ... 29
3.2 General Protocol Model for UTRAN interfaces [3GP02c] ... 31
3.3 Control plane protocol stack in UTRAN ... 33
3.4 User plane protocol stack in UTRAN (Packet-Switched Domain) ... 35
3.5 User plane protocol stack in UTRAN (Circuit-Switched Domain) ... 35
3.6 Channel types in the UTRAN (UMTS Rel99) ... 36
3.7 UMTS Rel99: Channel mapping in the uplink and downlink ... 37
3.8 Iub interface - user plane protocol stack ... 39
3.9 AAL2 multiplexing ... 42
3.10 Evaluating the Buffer Utilization ... 48
3.11 Protocol stack of the IP-based UTRAN ... 51
3.12 IP DiffServ Router Building Blocks ... 54
3.13 DiffServ-based QoS structure in the IP-based UTRAN ... 56
4.1 An overview of dimensioning issues ... 63
4.2 Basic network topology configurations for UTRAN ... 65
4.3 UTRAN network topology example ... 66
4.4 UMTS network dimensioning framework ... 69
4.5 An overview of the dimensioning procedure ... 71
4.6 Dimensioning process to derive the minimum required link capacities ... 72
4.7 Overview of analytical dimensioning models ... 73
5.1 Simulation Model Framework ... 78
5.2 UMTS Rel99 Simulation Model Structure ... 81
5.3 Modeling of UMTS Rel99 (ATM-based UTRAN) in OPNET ... 82
5.4 Implemented AAL2 QoS Structure ... 85
5.5 Measuring the CAC guaranteed bit rate ... 87
5.6 Modeling of IP-based UTRAN in OPNET ... 88
5.7 Modeling of combined UMTS Rel99 and HSPA ... 91
5.8 Voice traffic model ... 92
5.9 General web traffic model ... 94
5.10 File transfer (ftp) traffic model ... 96
6.1 State transition diagram for a system with a limited number of channels (n), Poisson arrival process (λ), and exponential service times (μ). ... 101
6.2 Blocking probability p_b as a function of the offered traffic A in Erlang for various values of the number of channels n with equation (6.6) ... 103

6.3 The average utilization per channel a equation (6.8) as a function of the number of channels for given values of the blocking ... 105
6.4 Two-dimensional state transition diagram for a loss system with n channels offered to two traffic streams [Ive04] ... 106
6.5 Two-dimensional state transition diagram for a loss system with class limitations: n_1 for the traffic stream 1 and n_2 for the traffic stream 2[Ive04] ... 108
6.6 Two-dimensional state transition diagram for a loss system with class limitations: (n_1 =10, λ_1=5, μ_1=1); (n_2 =3, λ_2=1, μ_2=1) ... 108
6.7 State Probability of two-dimensional Erlang-B formula with class limitations: (n1=10, λ1=5, μ1=1); (n2 =3, λ2=1, μ2=1) ... 109
6.8 Validation of Erlang-B formula for the voice scenario: CAC reject ratio and link utilization over offered voice traffic ... 112
6.9 Validation of Erlang-B formula for the voice scenario: state probabilities ... 113
6.10 Insensitivity of Erlang-B model (different service time distributions) ... 113
6.11 Validation of Erlang-B formula: dimensioning for 1% CAC reject ratio ... 115
6.12 Validation of Erlang-B formula: dimensioning for 5% CAC reject ratio ... 115
6.13 Validation of Erlang-B formula for voice scenario: CAC reject ratio and link utilization over offered voice traffic ... 115
6.14 TCP slow start and congestion avoidance [Rie04] ... 118
6.15 Illustration of rate sharing ... 122
6.16 Flow Level Model for Iub Interface ... 123
6.17 Illustration of M/G/R-PS Model ... 125
6.18 Delay factor and normalized throughput vs. load - M/G/R-PS Model ... 126
6.19 Data transfer in TCP with slow start ... 128
6.20 TCP Procedure ... 132
6.21 Flow level model for multiple RAB scenarios ... 134
6.22 Processor Sharing ... 134
6.23 Elastic traffic shares the Iub link with circuit-switched traffic ... 138
6.24 Average file transfer delay over load: ftp scenario, RAB 128 kbps, no CAC, constant file size of 12 kbyte, Iub link = 1 E1 line ... 140
6.25 Average file transfer delay over load: ftp scenario, RAB 128 kbps, no CAC, constant file size (5; 100 kbyte), Iub link = 1 E1 line ... 141
6.26 Different file size distributions: ftp scenario, RAB 128 kbps, no CAC, constant file size of12 kbyte, Iub link = 1 E1 line ... 142
6.27 Improvement for Pareto file size distribution with RTT_{est}: ftp scenario, RAB 128 kbps, no CAC, constant file size of 12 kbyte, Iub link = 1 E1 line ... 143
6.28 File transfer delay (left), blocking ratio (right): ftp scenario, RAB 128 kbps, CAC reserved rate = 96 kbps, constant file size of 12 kbyte, Iub link =1 E1 line . 144
6.29 File transfer delay (left), blocking ratio (right): ftp scenario, RAB 128 kbps, CAC reserved rate = 96 kbps, constant file size of 50 kbyte, Iub link =1 E1 line . 145
6.30 Average application delay over the load for multiple RABs: ftp scenario, no CAC, constant file size (12 kbyte) ... 146
6.31 The impact of different distributions of various RAB types: ftp scenario, no CAC, constant file size (50 kbyte) ... 147
6.32 Average application delay over the load for multiple RABs with different RAB distributions: web scenario, no CAC, constant page size (12 kbyte) ... 147
6.33 Average application delay over the load for BRA: web scenario, no CAC, page size =12 kbyte, Iub link = 1 E1 line ... 148

6.34 Average application delay over the load for web mixed with voice: RAB 128 kbps (for web traffic), no CAC, page size =50 kbyte, Iub link = 1 E1 line 150
6.35 Iub link dimensioning with processor sharing (PS) models 150
6.36 Bandwidth partitioning for elastic traffic & circuit-switched traffic 153
6.37 Iub Dimensioning: web scenario, no CAC, RAB 128 kbps, constant page size (50 kbyte), f=1.25 154
6.38 Normalized capacity and link utilization 155
6.39 Capacity vs. traffic load (with different delay factors): web scenario, constant page size of 12kbyte, RAB 128 kbps, no CAC 156
6.40 Capacity vs. traffic load (with different peak rates): web scenario, constant page size of 12kbyte, target delay factor f = 1.5 for all RABs, no CAC 156
6.41 Capacity vs. traffic load (different traffic mixes of web and voice traffic) for web traffic: RAB 128kbps, no CAC, page size =50kbyte, delay factor f=1.25 157
6.42 Delay in the UTRAN Iub 159
6.43 System Models (AAL2 packet level) 161
6.44 Squared CoV of packet interarrival time for RT voice traffic 162
6.45 Squared CoV of packet interarrival time for NRT traffic 163
6.46 CDF of FP PDU inter-arrival time 164
6.47 Self-similarity of the aggregated UTRAN traffic (Hurst Parameter) 164
6.48 Two State MMPP 165
6.49 Basic idea to study the delay distribution of a priority queue [HZ05] 170
6.50 Validation of MMPP/D/1 model (voice scenario) 174
6.51 Analytical models vs. simulation for Iub bandwidth (voice scenario) 174
6.52 Relative error of the analytical model (voice scenario) 175
6.53 Analytical Models vs. Simulation (web traffic with BRA) 176
6.54 Relative Error of the analytical model (web traffic with BRA) 176
6.55 Analytical Model vs. Simulation (90% web with BRA +10% Voice) 176
6.56 Packet arrival process (BMAP model) 177
6.57 CDF of FP PDU inter-arrival time – single RAB scenario 180
6.58 Measured user traffic (left) and customized BMAP (right) 180
6.59 R/S statistic of the measured traffic (left) and customized BMAP (right) 181
6.60 CDF of FP PDU Inter-arrival time – multiple RABs scenario 181
6.61 BMAP/D/1 Queue in OPNET 182
6.62 CDF of FP PDU Inter-arrival time (simulation vs. BMAP) 183
7.1 Dimensioning for IP DiffServ-based QoS Structure 190
7.2 Validation of M/G/R/N-PS Model – transfer delay 195
7.3 Validation of M/G/R/N-PS Model – IP bandwidth 196
7.4 Dimensioning of a single IP Iub link 197
7.5 Capacity vs. traffic load 198
7.6 Bandwidth Efficiency: ATM vs. IP 199
7.7 ATM vs. IP (without multiplexing) link bandwidth over traffic load 200
7.8 ATM vs. IP: web scenario, RAB 128 kbps, no CAC, constant page size of 50 kbyte, target delay factor f=1.25 200
7.9 ATM vs. IP (capacity over traffic load): web scenario, RAB 64 kbps (left) and RAB 384 kbps (right), no CAC, page size =50 kbyte, target delay factor f=1.25 201
8.1 Dimensioning for circuit-switched traffic in a multi-Iub RAN: with MD-Erlang model for the last mile and MMPP/D/1 model for the backbone link 208
8.2 Dimensioning the backbone link for 1% packet loss ratio 210

8.3 Overbooking the backbone link ... 210
8.4 System model for elastic traffic in the multi-Iub RAN ... 212
8.5 Delay estimations for case (1) with congested last mile link and ideal backbone link ... 215
8.6 Delay estimations for case (2) with ideal last mile link and congested backbone link ... 216
8.7 Delay estimations for case (3) with congested backbone and last mile links ... 217
8.8 Dimensioning for elastic traffic in a multi-Iub ATM RAN: with processor sharing model for both last mile and the backbone link ... 220
8.9 Validating approach (3): dimensioned last mile link bandwidths over different target delay factors ... 221
8.10 Validating approach (3): dimensioned backbone link bandwidths over different target delay factors ... 221
8.11 Validating Approach (3):E2E delays over different target delay factors ... 222
8.12 A star topology multi-Iub IP RAN using DiffServ QoS for IP transport ... 225
8.13 Dimensioning for elastic traffic in a multi-Iub IP RAN with **K** service classes using DiffServ QoS Structure ... 227
8.14 Dimensioned last mile link bandwidths over different target delay factors in the star-structured IP-based UTRAN with DiffServ QoS scheme ... 230
8.15 Dimensioned backbone link bandwidths over different target delay factors in the star-structured IP-based UTRAN with DiffServ QoS scheme ... 230
8.16 Overbooking factor over different target delay factors in the star-structured IP-based UTRAN with DiffServ QoS scheme ... 232
8.17 Validating the calculated Overbooking factor for a voice traffic scenario ... 235
8.18 Validating the calculated Overbooking factor for a web traffic scenario ... 235
8.19 Overbooking factors over different number of Node Bs (RAB=128, f=1.2) ... 236
8.20 Overbooking factors over different QoS requirements (RAB=128) ... 236
8.21 Validating overbooking factor vs. the number of Node Bs (web traffic) ... 238
8.22 Overbooking factor vs. last mile utilizations (4 Node Bs) - web traffic ... 238
8.23 Overbooking factor vs. the number of Node Bs (mixed traffic scenario) ... 239
8.24 Overbooking factor over number of Node Bs (with various last mile loads) ... 241
8.25 Overbooking factor over last mile loads (with different QoS requirements) ... 241
8.26 Overbooking factor over number of Node Bs (various last mile capacities) ... 242
8.27 Overbooking factor over number of Node Bs (with different voice portions) ... 242
8.28 ATM vs. IP- dimensioning of last mile link and backbone link ... 243
8.29 ATM vs. IP- overbooking factor over traffic load ... 244
A.1 Format of the ATM Cell ... 251
A.2 Format of the AAL2 CPS Packet ... 251
A.3 Format o the AAL2 CPS-PDU ... 253
A.4 RED Packet Drop Function ... 257
A.5 AMR Speech Codec Model for 12.2 kbps speech traffic ... 259
A.6 HSDPA protocol architecture ... 263
A.7 HSUPA protocol architecture ... 264
A.8 HSDPA Simulation Model Architecture [WL+06] ... 267
A.9 HSPA simulator overview [LZK+08] ... 268
A.10 CIP packet format ... 270
A.11 LIPE packet format ... 271
A.12 Bandwidth utilization efficiency over file sizes ... 273

A.13 TCP connection throughput of each RAB type – multiple RABs scenario 274
A.14 Average cell throughput 278
A.15 Average noise rise 278
A.16 Overall average application throughput 279
A.17 COV of cell throughput 280
A.18 Dimensioning factor 280
A.19 Iub link utilization 281
A.20 Dimensioning factor vs. QoS target 281
A.21 UMTS Network supporting Rel99 and HSPA traffic 283
A.22 Concept of Traffic Separation 285
A.23 Performance comparisons: with TS and without TS 288
A.24 Dimensioning for different UTRAN load 291
A.25 Total ATM cell discards on downlink and uplink 293
A.26 HSDPA and HSUPA TCP retransmissions 294
A.27 HSDPA application throughput 294
A.28 Downlink and uplink ATM link throughput 295
A.29 Validating the dimensioning approach for the mixed traffic scenario 296
A.30 Validating Approach (1):E2E delays over different target delay factors 297
A.31 Validating Approach (2): E2E delays over different offered last mile loads 298
A.32 Impact of overbooking 300
A.33 Optimum OB_factor (left), link utilization (right) vs. last mile utilizations 301
A.34 Optimum OB_factor vs. last mile utilization (various last mile links) 302
A.35 Optimum OB_factor vs. last mile link capacity (packet-switched dominant) 303
A.36 Impact of number of Node Bs 304
A.37 Average application delay with 95% CI: web scenario, RAB 128 kbps, no CAC, constant file size of 50 kbyte, Iub link = 1 E1 line (ATM) 305
A.38 Packet loss ratio with 95% CI: web scenario, RAB 128 kbps, no CAC, constant file size of 50 kbyte, Iub link = 1 E1 line (ATM) 306
A.39 CAC reject ratio with 95% CI: voice scenario, Iub link =1 Mbps (ATM) 307
A.40 LRE distribution function F(x) and relative error d(x) for queuing delays of circuit-switched voice: voice scenario, Iub link =1.7 Mbps (ATM) 308
A.41 Calculated bandwidth efficiency - ATM vs. IP-based UTRAN 309
A.42 Simulated bandwidth efficiency - ATM vs. IP-based UTRAN (voice traffic) 310
A.43 Simulated FP PDU delay - ATM vs. IP-based UTRAN (voice traffic) 310
A.44 Simulated bandwidth efficiency - ATM vs. IP-based UTRAN (web traffic) 311
A.45 Simulated FP PDU delay - ATM vs. IP-based UTRAN (web traffic) 312

List of Tables

3.1 Traffic classes [3GP06c] ... 45
3.2 RABs defined in 3GPP TS 34.108 [3GP04e] ... 46
5.1 Measured average ATM link throughput for different RABs ... 87
5.2 Traffic model for voice [TWP+02] ... 93
5.3 Traffic models for web application ... 95
6.1 State probabilities of two-dimensional Erlang-B formula with class limitations: (n_1=10, λ_1=5, μ_1=1); (n_2 =3, λ_2=1, μ_2=1) ... 109
6.2 Dimensioning with Erlang-B formula (Traffic Parameters) ... 110
6.3 For a fixed CAC reject ratio (1% and 5%), the required number of channels n given the offered traffic A in Erlang ... 114
6.4 Defined variables for general M/G/R-PS Model ... 135
6.5 UTRAN Delay Components ... 159
7.1 Traffic classes and their mapped PHBs ... 193
7.2 WRED and WFQ Parameters ... 194
8.1 Dimensioning results for different target delay factors with approach (3) ... 222
8.2 Service classes and their mapped DiffServ PHBs ... 228
8.3 WRED and WFQ Parameters ... 228
8.4 QoS requirements per service class for IP RAN dimensioning ... 229
8.5 Target end-to-end delay QoS vs. simulated delay ... 231
8.6 Settings for different service classes (overbooking) ... 237
A.1 Ethernet Physical Link Rate and Standard ... 255
A.2 Source codec bit-rates for the AMR codec [3GP99a] ... 258
A.3 Bit allocation of the AMR coding algorithm for 20 ms frame ... 259
A.4 Traffic Models for HSUPA ... 277
A.5 System Parameter Configurations for HSUPA ... 277
A.6 ftp Traffic Model for HSPA ... 292
A.7 Simulation Scenarios ... 292
A.8 Dimensioning results for different target delay factors with approach (1) ... 297
A.9 Dimensioning results for different offered traffic loads with approach (2) ... 299
A.10 95% CI for application delays of a 50kbyte web page (RAB 128kbps) ... 306

List of Abbreviations

2G	The Second Generation Mobile Communication Systems
3G	The Third Generation Mobile Communication Systems
3GPP	3rd Generation Partnership Project
AAL2	ATM Adaptation Layer 2
AAL5	ATM Adaptation Layer 5
ABR	Available Bit Rate
ACF	Autocorrelation Function
ADSL	Asymmetric Digital Subscriber Line
AF	Assured Forwarding
ALCAP	Access Link Control Application Part
AM	Acknowledged Mode
AMC	Adaptive Modulation and Coding
AMR	Adaptive Multi-Rate
ARIB	Association of Radio Industries and Businesses
ARP	Address Resolution Protocol
ATIS	Alliance for Telecommunications Industry Solutions
ATM	Asynchronous Transfer Mode
AUC	Authentication Center
BCH	Broadcast Channel
BE	Best Effort
BER	Bit Error Rate
BLER	Block Error Rate
BMAP	Batch Markovian Arrival Process
BPSK	Binary Phase Shift Keying
BRA	Bit Rate Adaptation
BS	Base Station
BTS	Base Transceiver Station
CAC	Connection Admission Control
CBR	Constant Bit Rate
CC	Congestion Control
CCSA	China Communications Standards Association
CDVT	Cell Delay Variation Tolerance
CE	Customer Edge
CN	Core Network

CID	Channel Identifier
CLP	Cell Loss Priority
CNCL	Communication Network Class Library
CI	Confidence Interval
CIP	Composite IP
COV	Coefficient of Variation
CPCH	Common Packet Channel
CPS	Common Part Sublayer
CTS	Channel Type Switching
DCH	Dedicated Channel
DCCH	Dedicated Control Channels
DiffServ	Differentiated Services
DL	Downlink
DSCH	Downlink Shared Channel
DSCP	DiffServ Code Point
DTCH	Dedicated Traffic Channels
EF	Expedited Forwarding
EIR	Equipment Identity Register
EM	Expectation Maximization
EPC	Enhanced Packet Core
E-AGCH	E-DCH Absolute Grant channel
E-RGCH	E-DCH Relative Grant channel
ETSI	European Telecommunication Standards Institute
FACH	Forward Access Channel
FC	Flow Control
FDD	Frequency Division Duplex
FDMA	Frequency Division Multiple Access
FP	Frame Protocol
FSM	Finite State Machine
GFC	Generic Flow Control
GGSN	Gateway GPRS Support Node
GPRS	General Packet Radio Service
GSM	Global System for Mobile Communications
GTP	GPRS Tunneling Protocol
HARQ	Hybrid Automatic Repeat Request
HDLC	High Level Link Control
HEC	Header Error Control
HLR	Home Location Register
HSDPA	High Speed Downlink Packet Access
HSPA	High Speed Packet Access

HSUPA	High Speed Uplink Packet Access
HS-DSCH	High-Speed Downlink Shared Channel
ID	Identifier
IETF	Internet Engineering Task Force
IntServ	Integrated Services
IMT-2000	International Mobile Telecommunications-2000
IPP	Interrupted Poisson Process
IP	Internet Protocol
ISDN	Integrated Services Digital Network
ISO	International Standards Organisation
ITU	International Telecommunication Union
ITU-T	Telecommunication Standardisation Sector of ITU
LIPE	Lightweight IP Encapsulation
LRD	Long Term Dependence
LRE	Limited Relative Error
LTE	Long Term Evolution
MAC	Medium Access Control
MBS	Maximum Burst Size
MCR	Minimum Cell Rate
MDCR	Minimum Desired Cell Rate
ME	Mobile Equipment
MM	Mobility Management
MMPP	Markov-Modulated Poisson Process
MIMO	Multiple Input Multiple Output
MPLS	Multi-Protocol Label Switching
MSC	Mobile Services Switching Center
MSS	Maximum Segment Size
MWIF	Mobile Wireless Internet Forum
NBAP	Node B Application Part
NGMN	Next Generation Mobile Network
NRT	Non Real Time
nrt-VBR	Non-Real-Time Variable Bit Rate
NS-RLS	Non-Serving Radio Link Set
NSS	Network and Switching Subsystem
OFDM	Orthogonal Frequency Division Multiplex
PASTA	Poisson Arrivals See Time Averages
PAG	Primary Absolute Grant
PCH	Paging Channel
PCR	Peak Cell Rate
PDCP	Packet Data Convergence Protocol

PDU	Protocol Data Unit
PE	Provider Edge
PHB	Per Hop Behavior
PPP	Point-to-Point Protocol
PS	Processor Sharing
PSTN	Public Switched Telephone Network
PVC	Permanent Virtual Circuit
PVP	Permanent Virtual Path
PWE	Pseudo Wire Emulation
QAM	Quadrature Amplitude Modulation
QoS	Quality of Service
QPSK	Quadrature Phase Shift Keying
RAB	Radio Access Bearer
RACH	Random Access Channel
RAN	Radio Access Network
RANAP	Radio Access Network Application Part
RED	Random Early Detection
RLC	Radio Link Control
RNC	Radio Network Controller
RNS	Radio Network Subsystem
RNSAP	Radio Network Subsystem Application Part
RRC	Radio Resource Control
RRM	Radio Resource Management
RSVP	Resource Reservation Protocol
RT	Real Time
rt-VBR	Real-Time Variable Bit Rate
RTT	Round Trip Time
RTP	Real-time Transport Protocol
SAE	System Architecture Evolution
SAG	Secondary Absolute Grant
SAP	Service Access Point
SCCP	Signaling Connection Control Part
SC-FDMA	Single-Carrier FDMA
SCR	Sustained Cell Rate
SDU	Service Data Unit
SGSN	Serving GPRS Support Node
SHO	Soft Handover
SIR	Signal-to-Interference Ratio
SLA	Service Level Agreement
SPP	Switched Poisson Process

SRB	Signaling Radio Bearer
SVC	Switched Virtual Circuit
TB	Transport Block
TCP	Transmission Control Protocol
TDD	Time Division Duplex
TDMA	Time Division Multiple Access
TM	Transparent Mode
TOS	Type of Service
TRM	Transport Resource Management
TS	Traffic Separation
TTA	Telecommunications Technology Association
TTC	Telecommunications Technology Committee
TTI	Transmission Time Interval
UBR	Unspecified Bit Rate
UDP	User Datagram Protocol
UE	User Equipment
UL	Uplink
UM	Unacknowledged Mode
UMTS	Universal Mobile Telecommunications System
UNI	User Network Interface
USIM	UMTS Subscriber Identity Module
UTRAN	UMTS Terrestrial Radio Access Network
UUI	User-to-User Indication
VC	Virtual Channels
VCI	Virtual Channel Identifier
VLR	Visitor Location Register
VoIP	Voice over Internet Protocol
VP	Virtual Path
VPI	Virtual Path Identifier
WCDMA	Wideband Code Division Multiple Access
WFQ	Weighted Fair Queuing
WiMAX	Worldwide Interoperability for Microwave Access
WRED	Weighted Random Early Detection

1 Introduction

1.1 Motivation

The *Universal Mobile Telecommunication System* (UMTS) is one of the most important and popular standards of the third generation (3G) mobile communication system specified by the *3rd Generation Partnership Project* (3GPP). It provides global high-speed mobile access for a great variety of services including voice, Internet access, video conferencing, and general multimedia applications. Each service has its own particular traffic characteristics and specific *Quality of Service* (QoS) requirements to be met by the network, for example end-to-end delay, jitter, packet loss and blocking ratio. In the public multi-service UMTS networks, the key objective is to support the required QoS for various services to deliver an improved end-user experience for mobile data applications while minimizing the network costs to attain efficient resource utilization.

Over the past few years UMTS has experienced an unprecedented development. Since the first UMTS networks were launched in 2002, there has been an intensive growth of the subscriber population and the number of operative networks all over the world. By the end of 2006, there were around 140 UMTS networks operational in more than 50 countries, and the worldwide subscriptions to UMTS networks reached 100 million, with more than 3 million new customers each month [UMTS06]. At present the number of UMTS subscribers and operative networks are still growing rapidly. On the technology evolution path, UMTS has unequivocally delivered on its promise to provide a compelling mobile multimedia experience with a wide range of new services. The evolved wireless technologies of UMTS have significantly increased the network capacity and throughput to improve the end-user experience. Along with the technology evolution, the UMTS terminal types and their capabilities are also changing quickly, and the portfolio of services that are offered by the network operators are expanding dramatically. All these aspects result in a drastic increase of the traffic volume, especially concerning the data traffic, and remarkable changes in the traffic pattern and characteristics in the current UMTS networks can be seen.

Despite the great development of UMTS networks, network operators and vendors have struggled hard in the last years to gain profit. At the beginning they had to spend huge investments in both licenses and infrastructure for UMTS. But unfortunately the downturn of the economy throughout the past years has considerably influenced the revenue of 3G mobile networks. As a consequence, it has been quite difficult for network operators to keep up revenue and until today they are still facing difficult times. In order for businesses to succeed in the current tense situation, network operators have to maximize their cost-efficiency in order to make maximum profit.

An important lever for improving the cost-efficiency of UMTS networks is efficient design and use of the existing network infrastructure, in particular the radio access network, referred to as *UMTS Terrestrial Radio Access Network* (UTRAN). The radio access network is considered as one of the most important economic factors for the network dimensioning. As it has to cover large suburban and rural areas, the transport resource within the UTRAN is considerably costly. With the rapid expansion of the radio access network and the fast growing number of mobile users and traffic volume, the expenses on the transport link capacities and the cost of infrastructure will increase significantly. To achieve a cost-efficient design of the UMTS network, the radio access networks will have to be dimensioned appropriately for the types of services which are to be offered. Excessive over-dimensioning will waste the expensive bandwidth resources unnecessarily, whereas under-dimensioning can lead to a less satisfactory quality of service perceived by the mobile users. A properly dimensioned UMTS radio access network can result in significant reductions in capital expenditure required for a desired target quality.

1.2 Scope of UMTS Network Dimensioning

In this thesis the scope of UMTS network dimensioning is focused on the radio access network, as an efficient design of the radio access network is the key process to guarantee the optimal use of the network resources and achieve the best cost-efficiency of a UMTS network. The specific task of dimensioning is to determine appropriate bandwidths for the transport links within the radio access network, with the objective of maximizing the utilization of the allocated transport resources while ensuring the QoS requirements of the UMTS network and the end-users.

Figure 1.1 illustrates a generic view of a UMTS network, which presents a hierarchical network structure. The core network is the highest level of the hierarchy, which is the backbone of the UMTS network providing the gateways and connections to the external networks. The mobile users (UEs) are the lowest level of this hierarchy. They are connected through the base station (Node B) to the radio access network, i.e. UTRAN. In the radio access network, groups of Node Bs are connected to one *Radio Network Controller* (RNC) via switches or routers. From there, the traffic from different Node Bs are aggregated and routed to the RNC.

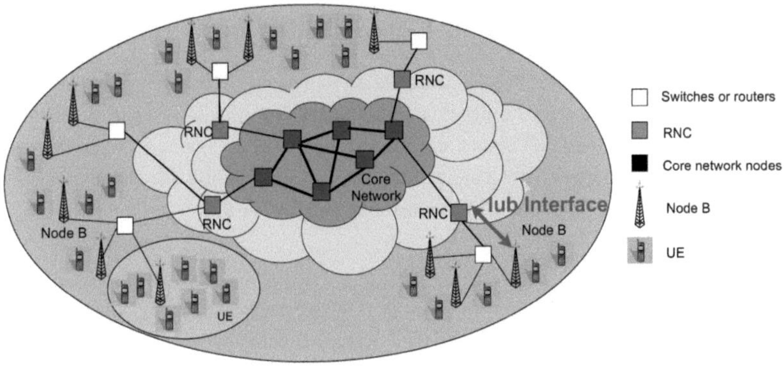

Figure 1.1: *Generic view of a UMTS network*

Typically, a large UMTS radio access network consists of a number of RNCs each controlling hundreds of Node Bs. Such a large radio access network often involves considerable network costs, as the required transport resources and the resulting capital expenditures of installing transport links and assigning capacities will be considerably high, especially on the Iub interface which connects a Node B to its serving RNC. Each Iub interface carries the traffic of all UEs which are served by the connecting Node B. In order to save the overall infrastructure cost, the Iub interface is usually assigned with a limited bandwidth. Thus, the Iub interface is easy to be under-provisioned and hence may cause performance degradation to the user traffic. In this context, the Iub interface is regarded as one of the major capacity bottlenecks in the UMTS radio access network. According to the one-way delay estimation from the UE to the core network given in 3GPP TR 25.853 [3GP01c], it can be seen that the Iub delay contributes a considerable portion. On the other hand, the air interface and the related radio control functions like power control or channel allocation also define strict delay requirements for the Iub interface. These delay requirements should be guaranteed over the Iub interface for all types of traffic. The Iub must deliver the content of a frame on time to the base stations for transmission over the air. Excessively delayed frames are discarded [3GP02e]. In such a network environment, the ability to appropriately dimension the Iub interface is a fundamental pillar of the UMTS network design and optimization processes. Therefore, the ultimate goal of this thesis is to investigate important aspects related to network dimensioning, and develop analytical methods to accurately estimate the required minimum capacity for the Iub interface, based on the amount as well as the characteristic of the aggregated traffic on the Iub interface and the desired QoS.

1.3 Important Challenges for UMTS Network Dimensioning

The evolution of UMTS wireless technologies as well as the derived new applications and services will greatly influence the development and the use of UMTS, and further cause considerable impact on the UMTS network dimensioning. The following paragraphs present the most important challenges, which are expected to significantly influence the dimensioning of a UMTS network, and point out their implications for network dimensioning.

Evolution of UMTS

With the fast development of UMTS in the past few years, evolved cell phones and smart-phones already support a broad range of data services, including video streaming, typical Internet applications like e-mail, file download, web browsing etc. In 2004, 3G UMTS interface cards for laptops or mobile computers were launched in the market, often coupled with volume based or flat-rate connectivity contracts, and met a considerable market success. In 2008 the iPhone 3G was released and it represents a new trend of UMTS terminals. It not only provides 3G data speeds and offers a wide range of Internet services, but also offers functionality of mobile music (music on the move) that borrows from "the iPod" experience. With these evolved UMTS terminals, the range of the delivered services in UMTS has rapidly expanded from voice-centric to a variety of appealing data services, and the volume of traffic generated by UMTS terminals and directed to the Internet has registered a rapid increase. The growing popularity of UMTS terminals and services has notably extended the coverage of Internet access, and the UMTS networks are now becoming key components of the Internet in Europe.

In addition to great improvements on UMTS terminals, the UMTS radio access technology is as well under evolution. After the operation of UMTS Release 99 (Rel99), which is the first UMTS release, *High Speed Packet Access* (HSPA) was introduced as a major step along the path of technology evolution of UMTS. It was launched in two phases, the first offering improvements in the downlink (HSDPA), while in the second phase the uplink offers enhanced performance (HSUPA). The launch of HSPA brought a major development in improving the end-user experience with the superior throughput and latency. In addition, operators will benefit from a significant increase in network capacity for data services. Beyond HSPA, further improvements are already being proposed, i.e. HSPA+ and Long Term Evolution (LTE) are currently making their way through the standardization process. A detailed introduction of the UMTS network evolutions is presented in Chapter 2. These evolutions will not only offer a much improved service experience for UMTS users but also enable new services such as IPTV, Voice over IP, multi-player gaming, high resolution video, and also provide high speed for popular Internet applications like web browsing, Skype, YouTube, etc.

The UMTS evolution and the resultant new applications and services are changing the nature of traffic in the current and future UMTS networks significantly, having a huge impact on the amount of traffic and its distribution as well as on the flow and packet-level characteristics. As a consequence, the evolution of the traffic pattern, traffic quantity and traffic flow characteristics will have considerable impact on the UMTS network dimensioning. Especially the "bandwidth greedy" applications pose great problems for the network dimensioning, as they often cause severe traffic congestions within the network. In order to avoid such congestion, network capacities have to be expanded. More detailed explanations on the implications of traffic and network dimensioning are given in Chapter 4. Nevertheless, it is not yet possible to predict typical traffic patterns for future UMTS networks, as these will depend on the popularity of future services and the prevalent application mix (telecommunications vs. data). The fact that traffic cannot be accurately estimated before the network is in operation stresses the need for a continuous network dimensioning throughout network operation phase. Once the amount, distribution and characteristics of the offered traffic are determined by means of monitoring, the dimensioning can be used to adapt the network capacities to support the demanded traffic in a cost-efficient way.

Advanced Transport Technology

The WCDMA radio control functions such as soft-handover, power control, scheduling, radio channel allocation, etc., are imposing strict delay and delay variation requirements for both *Real Time* (RT) and *Non Real Time* (NRT) services over the Iub interface between the RNC and Node B. These requirements should be guaranteed in the transport network of the Iub interface with the maximum utilization of the transport capacity which is a limited resource. Corresponding to the imposed requirements, the *Asynchronous Transfer Mode* (ATM) was selected as the transport technology for the UTRAN transport network in UMTS Release 99 (3GPP TS 25.434 and TS 25.426), due to its ubiquitous nature for the heterogeneous traffic types, quality of service guarantee and its widespread deployment in public networks [LK05].

With the evolutions of UMTS radio access technologies like HSPA, HSPA+ and LTE, which are aiming to increase the system capacity and the user peak throughput for packet based services, the capacity bottleneck mainly lies in the transport network as the transport resources on the radio access side are scarce. These high speed radio access technologies are optimized only regarding the radio resource usage and are mostly defining strict delay requirements for the transport network. In order to achieve optimal radio resource utilization, the congestion in the transport network has to be avoided. Moreover, the transport network is not dedicated to high speed radio access traffic only, it is a shared resource used to carry fixed traffic (fixed mobile convergence) as well as Rel99 traffic. Based on these considerations, the transport network should be dimensioned carefully in order to support different traffic and as well guarantee their strict delay requirements over the transport network and to reduce the probability of network congestion. Usually, in order to guarantee the stringent delay and loss requirements over the transport network, various functions can be taken within the transport network for managing the transport resources: traffic control, resource control, routing optimization and traffic engineering. These functions will cause considerable impact on the performance of the transport network, which are discussed in detail in Chapter 4. They also need to be carefully considered for the dimensioning of the transport network.

In order to achieve optimal transport resource usage various transport solutions have been adopted: service prioritization, hybrid backhaul, migration from ATM to *Internet Protocol* (IP) transport, etc. These solutions are aiming to increase system resource usage and at the same time decrease the transport costs by improving the multiplexing gain on the links, e.g. with service prioritization and path separation, and by migrating to IP transport, like changing to full IP UTRAN (see section 2.3.4.1) or using *Pseudo Wire Emulation Edge-to-Edge* (PWE3) technology to emulate ATM over *Ethernet* or other packet switched networks like *Asymmetric Digital Subscriber Line* (ADSL) (see section 2.3.4.2). The resulting system requires a modified dimensioning paradigm. The Iub transport cannot be modeled as a set of dedicated point-to-point links between RNC and Node B pairs anymore, but as a complex topology, where traffic going to different destinations is competing for the system resources with connections having mutual impact on each other.

Quality of Service Requirements

One of the most essential features of UMTS is its multi-service support to provide adequate service-specific QoS for a wide range of applications and services. The following fundamental QoS functionalities will have to be realized in the UMTS networks:

- *Differentiated service treatment*: in order to offer several classes of QoS, UMTS has to be able to distinguish different service classes and handle them in a differentiated way, corresponding to the class of service they belong to.
- *Resource reservation and admission control*: whenever network resources are limited and strict QoS requirements have to be met, the usage of resources needs to be controlled, monitored, and regulated.
- *End-to-end service provisioning*: QoS has to be provided on an end-to-end basis, making it necessary to coordinate service provisioning across multiple domains and provider networks.

Over the past years, numerous QoS mechanisms have been developed, which provide fundamental building blocks for the overall QoS framework [LF03]. Furthermore, various concepts have been introduced, which incorporate fundamental QoS schemes within one network architecture to realize comprehensive QoS-enabled networks. For the ATM transport networks, a number of QoS categories are defined by the ATM forum [ATM96] and various QoS mechanisms are suggested such as call admission control. For the IP transport networks, examples are the *Internet Engineering Task Force* (IETF)'s *Integrated Services* [BCS94] and the *Differentiated Services* [BBC+98] frameworks.

It is clear that QoS aspects, i.e. the various frameworks for QoS as well as the enabling technologies, need to be considered for network dimensioning. They determine the constraints and the objective of the relevant dimensioning procedures. As soon as QoS requirements are defined, the networks need to be dimensioned appropriately to meet the requirements.

1.4 Contributions of this Thesis

In this thesis dimensioning is investigated for the UMTS radio access network with specific focus on the transport network of the Iub interface, which is connecting the Node B with the RNC. The main contribution of this thesis is an elaborate analysis of the important aspects that have significant influence on dimensioning the Iub interface and evaluating their particular impacts; and secondly developing novel analytical methods to accurately calculate the required transport capacity for the Iub interface for the maximum utilization of the transport resources.

In order to give a comprehensive investigation of the dimensioning of the Iub interface, different traffic scenarios, evolutions of UMTS radio access networks, transport solutions, QoS mechanisms and network topologies are investigated. For performance analysis and evaluations, as well as for validating the developed analytical dimensioning models, several dedicated simulation models were developed to model different UMTS radio access networks, including modeling of their specific protocol structures, channels, radio functions, and transport technologies (see Chapter 5). By performing extensive simulations and analysis of the simulation results, important dimensioning rules are derived and the proposed analytical dimensioning models are demonstrated. At the end, to summarize the proposed analytical dimensioning approaches in this work, a dimensioning tool is implemented including all the proposed analytical models. The developed dimensioning tool can perform dimensioning for various traffic scenarios and UMTS radio access networks, considering certain transport solutions, QoS mechanisms and network topologies in use.

Traffic Scenarios

When dimensioning multi-service UMTS networks, it is not practicable to consider every characteristic of each application. Usually, only a few typical traffic types (or traffic classes) are defined and various applications are mapped upon them. Each traffic type has its own specific QoS requirements and traffic characteristics. In this thesis two fundamental types of traffic are distinguished: elastic and circuit-switched traffic.

Elastic traffic is generated by *non real time* applications and is typically carried by the *Transmission Control Protocol* (TCP). Typical applications are Internet services like web browsing or file download.

Circuit-switched traffic is associated with *real time* applications, which is delay-sensitive and have strict QoS requirements concerning the transport of packets over networks such as packet delay, delay jitter and packet loss. Typical applications in this traffic class comprise voice and audio or video conferencing. In order to guarantee the required stringent

QoS, they can reserve a certain bandwidth though the network in order to ensure a minimum transport bandwidth. In this context, they are called as circuit-switched traffic and controlled by admission control function.

These two traffic types are identified as the two main traffic classes served in the current UMTS networks. Voice, which is typically considered as circuit-switched traffic, has been the most important service in mobile networks. In UMTS Rel99, voice is still the main service. With the evolution towards a high-speed UMTS networks, elastic traffic is growing rapidly in UMTS and becoming prevailing. Due to the importance of these two traffic types in the UMTS networks, they are chosen to be considered for network dimensioning in the framework of this thesis.

Based on the two basic traffic classes, three different traffic scenarios are considered: pure elastic traffic, pure circuit-switched traffic, or a mix of both traffic types.

In this thesis, for each traffic type, different QoS requirements are investigated. The QoS requirements determine the objective of dimensioning and the selection of suitable analytical dimensioning models. Two kinds of QoS are considered: *user-relevant QoS* and *network-relevant QoS*. User-relevant QoS refers to the QoS related to the individual user flow; while network-relevant QoS measures are used to evaluate the quality of the transport network such as packet delay, packet drop ratio, and link utilization (see Chapter 4 for a detailed description).

The relevant user QoS for the elastic traffic flows used in this thesis is *throughput*, which specifies the amount of data that can be transferred in a certain time period. Specifically, in this thesis, the notion of throughput is the average transaction time for transmitting a specific file size. The proposed dimensioning model for guaranteeing this QoS for the elastic traffic is based on the *processor sharing model*. It can be applied for dimensioning in UMTS networks to meet the desired per user application layer throughput. One of the main contributions of this thesis is proposing a number of extensions of the process sharing model to incorporate important UMTS radio and transport functions and network topologies, and to demonstrate the applicability of the theoretical model through extensive simulations. For circuit-switched traffic the considered user QoS is *connection reject ratio*, as this traffic type is subject to the admission control function. The presented corresponding dimensioning model is based on the *Erlang loss model*. In the mixed traffic scenario, it is assumed that there is complete bandwidth sharing between the two traffic types. A dimensioning approach is suggested for the mixed traffic scenarios in this thesis to consider the bandwidth sharing and its achieved multiplexing gain.

For dimensioning to meet the desired network-relevant QoS, the proposed analytical models are based on *queuing models*, which are focused on the packet level taking the aggregated traffic into account. The proposed queuing models define an appropriate arrival process to capture the important characteristics of the aggregated traffic on the link, and an accurate server process considering the packet length distribution and the given link capacity. In the mixed traffic scenario, packet scheduling is considered in the queuing models, taking into consideration the potential multiplexing gain. An overview of analytical models used for dimensioning is given in Chapter 4.

UMTS Radio Access Network Evolutions

Within this thesis a number of evolutions of UMTS radio access networks are investigated for dimensioning: UMTS Rel99 as the basic UMTS network and HSDPA/HSUPA which are the high speed radio access technologies. For each evolution of the UMTS radio access network, its specific protocol stacks and important features are studied, especially on the Iub interface. Their particular traffic- and resource control functions are carefully considered for dimensioning the Iub transport network, as they will cause significant impact on the usage of the transport resources. Examples for traffic control functions are *congestion control* and *flow control* algorithms (for HSPA), packet scheduling, buffering and shaping. Examples for resource control functions are the *Connection Admission Control* (CAC) and the *Bit Rate Adaptation* (BRA) algorithm (see Chapter 3 for a detailed introduction). Furthermore, various traffic scenarios and network structures are investigated for dimensioning of the Iub interface for each radio access network evolution.

Transport Solutions

Transport solutions have great impact on the utilization of the transport resource. In this thesis, two fundamental transport technologies used in the UTRAN are investigated: ATM and IP. ATM is the transport technology used in UMTS Release 99 and Release 4 to transfer user data and control traffic in the UTRAN. From Release 5 onward, IP transport is introduced as an alternative transport technology to replace ATM in the UTRAN to provide improved cost efficiency of the transport network. Migrating from ATM to IP transport requires some change of the deployed QoS schemes as well as traffic- and resource control functions in the transport network, which result in different transport efficiency.

In addition, when combing Rel99 traffic with high speed radio access traffic over the ATM transport on the Iub interface, service prioritization and path separation are applied to give higher priority for Rel99 traffic and at the same time to enhance the transport efficiency by improving the multiplexing gain on the links. The dimensioning results demonstrate the improved cost efficiency of applying these transport solutions.

QoS Mechanisms

In the transport network, various QoS mechanisms can be applied to provide distinguished service treatment and control the usage of resources in order to guarantee the desired level of QoS for a variety of services.

Resource reservation and admission control: the UTRAN transport network has to guarantee strict delay requirements and moreover its transport resources are limited. Therefore reservation of the transport resource for different flows and controlling the usage of the resources in the UTRAN are important [NRM+05].

Strict Priority: traffic is served in the order of its priority. Lower priority traffic can be served only when the higher priority traffic has been served. Usually Rel99 traffic has higher priority than the high speed radio access traffic. This is usually realized on the packet level in both ATM and IP transport networks.

Differentiated Services (DiffServ): DiffServ is an IP-based QoS framework. It provides different treatment of flows or aggregates flows by mapping multiple flows into a finite set of service levels. In this thesis, a DiffServ-based QoS architecture is used in the IP-based transport of the radio access network (see Chapter 3). The applied packet scheduling in this QoS framework is *Weighted Fair Queuing* (WFQ). It defines a weight associated to each service class, according to which the bandwidth is shared.

The applied QoS mechanisms will have impact on the utilization of the transport resources, and as a consequence influence the outcome of the dimensioning. In this thesis, the proposed dimensioning models consider the effect of the deployed QoS functions.

Network Topologies

In this thesis, two logical network structures are investigated for the dimensioning of the Iub interface, as shown in Figure 1.2. The first structure has only a single link between the Node B and the RNC, which is the most basic and simple structure for the Iub interface. The second structure considers a star network topology for the Iub interface, where a number of Node Bs are aggregated to an intermediate switch or router that is connected to the RNC over a *backbone link*. The link between the Node B and the intermediate switch or router is called *last mile link*.

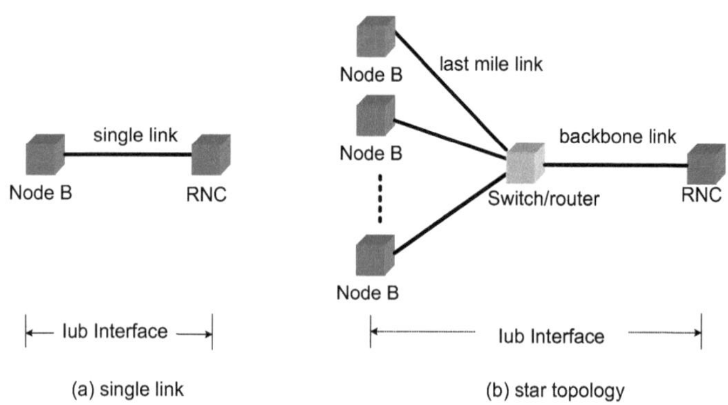

Figure 1.2: *Investigated network structures*

The single link case (a) is the most basic logical network structure. It is used to derive and demonstrate the fundamental analytical dimensioning models in this thesis. Furthermore, dimensioning of a single link can be used for dimensioning individual links in the networks.

However, the UTRAN is typically a tree-structured access network with several aggregation stages. The star network topology (b) represents the basic single-stage aggregation of a tree structure. In (b), the traffic is aggregated at the backbone link, which can thus achieve certain multiplexing gain as a consequence of aggregation. The purpose of dimensioning such a star-structured network is to investigate the potential multiplexing gain which can be achieved at one aggregation stage. Given the number of aggregation stages and the amount of traffic aggregates at each stage, the dimensioning of a complete tree-structured access network can be derived by performing dimensioning for the basic star network topology at each aggregation stage of the tree-structured access network.

In a star-structure access network as shown in (b), the dimensioning task is to determine a necessary bandwidth for each last mile link and the backbone link separately. The analytical models, which have been developed for case (a), are extended and applied in the scenario (b) to dimension the individual links in the network. One important issue for dimensioning the backbone link is *overbooking* (see Chapter 8), which is to allocate less bandwidth on the backbone link than the total bandwidth request by all the connected last mile links. Usually, the backbone link can be overbooked by taking advantage of the statistical multiplexing gain among traffic from different Node Bs. For a cost-efficient dimensioning of the backbone link, an optimum overbooking is desired to provide balance between the costs and the QoS objectives. In this thesis, overbooking is investigated for a wide variety of scenarios through extensive simulations, with special focus on determining the optimum overbooking for the backbone link. From the investigated results, important factors that have great impact on the overbooking are found. Furthermore, the analytical models are applied in various network scenarios to derive optimum overbooking factors (which represent the degree of overbooking), which are validated by simulations.

1.5 Thesis Overview

In chapter 2, the evolution of UMTS networks is introduced. The chapter provides an overview of development of UMTS and emphasizes the technologies of evolved UMTS radio access networks.

Chapter 3 gives a detailed introduction of UMTS Terrestrial Radio Access Network (UTRAN) for various UMTS radio access evolutions including their important network elements, internal and external interfaces, channels, and protocol stacks. Special focus on is put on the Iub interface, which is the key interface considered for dimensioning in this thesis. The related transport technologies, resource control and traffic control functions for the Iub transport network are introduced in detail. This chapter provides a profound understanding of the UMTS system and its important functions related to dimensioning.

In Chapter 4, a general framework for the UMTS network dimensioning is presented. In the proposed dimensioning framework, the objectives of UMTS network dimensioning are defined, and related important aspects which need to be particularly considered in the dimensioning process are discussed. Based on the proposed framework, a detailed dimensioning procedure which is used in this thesis is described.

Chapter 5 introduces in detail the UMTS simulation models which have been developed and used in this thesis, including the simulation models for HSPA and UMTS Rel99 with ATM and IP based transport. Besides, an overview of different traffic models is given. Among them several traffic models are selected in this thesis, which are described in detail.

In Chapter 6 the dimensioning for the ATM-based Iub transport is investigated in detail with dedicated focus on the single link scenario. In this thesis, both Rel99 and HSPA radio access networks are based on the ATM transport technology. Their individual dimensioning results are presented in this chapter. For the UMTS Rel99 network, novel and elaborate analytical models are developed for dimensioning of the Iub link. The proposed analytical models carefully consider different traffic scenarios, various QoS requirements, and important UMTS radio and transport functions, etc. Through simulations it is demonstrated that the proposed analytical models are able to capture relevant characteristics and provide accurate dimensioning results, and thus can be applied for network dimensioning. The dimensioning for HSPA networks are mainly based on simulations, from which general guidelines for dimensioning are derived in this thesis. In addition, dimensioning for combined traffic scenarios (transmitting HSPA traffic together with Rel99 traffic in the same radio access network) are also investigated and various traffic separation approaches are explored.

Chapter 7 presents the dimensioning for the IP-based Iub transport for the single link scenario. In the IP transport network, the DiffServ QoS architecture is considered. The proposed analytical models for the Rel99 ATM-based transport network are extended for the IP-based transport, taking into consideration of the IP specific QoS scheme and transport functions. The extended analytical models are validated for the IP-based Iub transport network through simulations. Additionally in this chapter the performance and the dimensioning of the IP-based Iub is compared with the ATM-based Iub.

In chapter 8, the dimensioning for multiple Iubs with a star network topology is investigated. For this case, dimensioning approaches are proposed to extend the analytical models, which are developed for the single link scenario, to dimension the capacity of the last mile link for each Node B and the backbone link individually. And the calculated link capacities are verified to be able to meet the desired QoS through simulations. In addition, overbooking is investigated in detail for dimensioning the backbone link. By applying the extended analytical models, the optimum overbooking factor can be derived. Through comparing the analytical results with simulation results in various network scenarios, the presented analytical models are proven to provide accurate estimation on the bandwidth dimensioning and the optimum overbooking.

Finally, Chapter 9 gives a summary of this thesis and summarizes the most important achievements and results. Additionally, interesting issues for further research are pointed out.

2 UMTS Evolution

With the dramatic development and expansion of 3G cellular networks all over the world, a profound evolution is taking place in the UMTS networks and continuing to unfold towards wider bandwidth, lower latency and packet-optimized radio access technology with highly improved peak data rate capability, in order to meet the rapidly growing demand for bandwidth and mobility of the global nomadic users. This chapter will give a brief introduction of the evolution of the UMTS standard, from its first release (Release 99) to the latest Release 8 (still being standardized). The purpose of this chapter is to provide background knowledge of the evolution of the UMTS to help understand the new requirements of the evolved UMTS radio access networks.

2.1 Background of UMTS

In parallel with the widespread deployment and evolution of Second Generation (2G) mobile communication systems during the 1990s, the *International Telecommunication Union* (ITU) started the process of defining the standard for Third Generation (3G) mobile systems, referred to as International Mobile Telecommunications 2000 (IMT-2000). *Universal Mobile Telecommunication System* (UMTS) is a key member of the global family of 3G mobile communication systems. It was initiated by the *European Telecommunications Standards Institute* (ETSI) in Europe. In 1998, the *3rd Generation Partnership Project* (3GPP) was formed by standards-developing organizations from all regions of the world to continue the technical specification work for UMTS. This solved the problem of trying to maintain parallel development of aligned specifications in multiple regions. The major standardization bodies of 3GPP are ARIB (Japan), CCSA (China), ETSI (Europe), ATIS (USA), TTA (Korea) and TTC (Japan).

UMTS is seen as the successor to 2G mobile communication systems such as *Global System for Mobile communications* (GSM) and to evolved 2G systems such as the *General Packet Radio Service* (GPRS). 3G UMTS brings an evolution in terms of greatly enhanced capacity, data rates and new service capabilities from 2G systems. It provides global mobility and high speed transmissions with a wide range of services including voice telephony, messaging, images, video, Internet access, and broadband data, whereas traditional 2G mobile systems were built mainly for speech service. The main improvements of 3G systems compared with 2G are [HT04]:

- Much higher peak data rates

- Variable data rates to allow bandwidth on demand
- Support of asymmetric data rates on uplink and downlink to transfer packet-switched traffic, e.g. web browsing, ftp download/upload
- Provisioning of *Quality of Service* (QoS) for various applications and services, e.g. guaranteed error rates and delays
- Support of QoS differentiation, e.g. between delay-sensitive real-time services and best-effort packet data
- High spectral efficiency

The air interface of UMTS is based on *Wideband Code Division Multiple Access* (WCDMA) technology that utilizes the direct-sequence spread spectrum method of asynchronous code division multiple access to achieve higher speeds and support more simultaneous users compared to the implementation of *Time Division Multiple Access* (TDMA) and *Frequency Division Multiple Access* (FDMA) used by the 2G GSM networks. Within 3GPP, WCDMA is called *Universal Terrestrial Radio Access* (UTRA). It has two modes of operation, namely UTRA-FDD (Frequency Division Duplex) and UTRA-TDD (Time Division Duplex). UTRA-FDD uses paired frequency bands for *Uplink* (UL) and *Downlink* (DL) data transmissions, whereas UTRA-TDD uses a common frequency band for both directions and adjusts the time domain portion assigned for UL and DL transmissions dynamically. In this thesis, only the UTRA-FDD mode is considered.

In 1999 the 3GPP produced the first version of the WCDMA standard, which is the basis of UMTS deployed in most of the world today. This release, called Release 99, contains all the basic elements to meet the requirements for IMT-2000 technologies. As WCDMA mobile penetration increases, it requires UMTS networks to carry a larger share of packet data traffic. Thus 3GPP Release 5 initiated the *High Speed Downlink Packet Access* (HSDPA) in March 2002. Further Release 6 followed with *High Speed Uplink Packet Access (*HSUPA) was standardized in 2004. HSDPA and HSUPA are commonly referred to as *High-Speed Packet Access* (HSPA). The first commercial HSDPA network started in fall of 2005, and the first HSUPA network was launched in February 2007. The latest step being studied and developed in 3GPP is an evolution of 3G into an evolved radio access referred to as the *Long-Term Evolution* (LTE) and an evolved packet access core network in the *System Architecture Evolution* (SAE). By 2009–2010, LTE and SAE are expected to be deployed.

In the following a more detailed introduction of the UMTS network architecture and key UMTS evolution technologies are presented.

2.2 UMTS Network Architecture

A UMTS network comprises three main parts: *User Equipment* (UE), *UMTS Terrestrial Radio Access Network* (UTRAN), and *Core Network* (CN). Their individual functionalities as well as the logical interfaces between them are specified in the 3GPP standards. More detailed introductions on the fundamentals of UMTS can be found in [KAL⁺01], [WSA03] and [HT04]. Figure 2.1 presents the schematic view of the structure of a UMTS network.

The UE includes two parts: the *Mobile Equipment* (ME), which is a radio terminal handling the communications on the Uu interface (radio interface), and the *UMTS Subscriber Identity Module* (USIM), which is a smart card that contains the subscriber's confidential data such as identity, authentication algorithm and subscription information. USIM and ME communicate over the internal Cu interface. The UE accesses the fixed network via the Uu radio interface.

The UTRAN contains one or more *Radio Network Subsystems* (RNS) each of which comprises one *Radio Network Controller* (RNC) and a number of *Node Bs* in a UMTS network, which make up the UMTS radio access network. The RNC is the central element of the UTRAN responsible for the control of the radio resources in all attached cells. It logically corresponds to the *Base Station Controller* (BSC) in GSM. The Node B is the counterpart of the *Base Transceiver Station* (BTS) in GSM. It supplies one or several cells. The UTRAN establishes connectivity between the UE and the core network, and is responsible for managing and operating the radio access to the UE, like radio resource management and handover control. The 3GPP standard defines four interfaces related to the UTRAN: Iu, Uu, Iub and Iur. The Iu interface is the interface that connects the UTRAN to the Core Network. Dependent on the type of traffic, i.e. packet switched or circuit switched traffic, the Iu is divided into Iu-CS (Iu-Circuit Switched) and Iu-PS (Iu-Packet Switched). The Uu is the radio interface of UMTS based on WCDMA technology, located between the UTRAN and the UE. It realizes the radio connection between the UE and the Node B. The Iub and Iur interfaces are used to connect the functional entities inside the UTRAN. The Iub interface connects RNC and Node B, while the Iur interface connects two RNCs with each other. A more extensive introduction of UTRAN interfaces is given in chapter 3.1.

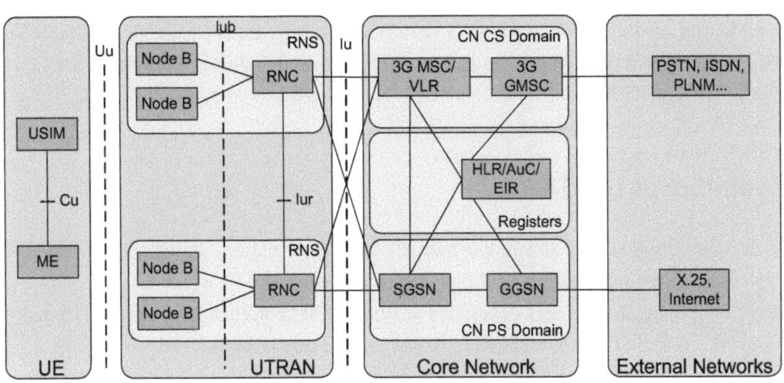

Figure 2.1: *UMTS Network Structure*

The CN is the backbone of UMTS, connecting the radio access network to other external networks like the *Public Switched Telephone Network* (PSTN), Internet or other data networks. The UMTS core network evolved from the current 2G networks based on GSM/GPRS, so that it provides backward compatibility and interoperability between GSM/GPRS and UMTS. In this way, it allows GSM mobile operators to directly proceed to UMTS cellular systems by simply replacing the *Radio Access Networks* (RAN) without significant changes in the core network architecture. Basically, the CN is divided into circuit switched and packet switched domains.

- The main functional entities in the circuit switched domain are the *Mobile service Switching Center* (MSC), *Visitor Location Register* (VLR) and *Gateway MSC* (GMSC). The MSC is responsible for the call handling of circuit switched calls, i.e. call setup, routing, inter-system mobility management, etc. The VLR is a database which stores copies of the service profiles and position information of subscribers who are currently being served by the MSC. The MSC/VLR is linked through the Iu-CS interface with the RNC and through the GMSC to the external mobile or fixed telecommunication networks.
- Packet switched domain consists of *Serving GPRS Support Node* (SGSN) and *Gateway GPRS Support Node* (GGSN). The SGSN is in charge of relaying packets from radio networks to the core network. Moreover, it also handles mobility management and session management. While the GGSN serves as a gateway to access outside networks. The SGSN is connected via the Iu-PS with the RNC and via the GGSN with the external packet switched networks.

Other network elements in the core network, i.e. *Home Location Register* (HLR), *Equipment Identity Register* (EIR), and *Authentication Center* (AuC) are shared by both packet switched and circuit switched domains. The HLR is a central database that contains detailed relevant information of each subscriber including associated telephone numbers, supplementary services, access priorities and security keys. The EIR is also a database that stores information about the identity of the mobile equipment that prevents calls from stolen, unauthorized or defective mobile stations. The AuC handles the authentication and encryption keys for every subscriber in HLR and VLR.

2.3 Evolution of UMTS

This section introduces in detail the path of UMTS evolution, starting from the first UMTS release (Release 99) up to the Long-Term Evolution (LTE) which is at present the latest step being standardized and developed in the 3GPP. For each evolved UMTS radio access network, its main features and key technologies are briefly introduced.

2.3.1 UMTS Release 99

Release 99 (Rel99) is the first UMTS release. It provides a smooth evolution path from GSM/GPRS to UMTS networks. The most important enhancement compared to GSM is the introduction of a new radio interface with WCDMA, which brings a great improvement on the spectrum efficiency and higher data rate to provide a better support for both packet data and circuit-switched data. WCDMA supports data rates up to 2 Mbps in indoor/small-cell-outdoor environments and up to 384 kbps with wide-area coverage, which is in full agreement with the IMT-2000 requirements.

WCDMA technology employed on radio interface naturally imposes new requirements to the radio access networks. As WCDMA and its radio access equipment is not compatible with the GSM equipment, it needs additional new network elements (namely RNC and Node B) and new interfaces, which form a new radio access network, i.e. UTRAN. In the Rel99 radio access network, most network intelligence resides in the RNC, which performs most of the *Radio Resource Management* (RRM) functions such as call admission control, handover, power control, radio access bearer set-up and release, channel allocation, packet scheduling, etc. The Node B manages one or several radio cells and is connected with the RNC over the Iub interface. The network architecture of UMTS Rel99 is according to the structure introduced in section 2.2. The corresponding radio interface protocol architecture and the related channels are standardized in [3GP99b]. Section 3.3.2 gives an overview of the channels defined in the 3GPP Rel99.

In addition to new network nodes and interfaces, the UTRAN employs a new transport technology. The transport within the radio access network is a very important and crucial aspect. It must provide the support from the RAN for the most efficient utilization of the radio resources. In other words, it should not be a bottleneck between the radio interface and core network. On the other hand, it should also achieve high cost-efficiency on the transport resources (also called backhaul resources) as well as supply certain QoS for various traffic types. In Rel99, *Asynchronous Transfer Mode* (ATM) has been chosen for the radio access network transport. In addition, the *ATM Adaptation Layer type 2* (AAL2) [ITU00] was selected by the 3GPP Rel99 to be used on top of the ATM to transport user data within the UTRAN (User Plane) because of its ability to perform multiplexing of low bit rate traffic such as voice traffic, and thus can efficiently support the QoS and link transmission utilization. A more detailed introduction on UTRAN of UMTS Rel99 is given in Chapter 3 including the defined network architecture, channels, protocol stacks, transport network and its related resource management functions.

An essential part of this thesis focuses on the UMTS Rel99 system. It is the basic UMTS release and considered as the reference system for the other evolved UMTS networks.

2.3.2 High Speed Downlink Packet Access (HSDPA)

With the gradual development of UMTS, the offered services by UMTS rapidly extend from primarily voice telephony to a variety of appealing data and multimedia-based applications. The offered data services, such as Internet access, remote database access, email or ftp, are now contributing a substantial traffic share in the UMTS networks. In order to greatly enhance the capability of transmitting such delay-tolerant data services with improved resource efficiency and service quality, 3GPP Release 5 specifications introduce the *High Speed Downlink Packet Access* (HSDPA) as a key evolution of the WCDMA radio interface to significantly improve the downlink packet data transmissions. Compared to the UMTS Rel99, HSDPA makes a significant improvement in peak data rate on the downlink for transmitting the packet data. It provides a maximum data transmission rate of up to 14.4 Mbps per user, the current HSDPA deployments support downlink speeds of 1.8, 3.6, 7.2 and 14.4 Mbps. The main features of HSDPA are its high data transmission rate, low latency, and high throughput, which facilitate a notably improved support for the downlink packet data transfer and enable the provision of various types of multimedia services [3GP01a].

A key characteristic of HSDPA is the use of *Shared-Channel Transmission,* which implies that a certain fraction of the total downlink radio resources, channel codes and transmission power belong to the common resource that is dynamically shared between users, primarily in the time domain. The use of shared-channel transmission in WCDMA is implemented through a new transport channel: *High-Speed Downlink Shared Channel* (HS-DSCH) that is shared by all HSDPA UEs within the cell. The HS-DSCH enables the possibility to rapidly allocate a large fraction of the downlink resources for transmission of data to a specific user. This is appropriate for packet data applications which in nature have bursty characteristics and thus rapidly varying resource requirements. The HS-DSCH excludes two basic features of WCDMA radio channels: variable spreading factor and fast power control. Instead, it introduces the following new features [HT06, DPSB07], which deliver the improved downlink performance in terms of higher peak data rate, lower latency and increased capacity.

1) Short *Transmission Time Interval* (TTI) of 2 ms. The use of such a short TTI for HSDPA reduces the overall delay and improves the tracking of fast channel variations exploited by the rate control and the channel dependent scheduling.
2) *Adaptive Modulation and Coding* (AMC): HSDPA is not power controlled instead it dynamically adjusts the modulation scheme between QPSK and 16QAM. The modulation scheme and coding is changed on a per user basis depending on signal quality and cell usage; and selected independently for each 2 ms TTI (*Transmission Time Interval*) by the Node B and can therefore track rapid channel variations.
3) *Fast packet scheduling* at the base station (Node B): by shifting the packet scheduler from the RNC in Rel99 to the Node B enables a more efficient implementation of the scheduling by allowing the scheduler to work with the most recent channel information, i.e. the scheduler can adapt the modulation to match the current channel conditions and

fading environment in a better way, and thus the channel adaptation can be performed in a more rapid and efficient manner.

4) Fast *Hybrid Automatic Repeat Request* (HARQ) with *soft combining*: HARQ is an effective technique to correct the erroneous transmission where the mobile terminal (UE) attempts to decode each transport block received and reports to the Node B upon the reception of the transport block, which allows for rapid retransmissions of unsuccessfully received data and significantly reduces the delays associated with retransmissions compared to Rel99. Additionally, soft combining is used, which means that the soft information from the previous transmission attempts is combined with the retransmissions to increase probability of successful decoding. In order to reduce the delay associated with HARQ, the entity of HARQ is located in the base station Node B without RNC involvement. Thus, in case of packet decoding failure, retransmission automatically takes place from the Node B. The RNC-based *radio link control* (RLC) layer retransmission may still be applied on top in case of the Node B exceeding the maximum HARQ retransmissions.

In Release 5, HSDPA is operated together with the Rel99 *Dedicated Channel* (DCH) where the uplink data are transmitted, whereas in Release 6 an alterative for transferring the uplink data is provided by the *Enhanced DCH* (E-DCH) with the introduction of HSUPA, as covered in the next section.

2.3.3 High Speed Uplink Packet Access (HSUPA)

After the successful introduction of HSDPA in 3GPP Release 5, the downlink capabilities are enhanced significantly, but the uplink capabilities did not match the HSDPA downlink. Therefore, *High Speed Uplink Packet Access* (HSUPA), also referred to as *Enhanced Uplink*, is promoted in 3GPP Release 6 as an enhancement on the uplink of UMTS [3GP06a]. It provides high-speed transmission with uplink speeds up to 5.76 Mbps for the transport of packet data traffic in the uplink to meet the requirements of the new data services and the vastly growing demand of user traffic in the 3G UMTS network. HSUPA improves UMTS uplink capabilities and performance in terms of higher data rates, reduced latency, and improved system capacity, and is therefore a natural complement to HSDPA.

In HSDPA, the shared resources are the transmission power and the code space, both of which are managed in one central node, the Node B. However, in the uplink the shared resource is the amount of allowed uplink interference [DPSB07], which depends on the transmission powers of the individual UEs. Since the uplink power resources are distributed among the users and not located in one place, it is physically impossible to schedule multiple distributed UEs to share one transport channel, thus for each UE a new uplink transport channel called *Enhanced Dedicated Channel* (E-DCH) is used in HSUPA that supports the following key techniques: fast Hybrid ARQ (HARQ) with soft combining, fast Node B scheduling, *soft handover* (SHO) and also optionally a short 2ms uplink TTI. The HARQ is used in HSUPA for the recovery of erroneous transmissions, which is the same as in HSDPA. Fast Node B scheduling is one of the fundamental technologies behind HSUPA, it is located at the Node B to control when and at what data rate a UE is allowed to transmit, thereby controlling the amount of interference affecting other users at the Node B and keeping the total interference below a maximum allowed level. Unlike in HSDPA where the scheduler and the transmission buffers are located in the same node, while in the uplink the scheduler is located in the Node B whereas the data buffers are distributed in the UEs. Hence in HSUPA, the UEs need to report the buffer status information to the scheduler at the Node B. Moreover, both soft and softer handover (the types of handover are introduced in section 3.3.4.2) are supported in HSUPA because receiving data from a terminal in multiple cells is fundamentally beneficial as it provides diversity. Another major difference between HSDPA and HSUPA is that the latter will not support adaptive modulation because it does not support any higher order modulation schemes and only uses the BPSK modulation scheme. The reason behind is that the higher modulation schemes requires more energy per bit to be transmitted than the BPSK modulation scheme.

The common resource of the enhanced uplink is the amount of tolerable interference. It is represented by the *Noise Rise* (NR) which is defined as $(I_0+N_0)/N_0$ where N_0 and I_0 are the noise and interference power spectral densities, respectively. A too large noise rise would cause some terminals not to have sufficient transmission power available to reach the required E_b/N_0, i.e. the ratio of Energy per Bit (E_b) to the Spectral Noise Density (N_0), at the base station. Hence, the uplink scheduler must keep the noise rise within acceptable limits. The basis for the HSUPA scheduling is *scheduling grants* (RG - *Relative Grant* and AG - *Absolute Grant*) sent by the Node B to the UE and limiting the E-DCH data rate, and *scheduling requests* sent by the UE to the Node B to request permission to transmit. The scheduler uses *E-DCH Absolute Grant channel* (E-AGCH) for transmitting *Primary AG* (PAG) or *Secondary AG* (SAG) information, and *E-DCH Relative Grant channel* (E-RGCH) to send relative grants (UP, Down, and Hold) to the UE. PAG or SAG are used for large changes in the data rate, while the relative grant is typically used for smaller adjustment of the data rate during an ongoing packet transmission. The HSUPA scheduler algorithm is vendor specific. More details can be found in [HT06]. More introductions of HSDPA/HSUPA and their UTRAN networks are given in Appendix A.9.

2.3.4 Introduction of "All-IP" UMTS Networks

It is now widely recognized that using *Internet Protocol* (IP) as the foundation for next generation mobile networks enables strong economic and technical profits, since it

(1) takes advantage of the ubiquitously installed IP infrastructure which is widely deployed and has high flexibility and scalability;
(2) capitalizes on the IETF standardization process;
(3) brings flexibility to an operator in choosing a layer 2 transport technology;

(4) profits from the low cost IP equipments and IP bandwidths based on a cheap layer 2 transport technology, e.g. Ethernet, which can significantly reduce the overall infrastructure costs;

(5) benefits from both existing and emerging IP-related technologies and services;

(6) and facilitates the integration of different radio access technologies operating over a common IP backbone and therefore provide a flexible and reliable way for enabling heterogeneous network access.

In this context, the concept of an *All-IP Network* (AIPN) is promoted in UMTS. It requires establishing a single unified UMTS network that is fully using IP as the underlying network architecture. That means all services, not only packet data but also control data and circuit-switched data like traditional speech services, will be delivered over IP. The evolution of the UMTS into all IP-based system is considered in the recent releases of the 3GPP specifications. The feasibility study for an All-IP network within 3GPP is given in 3GPP TR 22.978 [3GP05a]. The architecture of an All-IP network is specified in 3GPP TR 23.922 [3GP01b].

With the scope of building an All-IP UMTS network, the UTRAN is certainly expected to evolve towards an IP-based network. The following subsections introduce different alternative approaches to use IP as the transport technology in the UTRAN.

2.3.4.1 IP-Based UTRAN

The most straightforward approach is to completely replace the legacy ATM transport networks with the IP networks in the UTRAN. The IP-based transport should meet the delay requirements of the UTRAN in a cost effective way in terms of efficiency and maximal resource utilization. To achieve this, multiplexing schemes and UDP/IP header compression are adopted to reduce the IP layer overhead. Multiplexing schemes seek to reduce the overhead to payload ratio by including multiple payloads within the same IP packet. 3GPP TR 25.933 [3GP04a] and [MWIF01] propose several multiplexing schemes for the IP-based UTRAN with different protocol stacks. The four main proposed schemes are briefly described below. Their corresponding protocol stacks are shown in Figure 2.2.

1) *Composite **IP*** **(CIP),** originally proposed by Alcatel, adopts the aggregation of small packets into one IP packet and the segmentation of large packets into smaller chunks in order to keep transmission delays low and avoid blocking of small time-critical packets by large packets. Several CIP packets are transported into one UDP/IP packet.

2) *Lightweight **IP** Encapsulation* **(LIPE),** proposed as a draft document at the IETF Audio/Video Transport working group, multiplexes various packets of low bit rate in one UDP/IP packet.

3) ***AAL2** over **IP**,* proposes to apply AAL2 user plane protocol over UDP/IP to achieve multiplexing. The protocol stack is then AAL2/UDP/IP. This solution has the

advantages of simplifying interoperability between IP and AAL2/ATM nodes and reusing already existing fragmentation and multiplexing standards.

4) ***M**ultiplexing with **P**oint-to-**P**oint **P**rotocol (PPP)*: ***PPP M**ultiplexing* (**PPPmux**), proposed by RFC 3153, provides a method to reduce the PPP framing overhead used to transport small packets, e.g. voice frames, over slow links. PPPmux sends multiple PPP encapsulated packets in a single PPP frame. As a result, the PPP overhead per packet is reduced. When combined with a link layer protocol, such as *High Level Link Control* (HDLC), this offers an efficient transport for point-to-point links.

In the above proposed multiplexing schemes, PPPMux is a Layer 2 multiplexing proposal, where the multiplexing gain is mainly achieved from multiplexing multiple PPP frames (each PPP frame encapsulates one IP packet) into a single PPP multiplexed frame. While LIPE and CIP are two proposals that apply multiplexing above Layer 3. They propose to create a container to encapsulate multiple packets into one UDP/IP packet. The AAL2 over IP scheme is also above Layer 3, but utilizing the existing fragmentation and multiplexing schemes from the AAL2 protocol. Its main benefit is easier interworking with ATM legacy transport networks. In this thesis, LIPE and CIP proposals are considered for the IP-based UTRAN. They are introduced in detail in Appendix A.11.

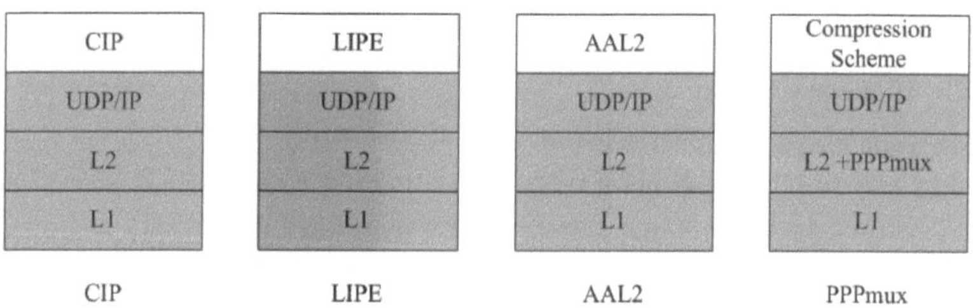

Figure 2.2: *IP-based UNTRAN Protocol Stack (CIP, LIPE, AAL2, PPPmux)*

In 3GPP TR 25.933 [3GP04a], ***M**ulti-**P**rotocol **L**abel **S**witching* (**MPLS**) protocol is also proposed as a promising solution for IP transport in the UTRAN. It is an interstitial, Layer 2.5 protocol which complements and enhances the IP protocol, in that it offers a complementary method of forwarding IP packets, while reusing the existing IP routing protocols. The main idea of this approach is to establish and manage *Label Switched Paths* (LSPs) for interconnecting Nodes B and RNC.

When the transport technology in the UTRAN is evolved from ATM to IP, this comes along with new QoS schemes, network configurations and network topologies (number of hops, star/chain connections, etc.). One of the most challenges of using the IP transport technology is how to properly differentiate the services and to provide guaranteed QoS for a wide range of services each with different QoS requirements, since IP by itself is designed for the "best effort" Internet where there is no guarantee for the QoS. Therefore, special attentions are paid in this thesis on setting up an appropriate QoS architecture for the IP-based transport in the UTRAN and then dimensioning such IP transport network by taking careful considerations of the applied QoS framework. In section 3.4.2, a short overview of different developed IP QoS schemes is given, and in particular a *Differentiated Services* (DiffServ) based QoS framework is presented, which is used in this thesis, for the IP transport in the UTRAN.

2.3.4.2 Pseudo-Wire Emulation (PWE) Technology

The above introduced fully IP transport solution requires a complete change of the transport network from ATM to IP in the UTRAN. However, substantial expenditures have been invested for the legacy ATM transport networks for numerous Node Bs and ATM-based interfaces. In order to maximize the usage of the existing ATM based infrastructure in the UTRAN for making the most profits for the network providers, an intermediate migration solution is required meanwhile for a gradual evolution towards fully IP transport. The intermediate migration solution should integrate the use of cost-efficient IP transport to reduce the transport cost while at the same time allow backward compatibility and interworking of RANs with different transport technologies. In this context, *Carrier Ethernet* [Sie06] is proposed as a cost-effective way for providing the migration solution. It is also a viable solution of converged fixed-mobile access networks, as well as a flexible and reliable way for enabling heterogeneous access networks towards an All-IP network.

The deployment of Carrier Ethernet for UTRAN is realized by establishing *Pseudo-Wires* in the backhaul network. This technique is standardized by the IETF's *Pseudo Wire Emulation Edge-to-Edge* (PWE3) working group defining various types of Pseudo-Wires to emulate traditional and emerging services such as ATM or frame relay over *Packet Switched Network* (PSN) [RFC04, RFC05, RFC06]. Figure 2.3 depicts the integration of a Carrier Ethernet-based transport network into an ATM-based UTRAN by means of PWE technique. Both the NodeB and RNC are *Customer Edges* (CE). They are not aware of using an emulated ATM service over Ethernet, so they keep transmitting and receiving ATM cells same as in the conventional ATM transport network. The NodeB and RNC are connected to the transport Ethernet network via two intermediate PWE routers which contain dual interfaces for ATM and Ethernet. Such routers are usually located at the edge of the ATM network and the Ethernet network, hence also called as *Provider Edges* (PEs), which are responsible for providing a tunnel emulating ATM services over the Ethernet network for the corresponding CEs. Between these two routers, an Ethernet Pseudo-Wire is established through the Ethernet network (which is a Packet Switched Network). Thus, the ATM cells coming from the Node B or RNC will be encapsulated into Ethernet packets within the PWE routers and then carried over the emulated Ethernet circuit, i.e. the Ethernet Pseudo-Wire. After the Ethernet packets arrive at the egress of the Ethernet network, they are decapsulated to the ATM cells and forwarded to their destination. Usually Ethernet switches can be used within the Carrier Ethernet transport network as an intermediate switches to multiplex/de-multiplex outgoing/incoming Ethernet frames.

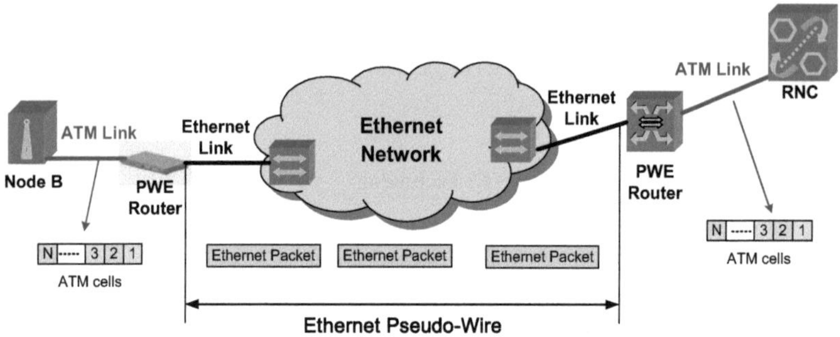

Figure 2.3: *Carrier Ethernet-based UTRAN using PWE3*

The Pseudo-Wire solution is becoming a popular approach for mobile wireless operators for the mobile network migration phase, because Pseudo-Wire is the enabling technology for transporting both mobile voice and data traffic over new high-capacity, lower-cost packet networks in the RAN. It gives operators the possibility to choose among multiple packet network technologies in the RAN, including Carrier Ethernet, MPLS, xDSL and even broadband packet radio (WiMAX). The Pseudo-Wire solution also provides support for all generations of mobile wireless services and ensures a smooth transition from one generation to the next. A detailed investigation on the performances of a PWE-based UTRAN by the author was published in [LZK+08].

2.3.5 LTE and SAE

The roadmap of *Next Generation Mobile Network* (NGMN) is to provide mobile broadband services. Services like high-speed internet access, distribution of video content (Mobile TV), fast interactive gaming, wireless DSL, fixed-mobile convergence will produce tremendous traffic in the future mobile networks. To make this happen, goals for the future evolved system include support for improved system capacity and coverage, high peak data rates, low latency, reduced network costs, multi-antenna support, flexible bandwidth operations and seamless integration with existing systems.

As a significant step for achieving such a system beyond 3G mobile communication systems, 3GPP is specifying a new radio access technology, known as *Long Term Evolution* (LTE) to ensure the competitiveness of the 3GPP technology family for the long term. LTE air interface and its radio access as called *Evolved UMTS Terrestrial Radio Access Network*

(E-UTRAN) are specified in the new evolved 3GPP Release 8 of the UMTS standards. To support the LTE radio interfaces and the E-UTRAN, 3GPP is also specifying a new Packet Core, the *Enhanced Packet Core* (EPC) network architecture. The work specifying the core network is commonly known as *System Architecture Evolution* (SAE).

LTE employs a completely new air interface technology which is different to the existing 3G technologies. It uses *Orthogonal Frequency Division Multiplex* (OFDM) technologies for the downlink and *Single-Carrier FDMA* (SC-FDMA) for the uplink. It employs *Multiple-Input / Multiple-Output* (MIMO) with up to four antennas per station which significantly increases throughput rates. With the new air interface technologies, LTE can achieve up to 100Mbps on the downlink and up to 50Mbps on the uplink. The spectral efficiency and the average user throughput in the LTE downlink will be 3 to 4 times of that of HSDPA while in the uplink, it will be 2 to 3 times that of HSUPA. More details on the LTE air interface technology can be found in [Mot07].

E-UTRAN is expected to substantially improve end-user throughputs and reduce user plane latency, bringing significantly improved user experience with full mobility. This can be achieved by introducing new, fully IP-based networks which will have a flat architecture with the *enhanced Node B* (eNode B) directly connected to the *access gateway* (aGW), where the Iub interface with its stringent delay constraints is no longer needed. The architecture of the LTE radio access network is illustrated in Figure 2.4. The eNode B is in charge of single cell *Radio Resource Management* (RRM) decision, handover decision, scheduling of users in UL/DL, etc. The aGW provides termination of the LTE bearer and acts as a mobility anchor point for the user plane. The eNode B is connected to the aGW with the *S1* interface. Between the eNode Bs the *X2* interface is defined, which is used to connect the eNode Bs with each other in the network.

Unlike HSPA, which was accommodated within the Release 99 UMTS architecture, 3GPP defines a new Packet Core, the EPC network architecture, to support the E-UTRAN. The EPC is designed to have a simplified network architecture with a reduction in the number of network elements, simpler functionality and improved redundancy. And most importantly it allows for connections and handover to other 3G networks and non-3GPP fixed line and wireless access technologies (e.g. WLAN, WiMAX), giving the service providers the ability to deliver a seamless mobility experience. Additionally, to reduce the operator system cost LTE/SAE will provide a smooth evolution path for the existing 2G and 3G networks, so that the inter-working with UMTS Rel99/HSDPA/HSUPA and GSM/GPRS is possible.

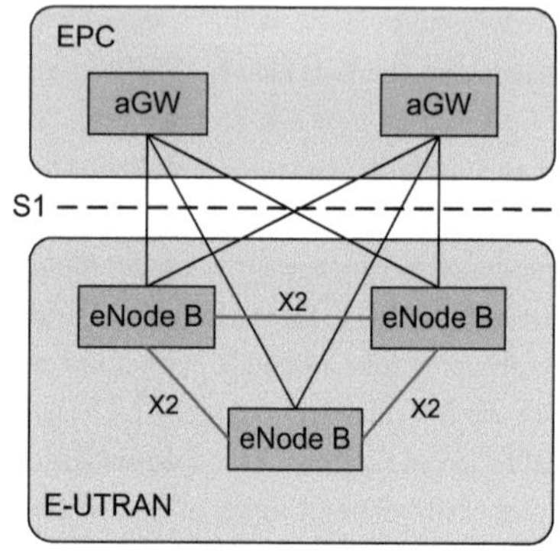

Figure 2.4: *LTE radio access network*

Though LTE/SAE is not a major concern of this thesis, it is worthy to mention here as it is so far the latest evolution of UMTS which is going to receive a lot of attention in the upcoming years. By understanding its goals and targets on increasing efficiency, reducing costs, improving services, making use of new spectrum opportunities, and better integration with other open standards, it helps to recognize the new requirements imposed by the LTE/SAE in the transport network and in the core network in the future.

3 UMTS Terrestrial Radio Access Network (UTRAN)

This chapter gives a more detailed and extensive introduction of the *UMTS terrestrial radio access network* (UTRAN) including its basic network elements, interfaces, transport channels, protocol stacks, and key management functions. Along with the evolution of UMTS, the UTRAN is under development aiming for an efficient network architecture, much higher transport bandwidths, improved transport network cost-efficiency, and enhanced support for the packet data services. This chapter starts with a detailed introduction of the UTRAN from UMTS Rel99 and then proceeds further towards evolved UTRAN networks along the UMTS evolution path.

3.1 UTRAN Architecture and Interfaces

The UTRAN has been introduced in chapter 2.2. Its architecture is highlighted in Figure 3.1. UTRAN consists of one ore more *Radio Network Subsystems* (RNS) each containing a *Radio Network Controller* (RNC) and a group of Node Bs.

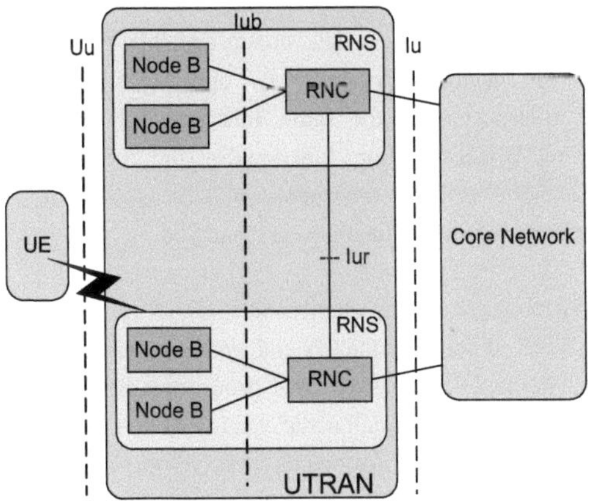

Figure 3.1: *UMTS terrestrial radio access network: nodes and interfaces*

As already introduced in chapter 2.2, there are two internal interfaces defined within the UTRAN:

- The Iub interface is used for communication between the Node B and the RNC. Each Node B is connected with its controlling RNC via an Iub interface.
- The Iur interface is used for communication between different RNCs. It is new to UMTS as compared to GSM, and is typically (but not exclusively) used to support mobiles that are in soft handover to Node Bs controlled by different RNCs without going through the *Core Network* (CN).

In addition the UTRAN defines two external interfaces towards the CN and towards the UE respectively:

- The Uu interface realizes the radio connection between the UE and the Node B. It is the radio interface between the UTRAN and the UE, based on WCDMA technology.
- The Iu interface establishes the communications between the RNC and the CN. The Iu divides the system into radio-specific UTRAN and CN which handles switching, routing and service control. The Iu is logically divided into two different instances: Iu-CS (Iu-Circuit Switched) for connecting UTRAN to the CN circuit switched domain and Iu-PS (Iu-Packet Switched) for connecting UTRAN to the CN packet switched domain.

All these interfaces defined in the 3GPP specification are open interfaces, this ensure the compatibility between equipment of different manufacturers and allows multivendor scenarios in the network configuration.

A Node B is handling transmission and reception in one or several cells. It is logically correspond to the *Base Transceiver Station* (BTS) in GSM. In contrast to GSM base station, Node B uses WCDMA as the air transport technology. The Node B converts the radio frames received from the radio interface into a data stream and then forwards it to the RNC via the transport channels over the Iub interface. In the opposite direction, the Node B prepares incoming data for transport over the radio interface and transmits them to the UE. Traditionally, the Node B has minimum functionality and is mainly controlled by an RNC. However, this is changing with the emergence of HSPA (HSDPA/HSUPA), where some control and resource management functions are handled by the Node B to achieve lower system latency.

The RNC is the central controlling element in UTRAN responsible for the control of the radio resources of UMTS. It connects the CN and also terminates the *Radio Resource Control* (RRC) protocol that defines the messages and procedures between UE and UTRAN. The RNC is essentially in responsible for call setup, quality of service handling, and management of the radio resources in the cells for which it is responsible by performing the *Radio Resource Management* (RRM) functions like admission control, power control, code allocation, radio bearer setup and release, handover, etc. The RNC usually controls multiple Node Bs. The number of Node Bs connected to one RNC varies depending on the implementation and deployment.

3.2 Generic Protocol Model for UTRAN Interfaces

Figure 3.2 presents the general UTRAN interface protocol model. It horizontally separates all the UTRAN related functionalities from the underlying terrestrial transport technology into two layers: *Radio Network Layer* (RNL) and *Transport Network Layer* (TNL), respectively.

All UTRAN related issues are visible only in the Radio Network Layer. The Radio Network Layer comprises the protocols that are especially designed for the UMTS system, dealing with the management and use of *Radio Access Bearers* (RABs) across the UTRAN. It is independent from the underlying transport technology.

The Transport Network Layer represents standard transport technology that is selected to be used for UTRAN but without any UTRAN-specific changes. It is responsible for the management and use of *transport bearers*, including the signaling, establishment, maintenance and release of transport bearers.

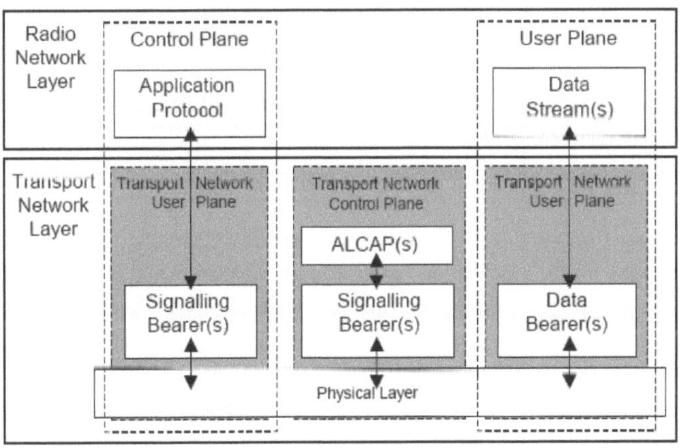

Figure 3.2: *General Protocol Model for UTRAN interfaces [3GP02c]*

In addition to horizontal layers, the UTRAN protocol structure is vertically divided into two planes: *Control Plane* and *User Plane*. The Control Plane consists of *Application Protocols* and *Signaling Bearers*. The Application Protocols take care of all signaling functionality in the UTRAN required for setting up and maintaining *Radio Access Bearers* and *Radio Links*, which include setting up dedicated signaling and user connections and exchanging other UTRAN-specific control information. The Signaling Bearers are used for transporting the application protocol messages. The involved application protocols in the control plane are:

- NBAP (*Node-**B** Application **Part***) is the signaling protocol responsible for the control of the Node B by the RNC. The NBAP protocol is carried out over the Iub interface.

The NBAP protocol is subdivided into Common and Dedicated NBAP (C-NBAP and D-NBAP), where C-NBAP controls the overall Node B functionality and D-NBAP controls separate cells or sectors of the Node B.

- ALCAP (*Access **L**ink **C**ontrol **A**pplication **P**art*) is the control plane protocol for the transport layer. It is used to set up the transport bearers (data bearer) for the user plane. It is also carried over the Iub interface.
- RNSAP (***R**adio **N**etwork **S**ubsystem **A**pplication **P**art*) is in charge of communications between RNCs and is carried on the Iur interface. Unlike NBAP, RNSAP is a symmetric protocol, thus run by two RNCs, of which one takes the role of SRNC and the other acts as DRNC.
- RANAP (***R**adio **A**ccess **N**etwork **A**pplication **P**art*) handles communication between RNC and the core network and is carried over the Iu interface. It is used to set up and maintain Control Plane and User Plane connections across the Iu interface thus handling communication between UTRAN and CN.

Whereas the Control Plane performs all control functionality, the actual user traffic, such as coded voice or IP packets, is conveyed through the User Plane. The User Plane consists of the *Data Streams* and *Data Bearers*. Data Streams contain the user data that are transparently transmitted in the UTRAN. The Data Steams are formed on the Radio Network Layer characterized by one or more frame protocols specified for that interface. The Data Bearers are used to transport user data (Data Streams) across the interface on the Transport Network Layer.

The *Transport Network User Plane* includes the Data Bearers in the User Plane and the Signaling Bearers for the Application Protocol in the Control Plane. The *Transport Network Control Plane*, which is located between the Control Plane and the User Plane, is used for all control signaling within the Transport Layer. It does not include any Radio Network Layer information. It is responsible for the transport bearer management, which includes the signaling, establishment, maintenance and release of transport bearers for the user plane. The Transport Network Control Plane includes the ALCAP protocol that is needed to set up the transport bearers (Data Bearer) for the User Plane, and also includes the Signaling Bearers needed for the ALCAP.

The following sections give a detailed introduction of the UTRAN network structure and protocol stacks of different UMTS releases. It starts from the first UMTS release, i.e. UMTS Rel99, which is considered as the reference model for the evolved UTRAN networks. The introduction of the HSPA (HSDPA / HSUPA) networks is given in Appendix A.9.

3.3 UMTS Release 99

UMTS Release 99 (Rel99) is established as the first UMTS version specified by 3GPP. It contains all the basic elements to meet the requirements for IMT-2000 technologies. As introduced in section 2.3.1, UMTS Rel99 selected the ATM as the transport technology for the UTRAN transport network.

3.3.1 UTRAN Protocol Stack

This subsection introduces the UTRAN protocol stack in the Control Plane and the User Plane respectively.

Protocol Stack for the Control Plane

Figure 3.3 shows the UTRAN protocol stack in the control plane.

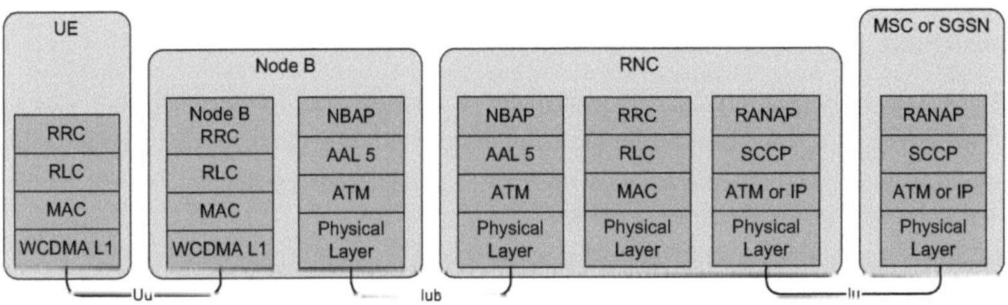

Figure 3.3: *Control plane protocol stack in UTRAN*

The signaling protocols in the control plane of UTRAN consists of NBAP on the Iub interface, RANAP at the Iu interface handling communication between RNC and CN, and RNSAP for the Iur interface in charge of communications between RNCs. NBAP uses ATM together with the AAL5 as the underlying transport network layer to transport signaling messages over the Iub interface. While at the Iu interface, the RANAP is conveyed by *Signaling Connection Control Part* (SCCP) and ATM or optionally IP as the underlying transport layer.

The Radio interface protocols establish, adapt, and free radio bearer services in the UTRAN platform. They have functions in Layer 1-3, i.e. physical (L1), data link (L2), and network layer (L3) according to the OSI model. The related radio interface protocols in the control plane are:

- The *Radio Resource Control* (RRC) is a signaling protocol in the control plane in L3. It is used to set up, and maintain the dedicated Control and User Planes radio specific connections between UE and RNC, i.e. the radio bearers, transport and physical channels between UE and RNC [3GP04d].
- The *Radio Link Control* (RLC) protocol performs the data link layer (L2) functionality [3GP04b]. It is in charge of flow control procedures, concatenation, segmentation and reassembly of higher-levels protocol messages and the reliable transmission of data. It

supports transfer of user data in three different modes: *transparent mode* (TM), *unacknowledged mode* (UM), and *acknowledged mode* (AM).

- The *Medium Access Control* (MAC) protocol is the second sublayer of the data link layer (L2). It has the responsibility for controlling the communications over the WCDMA transport channels provided by the physical layer [3GP04c]. The MAC layer maps *logical channels* to *transport channels* and performs the scheduling of radio bearers (or logical channels), as well as the selection of the data rates being used, i.e. selection of the transport format (TF) being applied. It should be noticed that both RLC and MAC protocols are also used in the user plane.
- WCDMA is the physical layer (L1) that is based on WCDMA technology. It is the air interface between the UE and the base station (Node B).

Protocol Stack for the User Plane

As introduced in section 3.2, the user plane is used for transferring user data. The user data transferred across the Iub can be basically divided into the circuit-switched data such as voice and packet-switched data such as ftp or web applications carried by TCP/IP transport protocols. Figure 3.4 presents the UTRAN protocol stack of the user plane for the packet-switched domain and Figure 3.5 for the circuit-switched domain.

The Radio Network Layer within the user plane consists of *Packet Data Convergence Protocol* (PDCP), RLC, MAC, and a set of *Frame Protocols* (FP) each for the corresponding transport channel.

- The *Packet Data Convergence Protocol (PDCP)* makes the UMTS radio interface capable to carry IP data packets [3GP02d]. It also performs IP header compression and decompression. The compression reduces the length of the header to a smaller number of bytes, thereby contributing to increased efficiency and a higher data rate, which is particularly important for small data packets. It is used specifically for the packet-switched domain.
- The RLC and MAC protocols have already been briefly introduced in the control plane protocol stack.
- The *Frame Protocol* (FP) is defined at the Iub interface responsible for the relaying of transport channels between the UE and the RNC via the Node B in order to extend the radio transport channels from Node B to RNC. The set of FP define the structure of the frames and the basic inband control procedures for each type of transport channel such as the Dedicated Channel, etc.

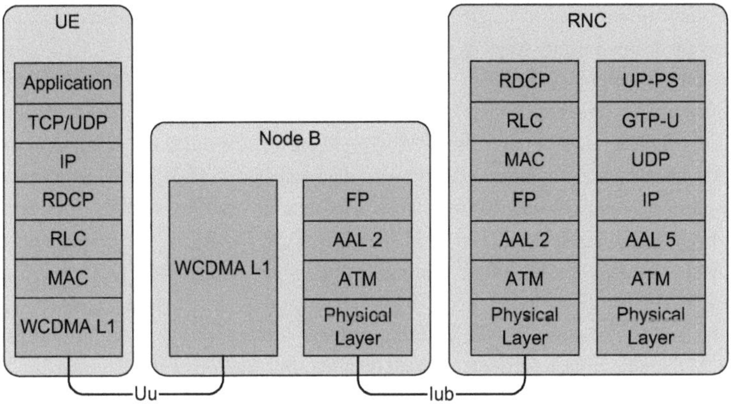

Figure 3.4: *User plane protocol stack in UTRAN (Packet-Switched Domain)*

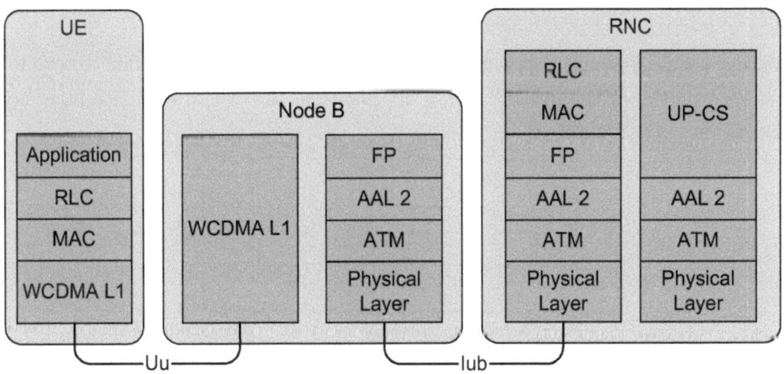

Figure 3.5: *User plane protocol stack in UTRAN (Circuit-Switched Domain)*

The transport network of UTRAN in Rel99 is built with *Asynchronous Transfer Mode* (ATM) transport technology. Since the ATM is suitable to carry traffic of different natures, it needs the adaptation layers for particular type of information flow. Therefore, above the ATM layer the *ATM adaptation layer* (AAL) protocol is employed. In the UTRAN transport network, *AAL Type 2* (AAL2) is chosen for the user plane of the Iub interface. The AAL Type 2 signaling protocol is defined by the *International Telecommunication Union Telecommunication Sector* (ITU-T) in recommendation Q.2630.1 (Capability Set 1) [ITU99] and recent enhancements in Q.2630.2 (Capability Set 2) [ITU00]. The reason for choosing AAL2 for the user plane at the Iub interface is that it allows the efficient multiplexing of several data flows over a single ATM virtual channel and is designed for efficient transmission of low-bit-rate services with stringent delay requirements.

The major difference between the packet-switched domain and the circuit-switched domain lies in the Transport Network Layers used for the Iu interface. In the Packet-Switched domain, the Iu interface employs the *GPRS Tunneling Protocol* (GTP), which is a transport network protocol that provides connectionless data transfer. GTP-U is one of the forms of GTP used in the core network for transfer of user data in separated tunnels. It is carried by UDP/IP and in turn transported over ATM. The *AAL Type 5* (AAL5) is selected since it is the usual AAL protocol for IP traffic transport. But in the case of the Iu Circuit-Switched interface, AAL2 is a good option for the efficient transport of circuit-switched data.

In the framework of this thesis, the Iub user plane is the major concern for dimensioning of the UMTS radio access network. A more detailed introduction of the Iub interface will be presented in section 3.3.3.

3.3.2 Channels in the UTRAN

When a user requests to establish a connection to transport data across the radio access network, the UTRAN need to allocate radio and transport resources for that connection. The resource allocations are handled by setting up channels. In UMTS, there are three layers of channels, as shown in Figure 3.6.

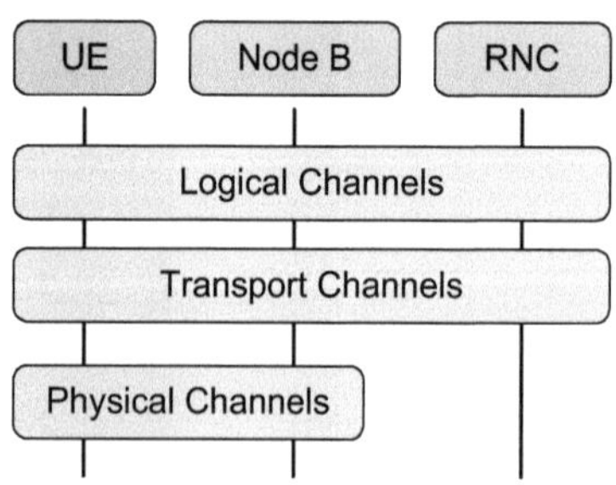

Figure 3.6: *Channel types in the UTRAN (UMTS Rel99)*

On the highest layer, the MAC layer provides data transfer services to the RLC layer by means of *Logical Channels*. A set of Logical Channel types is defined for different kinds of data transfer services as offered by MAC. Each Logical Channel type is defined by *what type of information* is transferred. It shall be noticed that Logical Channels are not actually channels. They can be interpreted as different logical tasks that the network and the terminal should carry out. Once the data is on the Logical Channel, the physical layer in turn offers information transfer services to the MAC layer via *Transport Channels* that are described by *how and with what characteristics* data is transferred over the radio interface. The Transport Channels performs actual information transfer between the UE and the access domain. Over the air interface, the Transport Channels are then mapped in the physical layer to different *Physical Channels*. As can be seen from Figure 3.6, Physical Channels are only present in the air interface whereas Logical Channels and Transport Channels are maintained between the UE and the RNC.

More information about the defined Logical Channels, Transport Channels, Physical Channels and the mapping between them can be referred to [3GP99b] and [3GP99c]. As mentioned in section 2.1, this thesis only considers UTRA-FDD mode. Figure 3.7 shows the corresponding Logical Channels, Transport Channels, and the mappings between them (refer to [3GP99b]).

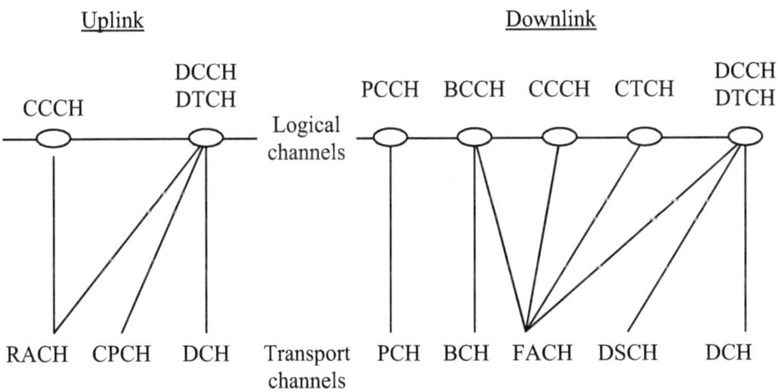

Figure 3.7: *UMTS Rel99: Channel mapping in the uplink and downlink*

According to the nature of transferred information, the transport channels can be divided into two groups: *common transport channels* shared by more than one terminal, and *dedicated transport channels* used for connecting individual terminals with the network. Each group includes different channels in uplink and downlink directions.

The Transport Channels in **downlink** direction are:

- ***Broadcast Channel*** (**BCH**) is a common channel used to transmit system information, e.g. code values used in the current and neighboring cells, allowed power levels, etc. Such data is broadcasted over the whole cell.
- ***Forward Access Channel*** (**FACH)** is a downlink low rate common channel used to transmit downlink signaling, e.g. for call setup and additionally small amounts of user data can be transmitted over this channel simultaneously.
- ***Paging Channel*** (**PCH**) is a common channel used by the network to reach a UE that is not currently maintaining an RRC connection with the radio access domain. The address

of the mobile terminal (UE) is paged once or several times. All terminals continuously listen to the PCH to pick up the paging request for them.

- ***Downlink Shared Channel*** (**DSCH**) is a common channel shared by several users, for whom the amount of information in downlink is small so that there is no need for the allocation of a dedicated channel. However, this channel is not mandatory for the operation of the network and has not been widely used commercially. With the introduction of HSPA in Release 5 and 6, it becomes obsolete.
- ***Dedicated Channel*** (**DCH**) is the only dedicated transport channel. It carries user data and higher layer control information coming from the *Dedicated Traffic Channels* (DTCH) and *Dedicated Control Channels* (DCCH) as logical channels. Moreover it can carry the information from several DTCHs. For instance a user may have several connections for different services (e.g. voice and video call). Each service requires an individual DTCH, but all of them share the same DCH. It should be noticed that DCH is the dynamically allocated resource, it can be set up on demand for one user and then released, whereas common channels basically exist permanently.

The traffic in the uplink direction is usually smaller, hence it requires less transport and physical resources. There are three Transport Channels available in **uplink**:

- ***Dedicated Channel*** (**DCH**) is a duplex channel. On the uplink it has the same function as a DCH in the downlink direction.
- ***Random Access Channel*** (**RACH**) is a common channel used to transmit control information from the mobile terminal to the network, e.g. a UE sends the initial access information via RACH to the network when it requests to set up a connection. RACH can also be used to send small amounts of packet data from the terminal to the network. For proper system operation the RACH must be heard from the entire cell coverage area, which also means that data rates have to be rather low [HT04].
- ***Common Packet Channel*** (**CPCH**) is a shared channel which carries small amounts of user packets on the uplink. Similar to DSCH on the downlink, it is also optional for the network operation and becomes obsolete with the emergence of HSPA.

3.3.3 Iub Interface

This section gives a detailed introduction of the Rel99 Iub interface, including the protocol stack used for the user plane, the underlying ATM transport technology and the involved protocols within the Transport Network Layer. At the end, the explicit transport requirements that the Iub interface has to satisfy are described.

3.3.3.1 Protocol Stack for the Iub User Plane

The major tasks of the Iub interface are to provide the means for transporting the uplink and downlink transport channel frames between RNC and Node B. This is performed by a set of data streams provided by the Iub interface: Iub DCH data stream (a bi-directional transport channel); Iub RACH data stream; Iub FACH data stream; Iub DSCH data stream; Iub CPCH data stream; and the Iub PCH data stream.

As shown in Figure 3.8, each of such Iub Data Streams is conveyed by the *Frame Protocol* (FP) and thus a set of FPs reside within the Iub user plane to provide the means to deliver the actual traffic on the transport channels. The Frame Protocol is in charge of putting the data stream to be transported on the corresponding transport channel over the Iub interface. Thus for each type of transport channel there is a certain type of FP. They all provide a basic service: the transport of *Transport Block Set* (**TBS**) between the Node B and RNC across the Iub interface. *Transport Block* (**TB**) is defined as the basic unit passed down to L1 (physical layer) from the MAC layer. An equivalent term for the Transport Block is *MAC PDU* (Protocol Data Unit). The Transport Block Set in turn is defined as a set of Transport Blocks which is passed to L1 from MAC at the same time instance using the transport channel. An equivalent term for Transport Block Set is *MAC PDU Set*. According to the UTRAN protocol structure, the MAC protocol is terminated at the RNC. That means MAC PDUs are not carried explicitly over the Iub interface. Instead, the information contained in the MAC PDUs is conveyed by the Frame Protocols from the RNC to the Node B in order to pass it further towards the UE across the air interface.

Frame Protocols are divided into two groups: FP for Dedicated Transport Channel data streams and FP for Common Transport Channels data streams. DCH is the single type of dedicated transport channel, so there is only one FP representing that group, DCH FP. For common transport channel data streams the following Frame Protocols exist: RACH FP, CPCH FP, FACH FP, and DSCH FP.

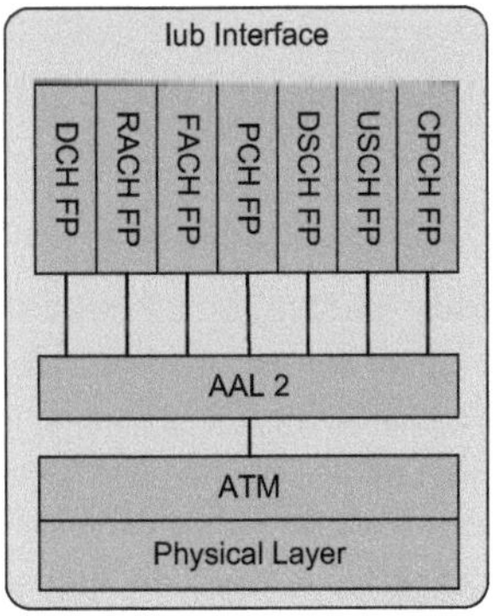

Figure 3.8: *Iub interface - user plane protocol stack*

As also shown in Figure 3.8, at the Iub user plane AAL2 and ATM protocols are used at the Transport Network Layer to transport MAC frames between Node B and its associated

RNC. Thereby, each Frame protocol stream is carried by an AAL2 connection over the ATM across the Iub. A detailed introduction of the ATM layer is presented in section 3.3.3.2, and the AAL2 protocol is given in section 3.3.3.3.

Higher layer PDUs (e.g. IP packets) as RLC SDUs (*Service Data Unit*) are segmented into a number of RLC blocks. After segmentation, RLC headers are added to form RLC PDUs. Then the RLC sends these RLC PDUs to the MAC layer over logical channels. In the MAC layer, each RLC PDU is put into a MAC PDU also called Transport Block. Here the MAC is transparent, i.e. without adding MAC headers. The MAC, in turn, performs scheduling and Transport Format selection for each specific transport channel. The Transport Format is defined as a format offered by L1 to the MAC (and vice versa) for the delivery of a Transport Block Set during a *Transmission Time Interval* (TTI) on a Transport Channel. The Transport Format defines the parameters for a specific transport channel, such as Transport Block Size and Transport Block Set Size, Transmission Time Interval which determines the time period between two consecutive TBS that arrives at the air interface. The feasible TTIs are 10, 20, 40, and 80 ms. Dependent on the transport format selection by the MAC layer, a set of MAC PDUs make a Transport Block Set which is sent at every TTI from the MAC layer down to the physical layer. In other words, the MAC layer groups a number of MAC PDUs from the same Transport Channel into one Transport Block Set and forwards it per TTI to the WCDMA air interface.

In the UTRAN Iub interface, the Transport Block Sets are handed over to the Frame Protocol layer. Thus, the information contained in the TBS is conveyed by the Frame Protocol to be transferred through the UTRAN. The FP layer adds its additional header to the TBS and in turn delivers it down to the Transport Network Layer in the form of FP PDUs. In the transport network layer, the AAL2 protocol segments the FP PDUs into small AAL2 packets. These AAL2 layer packets are then packed into ATM cells before being transmitted to the ATM link.

3.3.3.2 ATM

Asynchronous Transfer Mode (ATM) is a connection-oriented, packet-switched technology designed for the high-speed transfer of voice, video, and data through public and private networks using *cell relay* technology. It uses very short fixed-length (53 bytes) packets, called cells, to transfer data traffic. The first 5 bytes of the ATM cell contain header information, such as the connection identifier, while the remaining 48 bytes contain the data or payload (the ATM cell format is given in Appendix 0). For the fixed-size cells the ATM switch does not have to detect the size of a unit of data, therefore switching can be performed efficiently. Moreover, the small size of the ATM cell makes it well suited for the transfer of real-time data, such as voice and video, which is intolerant of delays.

ATM defines two levels of virtual connections: *Virtual Paths* (VP) and *Virtual Channels* (VC). The connection is identified by two values in the cell header: the *Virtual Path Identifier*

(VPI) and the *Virtual Channel Identifier* (VCI). A virtual path (VP) is a bundle of virtual channels, all of which are switched transparently across the ATM network based on the common VPI.

ATM provides several types of logical connections between two end users, which can be set up statically or dynamically. The *Permanent Virtual Circuits* (PVCs) or *Permanent Virtual Paths* (PVPs) are usually created long before it is used and remains in place until the connection is deprovisioned. Bandwidth is allocated for them whether it is used or not. In this way they are similar to leased lines. The *Switched Virtual Circuits* (SVCs) in contrary are dynamic connections. They are established and released on demand and remain in use only as long as data is being transferred.

One of the main advantages of ATM is that it provides a sophisticated QoS support for the transfer of various types of services with individual QoS requirements. The ATM forum defines five different service categories. Each class is designed to accommodate data bursts according to customer needs and provide the appropriate quality of service (QoS) for each service class. The five service categories are *Constant Bit Rate* (CBR), *Real-Time Variable Bit Rate* (rt-VBR), *Non-Real-Time Variable Bit Rate* (nrt-VBR), *Available Bit Rate* (ABR), and *Unspecified Bit Rate* (UBR). More detailed definitions for these service categories can be found in Appendix A.3. Each ATM connection contains a set of parameters that describe the traffic characteristics of the source (see Appendix A.3). The ATM network checks with the source to determine the specific traffic requirements according to the traffic characteristics. When each connection is set up, traffic parameters are determined and a traffic contract is made. Then the ATM network shall provide the QoS that falls in the agreed traffic contract.

3.3.3.3 AAL2 Protocol

The ATM Adaptation Layer (AAL) is designed to support different types of traffic and applications. The AAL performs functions required by the user, control and management planes and supports the mapping between the ATM layer and the next higher layer. The main services provided by the AAL are segmentation and reassembly, handling of transmission errors and losses, timing and flow control. AAL2 (AAL type 2), specified in ITU-T recommendation I.363.2, has two main advantages:

(1) AAL2 allows variable packet length within cells and across cells. The packet size is controlled by a system defined maximum AAL2 packet length and the duration of the AAL2 multiplexing timer (called CU Timer). Thus it makes an easy control of the packetization delay by letting the application choose the most convenient maximum packet size for its delay requirements. The lower the size, the lower is the delay. Nevertheless, lower sizes lead to an increase of the relative overhead as each packet has a dedicated header.

(2) AAL 2 provides for multiplexing several AAL2 connections over a single *Virtual Channel Connection* (VCC). That means, instead of having user data partially filling an ATM cell,

AAL2 will fill the cell with data from other several active AAL2 user connections. Thereby, high bandwidth efficiency is gained by efficient use of transport resources.

With the above features, the AAL2 protocol is suitable for bandwidth efficient transmission of low-rate, short and variable packet length in delay sensitive applications such as telephony in UMTS. The exact function of the AAL2 protocol is illustrated in Figure 3.9. The first step is generating an AAL2 *Common Part Sublayer* (CPS) packet flow from each user data flow (i.e. FP stream). Each AAL2 CPS packet contains a 3-octet header and a payload of up to 45 (or 64) octets. The header of the AAL2 CPS packet consists of an 8-bit *Channel Identifier* (CID), 6-bit *Length Indicator* (LI), 5-bit *User-to-User Indication* (UUI) and 5-bit *Header Error Control* (HEC). The CID field uniquely identifies the individual user channels within the AAL2, and allows up to 248 individual AAL2 user connections within each AAL2 structure. The LI field is used to indicate the actual length of the CPS packet payload associated with each individual user. Next, AAL2 CPS packets belonging to different flows are multiplexed and encapsulated into a 47-octet block. Each block is preceded with a 1-octet ***Start Field*** containing a pointer (useful for resynchronizing after a cell loss), to form a 48-octet AAL2 CPS PDU. Each CPS PDU is exactly an ATM cell payload of 48 octets, to which a 5-octet ATM header will be attached. The format of an AAL CPS packet and AAL2 CPS-PDU is given in Appendix A.2.

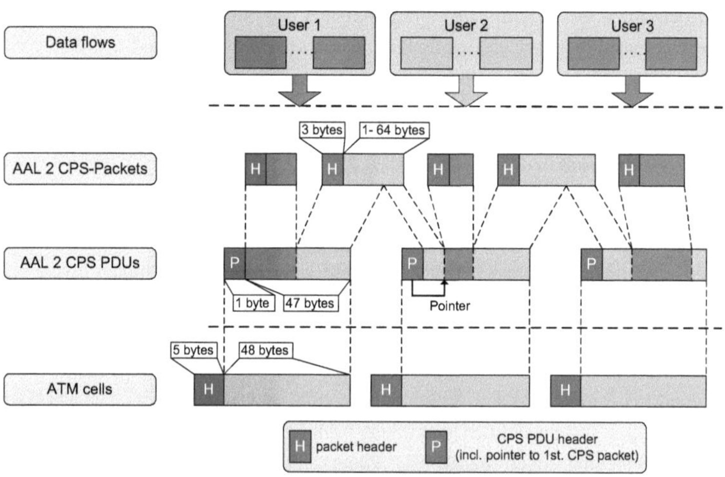

Figure 3.9: *AAL2 multiplexing*

3.3.3.4 Transport Requirements on the Iub Interface

The Iub interface is responsible for providing an efficient, seamless transport between the Node B and the RNC in order to meet the requirements of WCDMA radio interface. For fulfilling this task, the transport of the Iub interface needs to satisfy the following requirements:

1) Low delay and delay variations across the Iub
2) Low packet losses
3) Bandwidth efficient transport

As already addressed in Chapter 1, delay is an issue of importance for the UTRAN Iub interface. In order to support WCDMA radio functions such as soft handover, power control, packet scheduling, radio channel allocation, etc., it is necessary to ensure that the Node B receives downlink frames within an exact time window to assure this radio frame to be able to be delivered to the UE according to the time requirements of the air interface. Whenever a radio frame arrives too late at the Node B, i.e. excessively delayed frames which are behind the scheduled transmission time, then this radio frame is no longer valid and is immediately discarded by the Node B. This leads to strict delay and delay variation requirements on the Iub interface for the transport of all types of user traffic: real-time as well as non real-time. The Iub delay mainly consists of the AAL2 multiplexing/demultiplexing delay, ATM transmission delay, propagation delay over cabled networks, switching delay due to intermediate switching nodes along UTRAN terrestrial interfaces. 3GPP TR 25.853 [3GP01c] provides the detailed delay budgets within the access stratum.

In addition to a low delay and delay variation, there should be a moderately low packet losses rate on the Iub interface, e.g. less than 1%. Because the discarded packets result in a degraded user performance, and moreover it triggers RLC retransmissions if with RLC *Acknowledged Mode* (AM). The retransmissions of the radio frames will waste the limited transport resources in the UTRAN and lead to a higher delay and lower QoS for the end user. The main reasons for packet losses on the Iub are: (1) the excessively delayed frames that are discarded by the Node B; (2) long queuing delays due to the congestion on the transport link; (3) limited buffer space at the ATM switch. The packet losses on the Iub link need to be as low as possible. So in order to reduce the packet losses on the Iub interface, the transport resources such as transport link bandwidths, buffer space and network structure need to be properly designed.

As mentioned in Chapter 1, the transport resources of the Iub links are considerably costly. Therefore, effective bandwidth utilization is directly related to the transport costs. High bandwidth efficiency will significantly reduce the radio access network costs.

In summary, the key objective of the transport at the Iub interface is to provide transmission in a cost efficient way in terms of high transport efficiency and maximal utilization of the Iub bandwidth while guaranteeing stringent requirements on delay, jitter, and loss ratio. For achieving this goal, a proper dimensioning for the Iub interface is necessary.

3.3.4 Resource Management

In UMTS, the resources on the radio interface as well as the transport network are limited and expensive. In order to achieve an efficient utilization of the resources while at the same time provide guaranteed QoS, a proper resource management is required. The main function of resource management is to (1) allocate, establish/release, modify, maintain, manage and schedule the available resources dynamically for each connection upon their requests and demands; (2) control the total amount of resources according to the available system capacity and protect from overutilization of the resources; (3) provide QoS for various services and traffic types in the network, and guarantee their QoS requirements.

Within UTRAN, there are two major resources: radio resources at the air interface and transport resources in the terrestrial radio access network. Accordingly, there are specific resource management functions for organizing each resource. The following subsections introduce the associated resources management functions for each type of the resource in the UTRAN.

3.3.4.1 Radio Access Bearer (RAB)

Before addressing the radio resource management functions, it is necessary to introduce the *Radio Access Bearer* (RAB). When one user requests a connection, this connection requires resources to be dedicated in order to transfer the information flows across the UMTS network. The RABs are the UTRAN resources, which are dedicated to transport user information for particular user connections across the UTRAN network, and it can be viewed as a service, which UTRAN provides to CN and UE.

Radio Access Bearer Service

The *Radio Access Bearer Service* is characterized by a set of attributes such as traffic class, maximum bit rate, guaranteed bit rate, maximum SDU size, etc., which define that particular traffic aspect or Quality of Service profile of that particular application or service [3GP06c].

Traffic classes are explained in Table 3.1. The main distinguishing factor between these traffic classes is how delay sensitive the traffic is: The Conversational class is for traffic which is very delay sensitive, while the Background class is the most delay insensitive traffic class. Both the Conversational and Streaming RABs require a certain reservation of resources in the network, they are mainly intended for carrying *Real-Time* (RT) traffic flows, such as video telephony and audio/video streams. The Interactive and Background RABs are so-called *Best Effort* (BE), i.e. no resources are reserved and the throughput depends on the load in the cell. They are mainly used for carrying Internet applications like web, email, and ftp. Due to less stringent delay requirements, compared to Conversational and Streaming classes, both Interactive and Background RABs provide better error rate by means of channel coding and retransmission. But traffic in the Interactive class has higher priority in scheduling than Background class traffic, so background applications use transmission resources only when interactive applications do not need them.

Traffic class	Fundamental characteristics
Conversational class	– Preserve time relation (variation) between information entities of the stream – Conversational pattern (stringent and low delay) – Example: speech, video
Streaming class	– Preserve time relation (variation) between information entities of the stream (i.e. some but constant delay) – Example: streaming audio and video
Interactive class	– Request response pattern – Preserve payload content – Example: web browsing
Background	– Destination is not expecting the data within a certain time – Preserve payload content – Example: background download of emails or file downloads

Table 3.1: *Traffic classes [3GP06c]*

Maximum bitrate (kbps) is the peak data rate a user/application can transmit. It is defined as maximum number of bits delivered by UMTS and to UMTS at a *Service Access Point* (SAP) within a period of time, divided by the duration of the period.
Guaranteed bitrate (kbps) is defined as guaranteed number of bits delivered by UMTS at a SAP within a period of time (provided that there is data to deliver), divided by the duration of the period. It describes the bitrate the UMTS bearer service must guarantee to the user or application. That means, UMTS bearer service attributes such as delay and reliability attributes are guaranteed for traffic up to the Guaranteed bitrate. The Guaranteed bitrate may be used to facilitate admission control based on available resources, and for resource allocation within UMTS.
Maximum SDU Size (octets) defines the maximum allowed SDU size. It is used for admission control and policing, and/or optimizing transport.
Other important RAB service attributes such as SDU error ratio, residual bit error ratio, delivery of erroneous SDUs, transfer delay, traffic handling priority etc., are defined in 3GPP TS 23.107 [3GP06c].

Radio Access Bearer Setup and Management

Radio Access Bearer (RAB) is described by a set of parameters to transfer user data between UE (User Equipment) and CN (Core Network) for a specific traffic class. 3GPP TS 34.108 defines a set of RAB types, various RAB combinations on uplink and downlink for transferring different traffic classes, and the detailed transport channel parameters for each individual uplink and downlink RAB type such as RLC PDU payload size, RLC/MAC headers, TFS (Transport Format Set), TTI (Transmission Time Interval), etc. Based on the given transport channel parameters, the maximum or peak data rate of the RAB is determined. Table 3.2 gives some examples of RAB types defined in 3GPP TS 34.108 [3GP04e] and their detailed transport channel parameters.

	Service (3GPP 34.108)	Peak rate [kbps]	Payload [bit]	RLC [bit]	Number of Transport Blocks	TTI [ms]
CS 12.2	Conversational/Speech/UL 12.2 kbps/ CS RAB	12.2	256	0	1	20
CS 64	Conversational/unknown/ DL 64 kbps/ CS RAB	64	640	0	2	20
PS 64	Interactive or background/DL 64 kbps / PS RAB	64	320	16	4	20
PS 128	Interactive or background/DL 128 kbps / PS RAB	128	320	16	8	20
PS 384	Interactive or background/DL 384 kbps / PS RAB	384	320	16	12	10

Table 3.2: *RABs defined in 3GPP TS 34.108 [3GP04e]*
(In this table, CS refers to Circuit-Switched and PS refers to Packet-Switched)

The selection of an appropriate RAB for a particular user connection is mainly determined by *Radio Resource Management* (RRM) functions applied in the RNC and delimited by the parameters provided by the SGSN during the RAB establishment request. The core network will select a RAB with appropriate QoS based on the service request from the subscriber, and ask the RNC to provide such a RAB. The RAB is chosen depending on the user subscription, service type, requested QoS, the radio situation like pathloss, power utilization within the cell, interference power, etc. For example, the interactive web application can be assigned RAB 64kbps or higher. Conversational voice usually applies the AMR codec with CS 12.2 kbps. A higher RAB rate is suitable for the UEs which are near the mobile station in low loaded radio cells and lower RAB rates are assigned for far-off UEs in high loaded cells.

3.3.4.2 Radio Resource Management (RRM)

The Radio Resource Management (RRM) is responsible for an efficient utilization of the air interface resources and for provisioning and guaranteeing QoS [Tür07]. The UMTS RRM algorithms are separated into two categories: the network based functions and the connection based functions. Network based functions are *Admission Control*, *Load Control* and *Packet Scheduling*, while connection based functions are *Handover* and *Power Control* . The following briefly describes the role of each function:

- Admission Control decides whether a new call can be admitted and/or a current call can be modified, according to the available capacity of WCDMA network in terms of power and allowed interference. The admission control functionality is used to avoid system overload and to provide the planned coverage.
- Load Control resolves and prevents overload or congestion situations in the network, which may occur due to traffic fluctuations, user mobility or channel variations.
- Power Control manages the transmit power of the UE and base station, which results in less interference and allows more users on the same carrier.
- Handover is required in cellular networks to support the user mobility. There are in principle three types of handovers in UMTS networks.
  1. **Hard handover** also exists in GSM networks, that is, at any time only one physical connection is maintained. The handover takes place when the mobile user moves between a WCDMA network and a network of different radio technology (e.g. between UTRAN and GSM), or when the mobile user changes between different carriers of the same WCDMA network.
  2. **Soft handover** is when a mobile station communicates simultaneously with up to three cells from different Node Bs. On the downlink, the data is broadcast over the Node Bs and combined again in the mobile station. On the uplink, the data from all participating Node Bs is received and forwarded to the RNC. The RNC combines the data streams and transfers the data to the CN.
  3. **Softer handover** is a special case of soft handover where transmissions can run in parallel over different cells of the same Node B.

Besides, two Radio Network Layer resource management functions are considered in this thesis, which have direct impact on the UTRAN transport network.

Bit Rate Adaptation (BRA)

In order to improve the effective utilization of the Iub bandwidth on transferring the packet switched data traffic, *Bit Rate Adaptation* (BRA) is usually applied in the UMTS system, specifically for the best effort or interactive packet traffic. The main function of BRA is to dynamically assign the available RAB for user flows during their transmission according to their QoS requirements and their activity. For example, when the user activity is considerably low, a low RAB rate will be allocated. When the user has larger needs of bandwidth for transmitting a large amount of data, a higher RAB will be assigned if there is still adequate bandwidth left on the link. It has to be noted that BRA is applied only for the Dedicated Channel (DCH) transferring the packet switched traffic, mainly best effort and interactive traffic types. Because the packet switched traffic, not like circuit switched traffic, normally does not utilize the resource all the time, only on demand. Therefore, when the user does not have any traffic or little traffic to transmit, the utilization of the assigned resources will be reduced. By using the BRA mechanism, the resource for each user connection can be dynamically allocated according to the user's traffic demand. In this way the total bandwidth utilization can be improved. The data rate adaptation is realized by performing RAB upgrades to increase the data rate and RAB downgrades to decrease the data rate.

The most important issue of BRA is to define the triggering of the upgrade or downgrade for a user connection. The decision should be made based on the activity of the user connection, which can be observed through monitoring the utilization of the RLC buffer located at both RNC and UE sides. The measurement of buffer utilization is illustrated in Figure 3.10. The buffer utilization is monitored based on a configurable time interval (e.g. multiples of 500 ms), which is denoted as "Monitoring Period". The time scale of the monitoring period is divided into frames. In each frame the RLC buffer content is observed. If the RLC buffer is empty within a frame, then it is counted as empty, otherwise as used. Buffer utilization is defined as the ratio between the used frames and all frames.

The triggering of RAB upgrade or downgrade is determined by the thresholds of the buffer utilization: if the buffer utilization is above the upgrade threshold value, e.g. 80%, then the system assumes that the corresponding user's activity is high and therefore tends to upgrade its current RAB to the next higher RAB rate from the available RAB set given in the above table. If the buffer utilization is below a downgrade threshold, e.g. 20%, then the system assumes that the user's activity is low and then tends to downgrade to the next lower RAB rate.

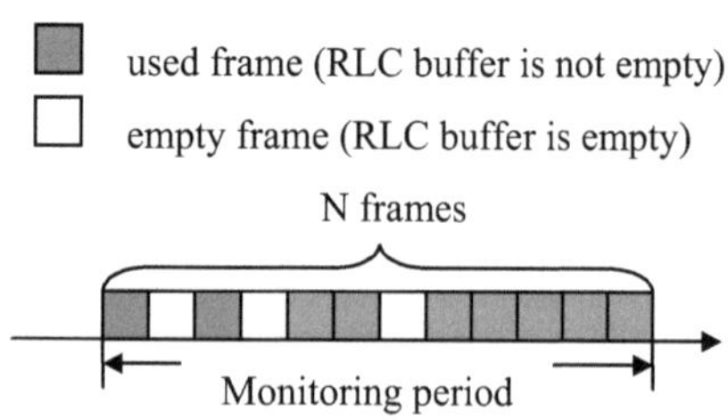

Figure 3.10: *Evaluating the Buffer Utilization*

Furthermore, in order to avoid frequent BRA triggering a "Waiting Time" is executed between two subsequent upgrades or downgrades. The waiting timer starts after an upgrade or downgrade, and before the timer expires no more upgrades or downgrades of the RAB rate are allowed.

Upgrade and downgrade events in uplink and downlink are triggered separately, i.e. there are independent RAB triggers on uplink and downlink. If upgrade in one link direction, either uplink or downlink, also requires an upgrade in the other link direction according to the defined RAB settings on both uplink and downlink given in the above table (although not triggered), then the upgrade is performed. But the downgrade on one link direction should not automatically affect the RAB rate on the other link direction.

Channel Type Switching (CTS)

Channel Type Switching (CTS) is to switch the user data transfer between DCH channel and the common channels, i.e. to FACH on the downlink and RACH on the uplink. The CTS mechanism is used to efficiently utilize the transport bandwidth for the packet switched data services. When one user is transmitting data over the common channels, it competes for the transmission resources with some other common channel users in the cell and only obtains a share of the bandwidth dynamically according to its demand. If the user does not transfer any data for an amount of time, it is assumed that this connection is inactive and does not need dedicated transport bandwidth anymore. Therefore, the system releases the DCH for this user and the corresponding transport connection, and switches it to the common channel. Then the reserved transport bandwidth for this user is released and given for other users to use. In this way, one user will not always occupy the available bandwidth even though it has no data to transfer. The released bandwidth can be reallocated for new user connections or shared by other existing user connections. On the other hand, the user in the common channel can be moved back to the DCH when the system detects this user is active again with increased traffic demand. Then this user will request to set up a DCH and a new transport channel in the transport network. The TNL CAC determines whether there are enough transport resources to allow the user to switch back to DCH. If not, the user has to remain in the common channel until there are enough transport resources again for setting up the transport connection for the requested DCH. With the use of the CTS function, the network resources can be allocated dynamically according to the user traffic demand. This can significantly improves the utilization of the transport resource and save the operation expenses while guaranteeing the desired QoS.

3.3.4.3 Transport Resource Management (TRM)

In contrast to the RRM, Transport Resource Management (TRM) is to control and manage the transport resources in the terrestrial radio access network in order to achieve an efficient use of the transport resources while provide guaranteed QoS. Following gives a detailed introduction of transport network admission control function, *Transport Network Layer Connection Admission Control* (TNL CAC), which is considered for the dimensioning in this thesis.

Unlike the admission control in the RRM group which controls the wireless resources in the WCDMA air interface where the resources are the transmission power and the interference, the Transport Network Layer (TNL) CAC is based on the transport resources supplied by the underlying wired transport network. TNL CAC is used to decide whether there are sufficient free resources on the requested link to allow a new transport connection. A connection can only be accepted if adequate resources are available to establish the connection with promised QoS while the agreed quality of service of the existing connections in the network must not be affected by the new connection. It shall be noted that in this thesis, the management and dimensioning of the transport resource is the major concern, so only TNL CAC is considered in this thesis (the CAC used in the radio interface is not considered in the thesis). In the rest of this thesis, the term CAC means the TNL CAC.

As the transport network within the terrestrial radio access network of UMTS Rel99 is based on ATM, the transport resource is the underlying ATM transport network with installed ATM links, leased ATM circuits or paths and their offered QoS, and intermediate ATM switches/routers. If an ATM VP or VC is used for the Iub interface in the UTRAN transport network, the total resources provided by this VP or VC is determined by the PCR (Peak Cell Rate) and this bandwidth will be reserved for the Iub permanently. When a user tries to set up a new transport connection (i.e. AAL2 connection) in the UMTS network, it requires a certain amount of transport bandwidth, at the Iub to transfer the user data. In order to ensure that the QoS of the ongoing transmissions of other users are not violated and the new connection can have enough bandwidth for guaranteeing its own QoS requirement, each AAL2 connection should request a certain bandwidth that can satisfy the QoS requested by the user. In this context, it is called *CAC guaranteed bit rate*. It is a configurable parameter in the system which is vendor specific. Their settings depend on a number of aspects like the applications, the applied traffic models, the QoS requirements, the RAB type, etc. For an efficient utilization of the offered transport bandwidth, the CAC guaranteed bit rate should be optimized for each traffic class and RAB type. A detailed implementation can be referred to section 5.4

The algorithm of the TNL CAC is described in the following. The TNL CAC accumulates the total reserved bandwidth of the current existing transport connections, named *Total Reserved Bandwidth*. When the new transport connection is requested to enter the network, the TNL CAC compares if the sum of the *Total Reserved Bandwidth* and the CAC guaranteed bit rate of the new connection is larger than the offered Iub bandwidth. If so, that means, there is not enough free bandwidth on the Iub link to set up this new transport connection offering the agreed QoS, this connection has to be rejected by the TNL CAC. Otherwise, this new transport connection will be accepted by the system as there is still sufficient bandwidth to provide for this new connection. In this case, the new connection can be established and the TNL CAC needs to update the *Total Reserved Bandwidth* by adding the allocated bandwidth for the new connection, i.e. the CAC guaranteed bit rate. When a transport connection is terminated, its previously occupied bandwidth shall be released. Thus, the TNL CAC needs to update the *Total Reserved Bandwidth* by decreasing the occupied bandwidth by this released connection. Another factor which is taken into account by the TNL CAC is the maximum number of transport connections supported by the ATM VCs or VPs. That means the TNL CAC also needs to make sure that each AAL2 PVC allows only up to 248 active connections simultaneously.

3.4 IP-Based UTRAN

ATM was selected as the underlying transport technology in the UTRAN transport network since UMTS Rel99. At present, there is a strong trend of replacing the current ATM with IP as the transport technologies in the UTRAN, as stated in section 2.3.4 and 2.3.4.1. 3GPP Release 5 firstly introduces the IP-based UTRAN transport network.

3.4.1 IP-Based UTRAN User Plane Protocol Architecture

The user plane protocol architecture of the IP-based UTRAN is shown in Figure 3.11. By comparing the ATM-based UTRAN (as shown in Figure 3.4 and Figure 3.5) and following the IP-based UTRAN protocol architecture, it can be seen that the radio network layer protocols, i.e. RLC, MAC and FP are kept unchanged, while the underlying transport network layer protocols replace the AAL2/ATM by UDP/IP layers (as highlighted in Figure 3.11). The UDP layer is responsible to establish/release the transport connections for each user flow. The IP layer is in charge of routing, addressing and QoS support. IP can either be IPv4 or IPv6. In the 3GPP standard for the IP-based UTRAN, the technology to be used in the layer 2 and layer 1 of the Node B and RNC is not specified. This provides the network provider freedom to choose a suitable implementation on the MAC and physical layer. Usually Ethernet is preferred to be used for commercial consideration due to its low cost, flexibility and scalability. In this thesis, Ethernet (see Appendix A.4) is considered to be the layer 2 transport technique on the IP-based Iub link.

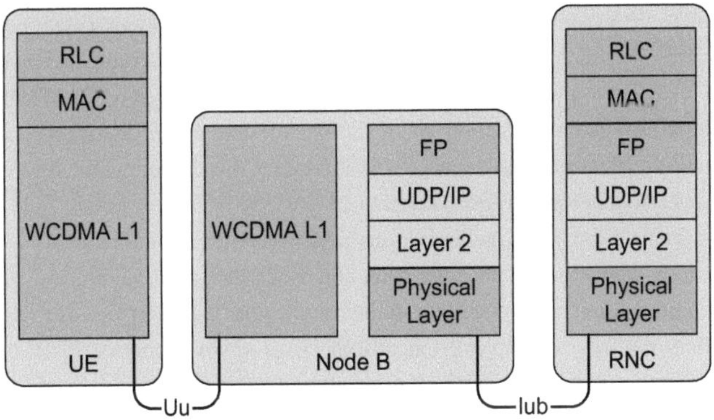

Figure 3.11: *Protocol stack of the IP-based UTRAN*

In addition, according to 3GPP TR 25.933 [3GP04a], the IP-based UTRAN can also apply multiplexing schemes in order to improve the transport efficiency by multiplexing several FP

PDU frames into one transport UDP/IP packet. The detailed introduction of these multiplexing schemes has been introduced in Chapter 2.3.4.1.

3.4.2 QoS Support in the IP-Based UTRAN

Despite prominent technique advantages and low costs of deploying IP and Ethernet for the transport, the most important challenge of an IP-based UTRAN network is how to provide and differentiate the QoS for a wide range of diverse services with various QoS requirements. The QoS challenge is mainly associated with the fact that IP is designed for the "best effort" Internet where there is no guarantee for the QoS. To provide appropriate QoS support in the IP-based UTRAN, an efficient IP QoS mechanism should be applied to differentiate different service flows and provide the desired level of service.

The *Internet Engineering Task Force* (IETF) has developed a number of IP QoS schemes to support multi-services, like *Integrated Services* (IntServ) [BCS94] with the *Resource Reservation Protocol* (RSVP) [BZB+97], *Differentiated Services* (DiffServ) [BBC+98], *Multi-Protocol Label Switching* (MPLS) [RVC01], etc. Among them, DiffServ is regarded as the most popular and notable one. It is considered as a practical and promising QoS scheme to be deployed in UMTS networks [AL02a]. Therefore, in this thesis DiffServ QoS scheme is considered for the IP-based UTRAN. The following sections give a detailed introduction of a DiffServ-based QoS structure to be deployed in the IP-based UTRAN network.

3.4.2.1 Introduction of the DiffServ QoS Scheme

DiffServ is an IP-based QoS support framework. It provides different treatment of flows or aggregates flows by mapping multiple flows into a finite set of service levels, namely *Per Hop Behavior* (PHB) groups. Each PHB is identified by a *DiffServ Code Point* (DSCP) in the IP header which provides information about the QoS requested for a packet, and defines a specific forwarding treatment that the packet will receive at each node. DSCPs are defined in IETF RFC 2474 [NBBB98]. It is specified in the 8-bit *Type of Service* (TOS) field in the IPv4 header (or in traffic class field in IPv6). The DSCP enables network routers to handle IP packets differently depending on the code point and hence their relative priority.

A DiffServ network is partitioned into domains, each having two types of routers: edge or core. A DiffServ domain is completely identified by its set of supported PHB, its edge and core routers, its policy in mapping QoS of an incoming packet to a pre-defined PHB, and the level of QoS associated with each PHB. Each packet belonging to a QoS class is marked with a DSCP in the IP header, and then each node in that domain forwards packet according to its DSCP.

DiffServ PHBs

The DiffServ working Group of the IETF has defined a number of different PHB groups for different applications. There are three most common PHBs defined in the IETF, *Best Effort* (BE) PHB, *Expedited Forwarding* (EF) PHB and *Assured Forwarding* (AF) PHB.

BE PHB is for the traditional Internet traffic and its usage implies that the nodes in the path will do their best to forward the packet in a fair manner, however, there is no guarantee on its delivery or its level of service. Everybody gets the service that the network is able to provide. Its recommended code point (DSCP) is 000000.

EF PHB is aimed for a low loss, low latency, low jitter, assured bandwidth edge-to-edge service through IP DiffServ domains. It can be understood as a virtual leased line service. Therefore, the bandwidth cannot be exceeded but the user can leave it idle or use it to the full extent of its capacity. The holder of this pipe should not be affected by the presence or absence of other users. An example of EF service is voice telephony. The DSCP of EF PHB is 101110.

AF PHB does not provide a bandwidth guarantee but packets are given a higher priority to be transmitted over the network than the packets from the BE PHB. In congestion situations the user of the Assured Forwarding service should encounter less bandwidth decrease than BE PHB users. The AF PHB group provides four independently forwarded AF classes. Each of these classes has three levels of dropping precedence: low, medium, and high. Therefore, 12 instances of AF with recommended DSCPs can exist in a DiffServ domain. Their defined DSCP codes are given following:

Low drop: 001010 010010 011010 100010

Medium drop: 001100 010100 011100 100100

High drop: 001110 010110 011110 100110

In each DiffServ node, each AF class is allocated a certain amount of forwarding resources (buffer space and bandwidth).

DiffServ Router

A DiffServ router consists of five components shown in Figure 3.12 [Man03]. On arrival, a packet is firstly classified by the classifier according to the bilateral *service level agreement* (SLA). Afterwards, the classifier forwards the packet to the traffic conditioner. The traffic conditioner may include a meter, a marker, a shaper, and a dropper. If accepted, the packet is enqueued into a corresponding buffer and then transmitted according to a specific scheduler policy.

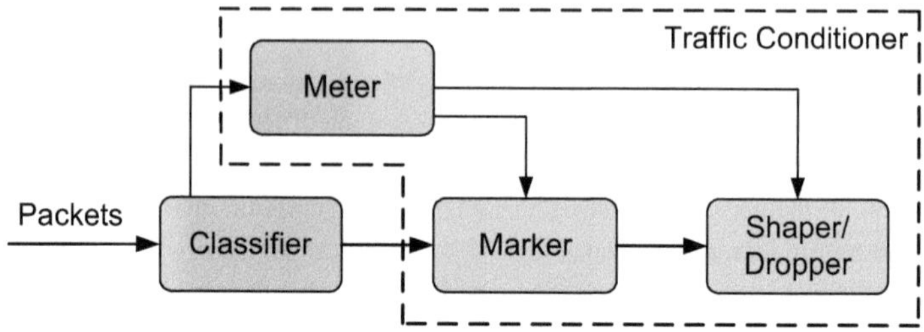

Figure 3.12: *IP DiffServ Router Building Blocks*

Mapping Function: Mapping is the function of translating the QoS of one system to the QoS parameters understood by another system. It is a necessary function for a UMTS packet entering the DiffServ-aware network, or when a packet coming from an external network enters the UMTS network. This function is usually integrated with the classifier into one functioning block. In the IP-based Iub domain (at the Edge Router), it is responsible for translating the QoS of the UMTS system to the corresponding IP DiffServ PHB, and vice versa.

Classifier: the packets are classified according to DSCP in the IP header. The classifier has a single input and splits the incoming stream into a number of outgoing ports in terms of different PHB aggregates.

Meters measure the temporal properties of the stream of packets selected by a classifier against a traffic profile. A meter passes state information to other conditioning functions to trigger a particular action for each packet, which is either in-profile or out-of-profile.

Markers set the packet with a particular code point, adding the marked packet to a particular DiffServ behavior aggregate. An existing marking may be changed. It may also use the statistics gathered by a meter to decide on the level of a packet's conformance to its allocated rate characteristics. For a non-conformant stream, it takes action by re-marking the packet in either a class with lower priority, or lower drop precedence.

Shapers delay some or all of the packets in a traffic stream in order to shape the stream into compliance with a traffic profile.

Droppers discard some or all of the packets in a traffic stream in order to bring the stream into compliance with a traffic profile. There are two major kinds of dropping elements in a DiffServ router: absolute and algorithmic.

- An absolute dropper is simply used to drop the overload traffic of a real-time application, e.g. for the EF PHB, because an overly delayed packet is no longer useful, and is better discarded. Therefore, when congestion happens, the real-time packets are forwarded to this block in the DiffServ node. When real-time traffic misbehaves, the overload traffic is sent to the absolute dropper.

- In contrast, an algorithmic dropper drops a packet according to a specific algorithm, usually well before the queue is full. *Random Early Detection* (RED) (see Appendix A.6) is such an algorithmic dropper, suggested for avoiding congestion in the network. It detects the incipient congestion and notifies the congestion manager of the source to decrease its flow rate. This way, congestion is avoided. To perform RED per traffic class, *Weighted Random Early Detection* (WRED) (see Appendix A.6) is used to drop packets selectively based on different priorities specified by IP precedence. Packets with a higher IP precedence are less likely to be dropped than packets with a lower precedence. Thus, higher priority traffic is delivered with a higher probability than lower priority traffic.

Scheduler: The scheduling has the most impact on the level of service a packet receives. It decides on which queue, among a set of queues, to service next, for example *Weighted Fair Queuing* (WFQ), priority queuing, etc.

Queuing: Additionally, there is also queuing function in the DiffServ router. They store the packets while they are waiting to receive service by the scheduler and depart to the next hop or DiffServ module.

Weighted Fair Queuing (WFQ)

WFQ is a popular scheduling algorithm to guarantee bounded delay, guaranteed throughput, and fairness [DKS89]. It is commonly used in a DiffServ node to provide different priorities for different traffic classes. The WFQ scheduling discipline offers QoS by adding a weight to the queues to assign different priority levels for different queues. Each queue is assigned a weight that determines the share of the available capacity that it will have and thus all queues share the bandwidth proportional to their weights. Traffic may be prioritized according to the packet information in the source and destination IP address fields, port numbers and information in the ToS field (e.g. DSCP). The WFQ discipline weights the traffic so that the low-bandwidth traffic gets a fair level of priority.

Let w_k be the weight of the kth queue and BW the total available IP bandwidth. If there are in total N queues and all queues are transmitting data, then the kth queue obtains a fraction of the total capacity BW_k as calculated in formula (3.1). But if one priority queue is empty (i.e. not utilizing its allocated bandwidth), then its spare bandwidth shall be fairly shared among the other queues according to their weights.

$$BW_k = \frac{w_k}{\sum_{i=1}^{N} w_i} \cdot BW \tag{3.1}$$

The WFQ can be also combined with other scheduling such as priority queuing. For example, the real time traffic mapped to the EF PHB, e.g. voice or video streaming traffic, is given the highest priority. While the interactive data traffic mapped to different AF PHBs are the middle priority, and among them WFQ is used. The lowest priority is assigned to the BE PHB, which utilizes the spare capacity which is not utilized by EF and AF PHBs.

3.4.2.2 Mapping of UMTS QoS to DiffServ PHBs

As introduced in section 3.3.4.1, UMTS defines four QoS classes: conversational, streaming, interactive and background class. Both the conversational and streaming class require a certain reservation of resources in the network, mainly intended for carrying *Real Time* (RT) traffic flows, such as video telephony and audio/video streams. The Interactive and Background class are so-called *Non Real Time* (NRT) traffic. They are mainly for carrying elastic Internet applications like web, email, and ftp.

For example, both the conversational class and streaming class are mapped to EF PHBs as they require low loss, low latency, low jitter, assured bandwidth, end-to-end service through IP transport domains. The interactive class is mapped to AF PHBs, e.g. web traffic. As mentioned in the last section, there are four AF Service classes and each of these classes can also have three levels of dropping precedence: low, medium, and high. In practice, different AF PHBs can be used for different RAB types which provide various QoS for the end users. The background class is mapped to BE PHB as it belongs to Best Effort traffic, which has low requirements on the QoS and utilizes the resources that the network is able to provide. For example, best effort HSPA traffic can be set to BE PHB. It should be noted that there can be other alternative mappings of UMTS QoS classes to DiffServ PHBs, depending on system requirements and traffic.

3.4.2.3 DiffServ-Based QoS Architecture in IP-Based UTRAN

Based on the DiffServ QoS scheme, the following QoS structure (Figure 3.13) is considered in this thesis for the IP-based UTRAN. In this proposed structure, there are in total six IP DiffServ PHBs: 1 EF PHB for sending signaling, conversational voice and video streaming traffic; 4 AF PHBs for Interactive traffic class with 4 different priorities (AF11, AF21, AF31 and AF41); 1 BE PHB for background traffic class. Each DiffServ PHB has its own dedicated queue.

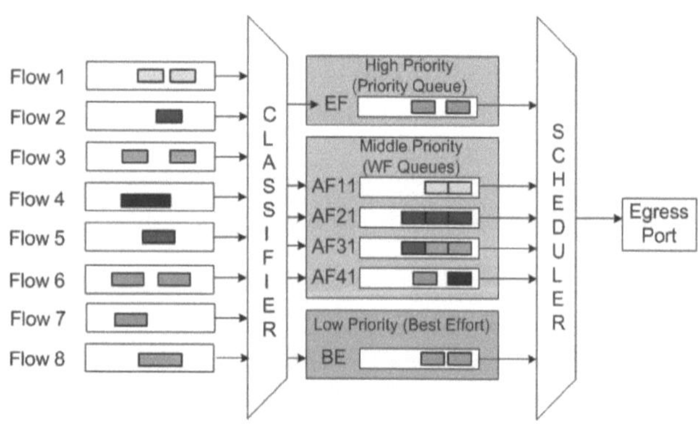

Figure 3.13: *DiffServ-based QoS structure in the IP-based UTRAN*

Typically in the IP DiffServ architecture, a combination of *Priority Queuing* (PQ) and *Weighted Fair Queuing* (WFQ) is often applied to support different service classes, e.g. [FPR00] [TNC+01]. PQ is used to provide premium service (like delay- and jitter- sensitive real time services) for the highest priority; and WFQ is used in addition to serve lower priority service classes with different priorities and fairness. In this thesis, the combined PQ and WFQ scheduling scheme is used in the DiffServ architecture of the IP-based UTRAN. As shown in Figure 3.13, the EF PHB is given the highest priority, therefore between EF and other PHBs is the priority queuing. While all AF PHBs and BE PHBs use the spare bandwidth which is not reserved by the EF PHB. The Weighted Fair (WF) Queues are used for AF and BE PHBs and for different AF PHBs and BE PHBs different weights are assigned which determine their priorities, and WFQ scheduling is used to allocate the bandwidth according to their weights among all AF and BE PHBs. WRED is applied for dropping packets for AF and BE PHBs. It is the probabilistic discard of packets as a function of queue fill before overflow conditions are reached. Random dropping takes place when the algorithm detects signs of congestion. The congestion variable is the average queue size that is calculated from the instantaneous queue size by using exponential averaging. More detailed description on WRED algorithm and its required parameters are given in Appendix A.5.

4 Framework of UMTS Network Dimensioning

This chapter provides an overview of dimensioning issues in the UMTS radio access network, which arises in planning and operation of UMTS networks. At first, the main objectives of the UMTS network dimensioning are defined. Then various important aspects which need to be particularly considered in the dimensioning process are addressed. Based on that, a general framework for UMTS network dimensioning is proposed in this thesis and the corresponding dimensioning procedure is presented. At last an overview of the dimensioning approaches used in this thesis is given.

4.1 Objectives of UMTS Network Dimensioning

In order to perform network dimensioning, a clear objective is necessary. In the framework of this thesis, *cost* and *quality of service* are the main objectives for dimensioning processes within the context of network planning and operation. The ultimate goal of network dimensioning is to minimize the cost while maximizing QoS. However, these two factors are usually negatively correlated, and thereby the service provider has to find a right balance and tradeoff between them.

4.1.1 Network Costs

Normally the network costs have two main categories: the expenditures associated with building a network and with running it. In this thesis, the network cost only considers the cost for leasing link bandwidths. For a given transport technology, usually the higher the link capacity, the higher is the network cost.

4.1.2 Quality of Service

The *European Telecommunications Standards Institute* (ETSI) and the *International Telecommunications Union* (ITU) define QoS as the quality perceived by the end user [ITU93]. The IETF network working group provides a more concrete definition where QoS is defined as the service requirements that need to be met by the network while transporting a traffic flow [CNRS98]. In this thesis, the term *quality of service* (QoS) is used based on the latter definition by the IETF. In the framework of this thesis, various QoS measures are taken into consideration for the UMTS network dimensioning and they are categorized into flow-based and network-based QoS measures. For further reference, they are called *user-relevant QoS* and *network-relevant QoS* (also called *network performance*) throughout this thesis. In the following paragraphs, the relevant measures for the quality perceived by the user as well as for network performance are presented.

4.1.2.1 User-Relevant QoS

User-relevant QoS refers to the QoS related to the individual user flow. From a user's perspective, typically user-relevant QoS criteria are application delay, throughput and delay jitter. If Connection Admission Control (CAC) algorithms are applied in the network, connection reject ratio (also referred to as blocking probability) is another important QoS aspect.

Application Delay The end-to-end application delay is defined as the total time of transferring a file (the data transaction is completed only after the last packet is received) or an individual application packet like a voice packet from the source to the destination. In addition to the delay for transmitting the entire data from the source to the destination, it may also include the extra time for setting up and releasing the transport connection, the possible retransmissions due to packet losses, and the delays caused by traffic shaping, flow control or congestion control from transport protocols like TCP or network functions.

Application Throughput Application throughput indicates the transaction speed, i.e. how long it takes to transfer a certain amount of data. It is directly related to the application delay and the volume of corresponding data transaction. The throughput is usually measured in bit per second (bit/s or bps), and sometimes in data packets per second or data packets per time slot.

Application Delay Jitter Application delay jitter refers to the variance of individual delay values. It is measured as the difference between the largest and smallest end-to-end delay. This QoS is mainly important for real-time traffic such as streaming audio or video.

Connection Reject Ratio As introduced in section 3.3.4.3, CAC algorithms are used to decide whether an incoming connection request should be accepted or rejected in a network in order to maintain the guaranteed QoS for the end users and networks. Connection reject ratio refers to the ratio of rejected connections to the total number of requested connections in the system. It is mainly used for reservation based services.

In the outline of this thesis, the user-relevant QoS is regarded as the main objective for the UMTS network dimensioning. The considered user-relevant QoS is the end-to-end application delay or throughput for the elastic traffic and connection reject ratio for the circuit-switched traffic.

4.1.2.2 Network-Relevant QoS (Network Performance)

For the objective of the network dimensioning and optimization problem, network-specific QoS measures are used to evaluate the quality of a network. The network-relevant QoS, measured on the packet level, is based on the aggregated traffic rather than individual flows. The most used network QoS measures are packet delay, packet loss ratio, and link utilization.

End to end Packet Delay In the context of this thesis, the packet delay refers to the delay of transmitting a packet through a network domain. It contains delay components from processing, segmentation and reassembly, transmission, propagation, and queuing, etc.

Packet Loss Ratio It is defined as the ratio of discarded packets to the total amount of transported packets. Within the UTRAN Iub transport network, the packet discards may be caused by excessively delayed Frame Protocol (FP) PDUs that are discarded by the Node B, or by the buffer overflow due to traffic overload, or as the result of a certain packet drop function to avoid long queuing delays and link congestion. As explained in section 3.3.3.4, the packet loss ratio is an important QoS requirement for the Iub transport network. It should be controlled to be a reasonably low ratio on the Iub interface in order to avoid too many retransmissions that will waste the expensive transport resources and also to protect the end user performance from severe degradation.

Link Utilization Link utilization serves as an appropriate link-specific QoS. It is expressed as a percentage of the achieved throughput over the link to the given link bandwidth. It is also known as bandwidth utilization efficiency.

In this thesis, the above defined network-relevant QoS are investigated for the UTRAN transport network. In Chapter 6.4, the dimensioning of the Iub interface to meet the transport network QoS, such as packet delay or packet loss ratio, is studied.

4.2 Important Issues for UMTS Network Dimensioning

Network dimensioning is used to determine the capacities of the links in the network, which is one crucial element of the network design process, of traffic engineering and network planning. Network dimensioning is strongly related to a number of issues such as the given traffic demand of the individual traffic classes, the designed network topology, and a certain routing pattern. This section discusses the important issues from the point of view of the network dimensioning in the network planning process.

Figure 4.1 illustrates the dimensioning issues within the context of network design, network management and network operation. The presented procedure generally applies to the planning of a UMTS network. First of all, it is necessary to initially set up a network. *Initial network design* contains the tasks to analyze the traffic that will be carried by the network, to design the network topology including node placement and link selection, to perform routing in order to find advantageous path patterns, and at last based on the traffic demand, the given network topology and a fixed routing pattern to carry out *initial dimensioning* to determine the capacities of the links for capacity assignment and bandwidth allocation. After the network has been designed and set up, *network operation* and *network management* come into play. Network operation includes processes and actions necessary to run a network and keep it in an operational state. Within the network operation, various actions can be taken to assure the desired QoS such as *traffic control*, *resource control*, and *rerouting*. Traffic control provides fundamental techniques on the packet level to differentiate and manage different traffic streams in order to achieve service-specific QoS for all traffic flows traversing the network. It includes technologies such as packet classification, scheduling, buffer management and traffic policing. Resource control refers to techniques which manage the access to available network resources at class, flow or connection level to keep the network from being overloaded and services from experiencing quality degradation. Rerouting takes place when the network topology changes due to temporary link or node failures. During the operation process, the operational state has to be continuously monitored in order to verify whether it conforms to the predefined requirements. As soon as a non-satisfactory state is indicated, countermeasures have to be taken. Usually such non-satisfactory states are caused by traffic overload situations, which might arise due to the increasing traffic demand or temporary traffic variations. The overload situations can be alleviated through *network management* procedures, such as *routing optimization* or *dimensioning update* (adaptation of bandwidth assignment). In either case the new configuration is uploaded to the network and network operation is continued. However, if the existing infrastructure is not sufficiently dimensioned to support traffic increases or new services, the network has to be extended. It often requires an adaptation of the network topology, which poses a redesign of the network. And therefore, the routing needs to be re-optimized, and *re-dimensioning* is needed for the extended network to add new links or extend the capacities for the existing links.

From this procedure, it is clearly seen that the dimensioning tasks arise at each phase throughout the network planning cycle. In the initial network design step, an initial dimensioning is required for setting up the network to assign appropriate capacities for the links. In the network operation phase, the dimensioning may be updated upon the indications of non-satisfactory states. If adapting the bandwidth allocation is not sufficient for keeping the QoS, the network needs to be expanded. In this case a re-dimensioning of the network is invoked. It can be concluded that the dimensioning should consider different essential issues like the current traffic demand, the existing network topology, the applied traffic control, resource control, routing and traffic engineering techniques in the network. The network dimensioning has to meet the objective of minimizing the overall network costs while satisfying the QoS requirements. The following sections give a more detailed discussion of dimensioning related issues in this framework and their implications for the dimensioning task.

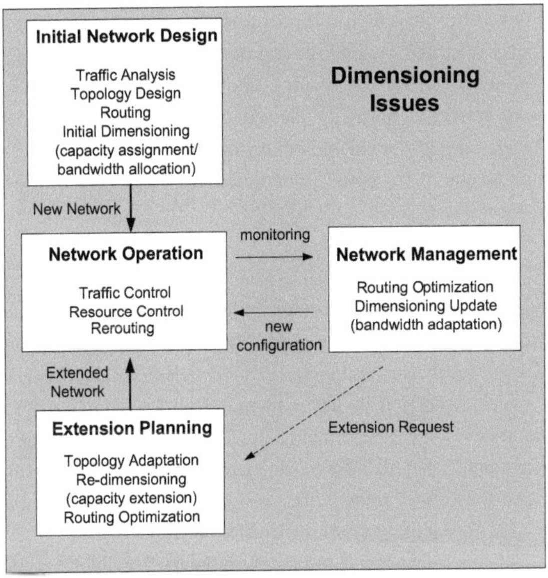

Figure 4.1: *An overview of dimensioning issues*

4.2.1 Traffic Analysis

The first fundamental issue to consider for network dimensioning is an accurate estimation for the traffic that is carried by the network. It comprises basically three aspects: *traffic classification*, *traffic distribution*, and *traffic characterization*.

Traffic Classification

The traffic, which will be offered in the network, needs to be classified with respect to the characteristics and required QoS. It is not practicable to consider every characteristic of each application, as this would make the dimensioning process too complex. Thus, it is recommended that only a few traffic classes should be defined and the overall set of applications and services mapped upon them. The traffic classes identify the required QoS that the network has to deliver for them. The implementation of mapping the applications to traffic classes in the network is technology dependent. In this thesis, two basic traffic classes are distinguished:

(1) elastic traffic, for which QoS correlates with the total time that it takes to transfer bulk data of a certain size, i.e. application delay or throughput;

(2) circuit-switched traffic, which reserves certain bandwidth in order to guarantee the end-to-end delay, delay jitter and packet loss, and thus is subject to admission control.

Elastic traffic is generated by data-centric applications (e.g. ftp, web, e-mail). In the Internet, this type of traffic is normally carried by the TCP protocol, which adapts its transmission rate to the currently available bandwidth. The circuit-switched traffic relates to delay-sensitive applications (e.g. voice, video conferencing) which use other transport protocols like *Real-time Transport Protocol* (RTP).

Traffic Distribution

It is an important task to predict traffic patterns, which depend on the popularity of services and the prevalent application mix. In the context of network dimensioning, traffic distribution refers to *traffic quantity*, *traffic mix*, and *geographic distributions*. Traffic quantity defines the amount of traffic which is carried in the network. With the ongoing development of the UMTS network, the traffic increases continuously. Thus for network dimensioning, not only the temporary traffic quantity but also the traffic growth and its growth rate are crucial factors to be considered. The traffic mix reflects the portions of different applications and services running in the network. In the early phase of UMTS networks, the voice services contribute the main portion of the overall traffic, but nowadays more "bandwidth greedy" applications like Internet traffic start to play a more important role in the UMTS network. The variation of the traffic mix, e.g. large percentage of Internet traffic, leads to the change of traffic properties, which results in different bandwidth demands for the dimensioning. Thus it is important to continuously observe the varying traffic pattern throughout the network dimensioning cycle. Finally, for dimensioning the overall network, the geographic distribution of traffic for each traffic class has to be taken into account. This is usually related to economic and demographic data, technological trends, user behavior, or relevant growth rates [KO02, SKO+02].

Traffic Characterization

Traffic characteristics are service-specific. They depend on the applications on the user flow level. For example, for web traffic small requests sent in one direction are followed by large data transfers in the opposite direction. It results in an asymmetric traffic property while the communication-oriented services like telephony and video conferencing typically establish bi-directional sessions and as a consequence the resulting traffic flows between the end systems are usually symmetric. Moreover, the web traffic usually has a much higher bursty nature than the voice or video traffic. On the network level, it is not only relevant to the applications, but also associated with the means of aggregation of various user flows. The traffic characteristics on both user and network level need to be considered carefully for network dimensioning. By taking into consideration a correct traffic property, the required capacities can be dimensioned more appropriately.

4.2.2 Network Topology

Network topology refers to the layout of the network. The topology of the UTRAN depends on the coverage of the access network, the location of Node Bs and RNCs, as well as the structure for interconnecting Node Bs and RNCs, and core network nodes. Figure 4.2 shows four basic network topologies for constructing the UTRAN: chain, star, ring, and tree structure.

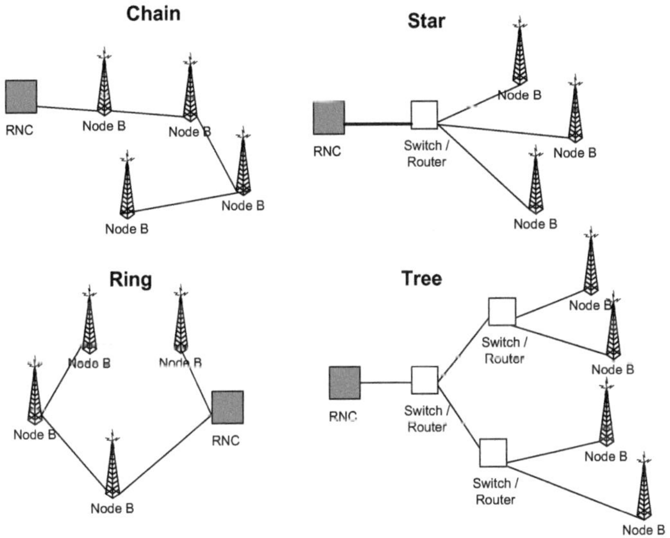

Figure 4.2: *Basic network topology configurations for UTRAN*

Usually the network is designed as a mixture of the above topologies. Figure 4.3 shows an example of a UTRAN network topology. A Node B can either be directly connected to the RNC, or act as a traffic concentrator for a set of Node Bs, or it can be connected to an intermediate router or switch. An intermediate router or switch can be used to achieve a higher degree of traffic concentration, which might be used to connect several Node Bs to a common and distant RNC, or to aggregate the traffic between groups of RNCs and the core network.

Choosing an appropriate network topology has a great impact on the traffic patterns in the network and as a consequence, affects the allocation of the link capacities, link costs and total network costs. Different network topologies indicate different degrees of traffic aggregates, and thus the resultant multiplexing gain and cost savings differ from one another. For example, the star topology can achieve high multiplexing gain as a result of high concentration of traffic on the backbone link, i.e. the link between RNC and the switch or router which connects all Node Bs.

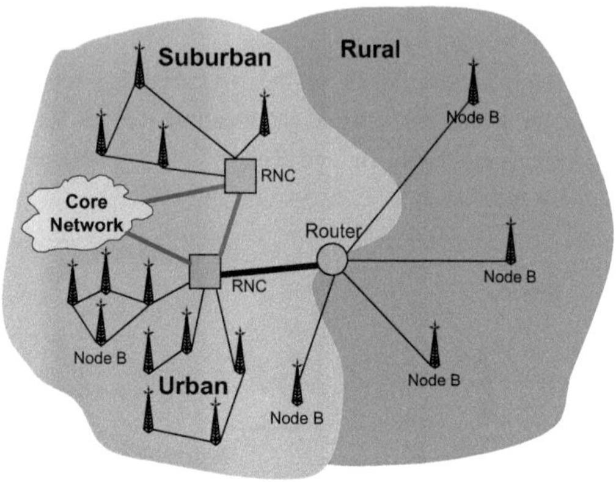

Figure 4.3: *UTRAN network topology example*

4.2.3 Traffic Control

Traffic control is used to distinguish different traffic streams and provide differentiated treatment to the corresponding packets such as packet buffering, scheduling, shaping, and forwarding. From a network dimensioning perspective, it is important to consider the impact of the traffic control schemes on the traffic carried in the network, and thus on the link dimensioning. For example, in the UMTS access network, to differentiate different traffic types with individual QoS requirements, typical scheduling strategies are prioritization where packets with higher priority are always sent before lower-priority packets (e.g. non-preemptive priority), or bandwidth fair sharing where certain shares of the link capacity are assigned to specific traffic flows or classes according to their configured priorities (e.g. WFQ). Different scheduling schemes determine different bandwidth allocations for different traffic types, and also results in different multiplexing gains that can be achieved, hence their influence on the dimensioning is significantly different. Several potential impacts of traffic control on network dimensioning are summarized as follows:

(1) With packet scheduling, each traffic class is allocated to one fraction of the total bandwidth, and thus receives a corresponding QoS guarantee. Usually the worst case (i.e. the lowest priority traffic stream) is mostly considered for the bandwidth dimensioning. Moreover, the potential multiplexing gains of sharing the bandwidth resources needs to be taken into account as well.

(2) Buffer management functions, such as buffer size, packet discarding, etc., have an impact on the achieved delay, delay jitter and packet losses, which are the relevant QoS measures for the dimensioning.

(3) Traffic shaping is a practical method for capacity assignment. It enables leasing bandwidth for transmitting UMTS traffic from fix network providers like ADSL or Ethernet to co-use the same physical links.

4.2.4 Resource Control

Resource control is responsible for controlling the access to the available network resources at flow or connection level to reduce or avoid the possibility of service degradation from traffic overload. For instance, *Connection Admission Control* (CAC) (see section 3.3.4.3) is used to control the number of traffic flows or connections entering the network, where individual flows or connections make bandwidth reservations along the data paths through the network. If there is no adequate bandwidth available on the path for a new traffic flow or connection, the request of the new connection has to be rejected. The relevant QoS is the connection reject ratio (see section 4.1.2.1). Additionally, a *Bit Rate Adaptation* (BRA) algorithm (see section 3.3.4.3) can be used for the transport of packet-switched traffic in the UTRAN for an effective utilization of the bandwidth resources. BRA dynamically adapts data rate (i.e. radio access bearer) per user flow according to its transmission activity and the available resources in the network. Another type of resource control approach is to employ policing entities such as rate limiting at the network edge to restrict the amount of traffic, which is entering the network. It should be noted that connection admission control works on a per-flow basis (a flow is either accepted or rejected), while policing is done within a traffic class on a per-packet basis (packets are dropped whenever the respective traffic exceeds a certain threshold).

The impact of the applied resource control functions in the network need to be considered for dimensioning. Whenever a type of per-flow CAC scheme is used, a blocking model shall be applied to dimension the network. The application of a BRA function can improve the utilization of the network resources, but also leads to a varying data rate per user flow and hence results in a bursty traffic property. With policing entities at the network edge, the traffic aggregates can be kept from exceeding per-class rate limits. Thus upper bounds can be given for the bandwidth requirements of such an aggregate within the network. But employing policing also leads to unwanted service degradation on the active flows.

4.2.5 Routing and Traffic Engineering

In order to achieve higher QoS, based on a given network infrastructure and a certain traffic load situation, a network service provider has the possibility to better distribute the traffic in the network by performing routing and traffic engineering. With traffic engineering techniques it is possible that traffic flows are either carried over single or over multiple paths between the source and the destination. Specifically, one core concept of traffic engineering is routing optimization. The idea of routing optimization is to find advantageous path patterns which can achieve the best QoS for a certain traffic demand and traffic load.

The applied routing and traffic engineering has significant influence on the distribution of the traffic in the network, i.e. control of the link loads on all links throughout the network, and thus their impacts on the overall traffic need to be considered essentially for the capacity assignments in the dimensioning process.

4.2.6 Summary

In section 4.2, several important issues and their impacts on the dimensioning are discussed. In the framework of this thesis, the dimensioning of the Iub interface will consider different traffic classes and traffic mix (elastic or circuit-switched traffic, or a mix of both), a certain network topology (either a single link or a star topology access network), traffic control functions like scheduling (e.g. strict priority, DiffServ), and different resource control functions such as CAC or BRA. However, routing and traffic engineering are not considered in this thesis. It is assumed that the network always keeps a fixed routing pattern.

4.3 Framework for UMTS Network Dimensioning

The preceding sections define the objectives for the UMTS network dimensioning, and analyze the related important issues which need to be particularly considered in the dimensioning process. This section proposes a general framework for the UMTS network dimensioning and presents the complete procedure for dimensioning which is used in this thesis including the necessary steps and the required inputs and outputs.

Figure 4.4 illustrates the framework for the UMTS network dimensioning. The aim of dimensioning in the context of this thesis is to decide the minimum required link capacities which should satisfy the desired QoS requirements. As mentioned at the beginning of this chapter, the goal of network dimensioning is to minimize the total network cost while maximizing the QoS to achieve a cost-effective radio access network. In this thesis, the dimensioning is specifically focused on the Iub interface within the UTRAN network.

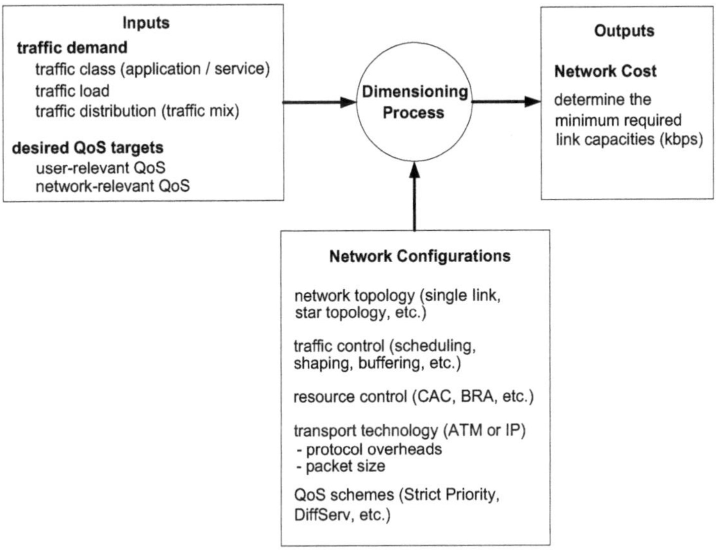

Figure 4.4: *UMTS network dimensioning framework*

As indicated in Figure 4.4, for carrying out the dimensioning, two types of input data are required:

Traffic Demand The traffic demand refers to the requirements of the traffic in the network, which specifies the total amount of the offered traffic, the traffic classes with respective to different services and applications as well as the traffic distribution in terms of a variation of traffic mix. The traffic demand is the outcome of traffic analysis in the framework of the network design process (see section 4.2.1).

Desired QoS Targets The QoS targets define the QoS requirements that need to be satisfied by the network. It describes the objective of the network dimensioning. As introduced in section 4.1.2, two types of QoS are considered: user-relevant QoS and network-relevant QoS.

Additionally, different **Network Configurations** need to be taken into consideration for the dimensioning. As discussed in section 4.2, they have considerable impact on the transmitted traffic in the transport network (traffic characteristics and load distributions), the performance of the networks and end users, the bandwidth utilizations, the achieved multiplexing gains, and in turn have significant influence on the dimensioning results.

The dimensioning process is the core of this dimensioning framework. It relies on the two inputs (i.e. traffic demand and desired QoS targets), and the given network configurations.

The output of the dimensioning process is the minimum required link capacity which should support the offered traffic demand while satisfying the desired QoS requirements. The detailed procedure for the network dimensioning is shown in Figure 4.5. It basically contains three steps.

Step 1: Check the input data and parameters. It includes three subtasks: analyze the traffic, check the network configurations, and identify the QoS targets as the objectives for the dimensioning.

Step 2: Select an appropriate dimensioning method. It can be either a simulation approach or an analytical approach. If using a simulation approach, a simulation model with UMTS functionalities needs to be set up and verified. The advantage of the simulation approach is that it can model detailed protocol behaviors, functions, traffic patterns, network topologies etc., which will give a rather accurate outcome. However, its main drawback is the high effort on implementing the simulation model and performing simulations. The analytical approach is usually preferred whenever it is available due to its low effort and easy implementation while it can achieve an acceptable accuracy on the results, though not all details can be modeled and more assumptions need to be taken. The selection of which analytical model to use for the dimensioning depends on the given traffic (i.e. traffic class, traffic mix), the network configurations and required QoS targets.

Step 3: This step is to process the dimensioning with the selected dimensioning method. The outcomes of the analysis on the traffic, network configurations and QoS requirements in step 1 are regarded as key parameters for the applied dimensioning method. A detailed view of how to derive the minimum required link capacities for the dimensioning output is given in Figure 4.6. It is an iterative process. The dimensioning starts with an initial link capacity, and then estimates its resultant QoS for the given traffic demand and network configurations using the selected dimensioning approach. The estimated QoS are compared with the required QoS. If the gap of the estimated QoS and the QoS targets is larger than a predefined convergence threshold, the link capacity need to be increased if the estimated QoS is worse than the required QoS, or decreased if the estimated QoS is better than the required QoS. This iteration process continues until the convergence threshold is reached, i.e. the estimated QoS is close to the required QoS. The outcome of this iteration process is the minimum required link capacity for meeting the QoS requirements of the given traffic demand and network configurations. If choosing analytical dimensioning approaches this process becomes a numerical calculation of the corresponding analytical models.

The above presented dimensioning framework and dimensioning procedure are applied in this thesis for the dimensioning of the UMTS Iub interface. In order to evaluate the cost-efficiency of the dimensioning results, in this thesis the following metrics are used.

- **Dimensioned bandwidth (kbps)**: it is the derived bandwidth for the Iub interface. It shall be noticed that in this thesis the derived dimensioning bandwidth is given in kbps. However, in practice the dimensioned bandwidth can be mapped to a number of physical lines, dependent on the underlying transport medium. For instance, in an ATM-based UTRAN, the dimensioned link bandwidth can be mapped to a number of E1 lines.

- **Normalized capacity**: it is defined as (link bandwidth / traffic load), where the traffic load is represented by the link throughput. It is also called as *overdimensioning factor* or *overprovisioning factor* [Rie04, GZFO07].
- **Link utilization**: it indicates the utilization of the link bandwidth as defined in section 4.1.2.2. It is calculated as (link utilization / link bandwidth). As seen, it is equal to (1/normalized capacity).

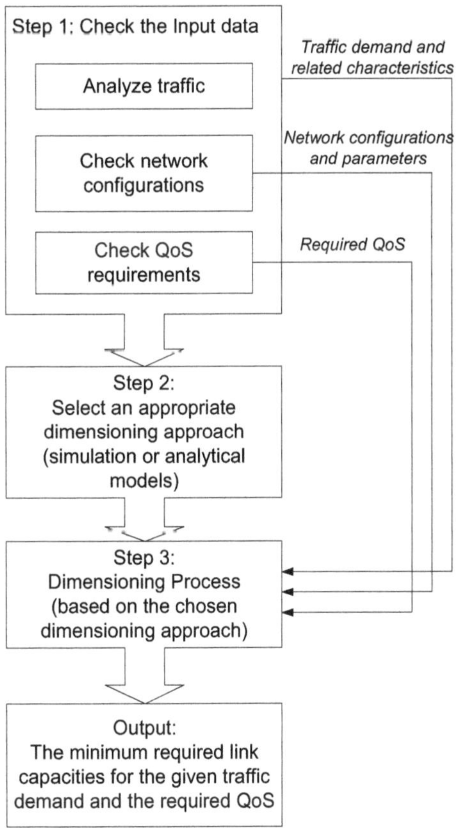

Figure 4.5: *An overview of the dimensioning procedure*

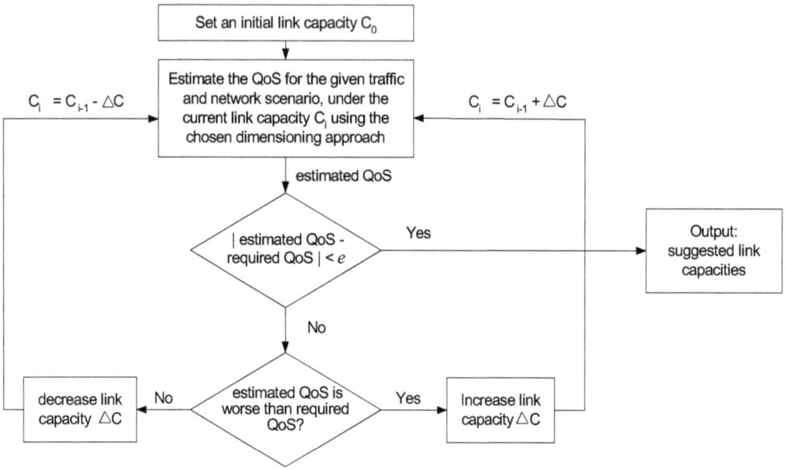

Figure 4.6: *Dimensioning process to derive the minimum required link capacities*

4.4 Dimensioning Approaches

This section presents an outline of dimensioning approaches which are considered in this thesis. As mentioned in the preceding section, the dimensioning approach can be basically categorized into simulation approach and analytical approach. The simulation approach relies on an established simulation model which can imitate the real system. The analytical approach is mainly based on queuing theory. In this thesis, both approaches are applied for the dimensioning of UMTS radio access network and compared with each other. The simulation approach usually can provide a higher accuracy than the analytical approach, as the simulation models can contain the detailed modeling of the UMTS network (i.e. modeling the required protocol layers, the main UMTS functionalities, the network structures, the traffic generation and management etc.), but the analytical models only make statistical assumptions and approximations. Therefore, in this thesis, the proposed analytical dimensioning models are validated through simulations. In general the analytical approach is preferred for the dimensioning due to its low effort and feasible implementation, and it can provide analytical expressions representing the dependency of the performance on the various model parameters. For this reason, one of the main goals of this thesis is to propose appropriate analytical models for the dimensioning.

The selection of an appropriate analytical model depends on several aspects.

1. The exactly modeled and the approximated elements of the considered network behavior: it is a common approach that the detailed models are applied only to the part

of the network that is under the scope of the analysis, but the overall network behavior is described with some reasonably simple model.

2. The required accuracy of the results: different analytical queuing models provide different level of accuracy. The applied model should be as simple as possible to minimize the computational complexity while providing the required accuracy.
3. The performance parameters of interest: the selected simplifying assumptions depend on the performance measure of interest. For example, when the performance measure of interest is the packet delay on the access link the detailed modeling of the individual users is usually not required. Instead, a simplified queuing model describing the aggregate traffic over the link and the transmission rate is sufficient.
4. The feasibility for the model: the applied analytical model should be feasible to implement and apply in the dimensioning process.

Figure 4.7 shows an overview of the considered analytical models for the dimensioning of the UTRAN networks in this thesis. Essentially, the analytical dimensioning models are divided into two main categories: models for estimating the user-relevant QoS and models for evaluating the network-relevant QoS. And for each of these categories, different traffic classes (i.e. belong to elastic traffic or circuit-switched traffic) and traffic mixes are considered separately.

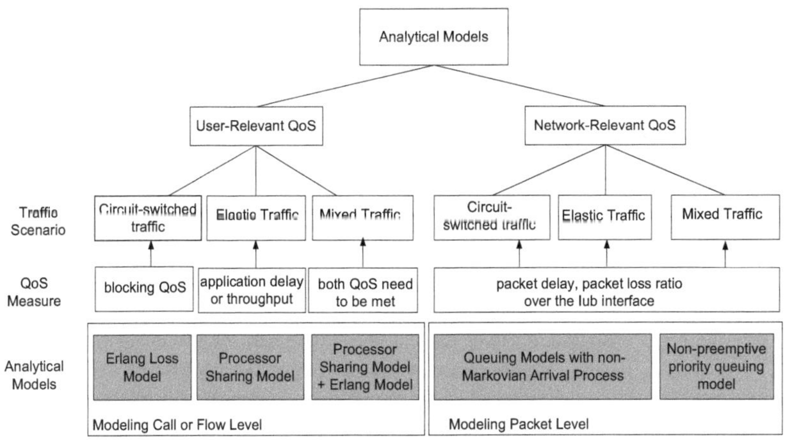

Figure 4.7: *Overview of analytical dimensioning models*

To obtain the user-relevant QoS, the applied analytical models should reflect the flow level behavior and predict per-flow performance. In this thesis, for the circuit-switched traffic a blocking model is used to estimate the blocking QoS (i.e. connection reject ratio). For the elastic traffic the dimensioning model is based on process sharing models on flow level, which is used to calculate the end-to-end application delay or throughput. In case of the mixed traffic scenarios, two cases can be considered: (1) bandwidth partitioning; (2) bandwidth sharing.

In the first case, it is assumed that technologies are employed in the network, which allow service-specific partitioning of link capacity. Thus a link can be divided up into bandwidth portions, which are assigned to the individual traffic classes. Traffic policing units (e.g. token buckets) at the network edges assure that individual traffic flows within the portion of each traffic type do not negatively interfere with each other. This assumption allows us to dimension the necessary transport capacity for elastic traffic and circuit-switched traffic separately with their individual dimensioning models. To obtain the total capacity of a link, the individual bandwidth shares are summed up.

In the second case, it is assumed that there is a complete bandwidth sharing among different traffic classes. That means a traffic class can use any bandwidth available in the network that is not used by others. In this thesis, strict priority scheduling is performed between the circuit-switched traffic (high priority) and the elastic traffic (low priority) in the transport network. At any time the low priority traffic is able to utilize the bandwidth which is not used by high priority traffic. For this case, the analytical dimensioning model needs to guarantee the QoS of both traffic types. Thus the processor sharing model for elastic traffic and the Erlang model for circuit-switched traffic need to be combined, but additionally the resultant multiplexing gain due to the bandwidth sharing has to be considered. This thesis will focus on this case with a full bandwidth sharing between the two traffic types.

For dimensioning to meet the network-relevant QoS such as packet delay or packet loss ratio over the Iub interface, the suggested analytical dimensioning models are queuing models which model the packet level. The arrival process of the queuing models needs to capture the important characteristics of the aggregated traffic on the link. They are usually Non-Markovian Arrival Processes. The server process depends on the packet length distribution and the given link capacity. In this context, the server refers to the link that needs to be dimensioned. In the mixed traffic scenario, a non-preemptive priority queuing model is considered.

Once the basic analytical queuing models are determined for certain QoS criteria and traffic scenarios, a number of extensions of the basic queuing models can be derived for adding specific resource control functions like BRA or CAC, traffic control functions such as DiffServ or deploying certain network topology. In this thesis, two network scenarios are considered for the Iub interface: single Iub link scenario and multiple links scenario with a star network topology. In the latter case, the gain on the bandwidth utilizations of multiplexing the traffic from the individual links should be considered for a proper dimensioning.

In this thesis, in order to validate the accuracy of the analytical models, *relative error* is used. Given a value a measured from simulation and b is the calculated value derived from the analytical model, and then the resultant relative error ∂ is calculated with equation (4.1).

$$\partial = |a-b|/a \tag{4.1}$$

4.5 Conclusion

In the framework of this thesis, the network cost and QoS serve as the main objectives for the dimensioning processes: costs have to be minimized while QoS needs to be maximized. In this thesis, the network cost only considers the cost for link capacities, and two types of QoS are considered for dimensioning: user-relevant QoS and network-relevant QoS. The dimensioning is strongly related to a number of important issues like traffic demand, network topology, traffic control and resource control functions as well as applied routing and traffic engineering techniques. Hence these issues need to be particularly considered in the dimensioning process. In this thesis, a general dimensioning framework and dimensioning procedure are proposed for the task of UMTS radio access network dimensioning. Different dimensioning approaches can be applied, which are basically categorized into simulation approach and analytical approach. The simulation approach relies on building simulation models and performing simulations, while the analytical approach is mainly based on queuing theory. Different queuing models can be employed to model flow or packet level behaviors in order to estimate the user or network related QoS that can be achieved under a given link capacity.

5 Introduction of Simulation Models

Simulation is a practical and scientific approach to analyze a complex system. In this thesis, simulation is used to investigate the dimension of the UMTS radio access networks. From the simulation results, important dimensioning rules can be derived and the proposed analytical dimensioning approaches can be verified. This chapter introduces in detail different UMTS simulation models which have been developed and used in this thesis. The simulation models of the UMTS Rel99 with the ATM-based UTRAN and the UMTS with the IP-based UTRAN were developed by the author in the framework of the MATURE project. The main objective of this project is the dimensioning of the UTRAN Iub interface. The simulations of HSPA (HSDPA and HSUPA) were using the HSPA simulation models established in the context of another project by Communication Networks working group of the University of Bremen. At the beginning of this chapter, the main requirements on the simulation models are defined. Then a basic framework for setting up the simulation models is described. Afterwards, the simulation environment and detailed introductions on the modeling of different UMTS networks are given. The last section introduces different traffic models.

5.1 Requirements of Simulation Models

In the framework of this thesis, the basic requirement of the simulation model is to model a complete UMTS system following the 3GPP specifications, which consists of the fundamental UMTS network components, the relevant UMTS functions and protocol stacks, individual user groups and the related external networks. The simulation model needs to provide end-to-end data communication paths between the UEs and their corresponding communication entities in the external networks.

However, to save the simulation effort it is not necessary to model every UMTS component and function in detail depending on the focus of the simulation task. As introduced in the first chapter, the scope of this thesis is to dimension the UMTS radio access network with main focus on the Iub interface (only considering the user plane). According to this scope, the simulation models can be designed to have a detailed and accurate modeling of the Iub interface (user plane) including its related network nodes, protocol stacks, channels, and transport technologies, whereas the modeling of the UMTS radio interface and the core network are simplified by considering their main impacts on the Iub dimensioning. In this way, the complexity of the simulation model can be reduced, which results in an improved simulation efficiency.

For modeling the UTRAN network, in addition to the modeling of the basic functions and protocol stacks of the Iub interface, several important aspects which are closely related with the network dimensioning as discussed in Chapter 4, i.e. traffic characterization, QoS mechanisms, traffic control, resource managements, and network topology, need to be implemented in the simulation model as well. Moreover, the network-relevant QoS can be defined and evaluated. On the user level, for a detailed modeling of the user behavior and its relevant QoS, the individual user entity needs to be modeled and each user has to be uniquely identified throughout the network for the network resources management and the QoS control. To model different user groups, the simulation models should allow defining various applications and services, different QoS classes, and a variety of transport protocols like UDP, TCP and IP. Another important requirement is that the simulation model should be flexible to choose a transport technology for constructing the transport network in the UTRAN, and easy to configure the transport link bandwidth and the network structure. To model a large UTRAN network, the simulation models should be able to support multiple Node Bs and feasible to configure the number of Node Bs in the network as well as choose the appropriate network topology to connect the Node Bs to the RNC.

5.2 Simulation Model Framework

Based on the above requirements, a basic and common framework is designed for building the simulation models for various UMTS networks as illustrated in Figure 5.1. The simulation model is designed as a layered network model. It consists of several main components: groups of UEs and their corresponding nodes serving as the traffic source and sink, the air interface, the core network, and the UTRAN network which comprises Node B, RNC, and the Iub interface.

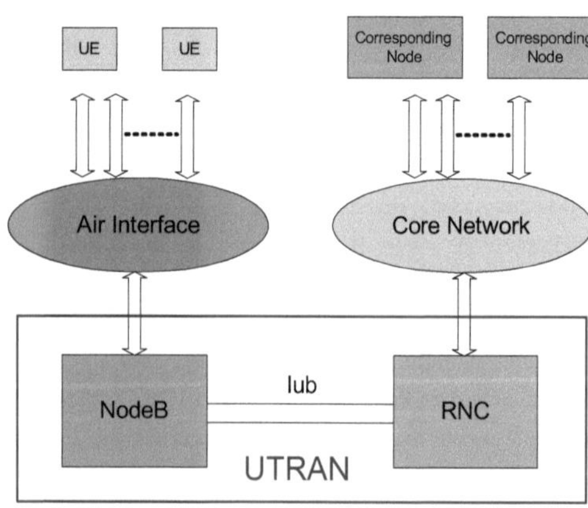

Figure 5.1: *Simulation Model Framework*

In this framework, for simplification of the simulation model, the UEs are not modeled as individual nodes distributing over the network, but instead all UE entities are created inside the same module where each UE connection is represented by an application session. Similarly, the corresponding nodes for individual UE connections are not modeled as separate servers or network nodes, but as the sink of individual UE applications. The UE entity includes all the protocol layers and functions related to applications, transport protocols, and UMTS radio access protocols. The applications are defined with traffic models, which define a set of parameters for characterizing the source traffic. The application traffic is carried by a specific transport protocol. The selection of transport protocol depends on the application type and the user preference. With UMTS radio access protocols, the user traffic is mapped to the UMTS transport channels. For the QoS managements, each UE entity can specify its QoS requirements and measure the obtained QoS on the user level.

The modeling of the UTRAN network including Node B, RNC and the Iub interface is the most important part of this framework. It requires a detailed and accurate modeling on the protocol stacks of the Iub user plane, the logical and transport channels for transmitting the user data, the required resource management and traffic control functions in the UTRAN, the applied QoS mechanisms, and various technologies for the building transport network.

The air interface model is implemented between the UEs and the Node B, and the core network is modeled between the corresponding nodes and the RNC. As addressed in section 5.1, the modeling of the air interface and the core network are simplified in the developed simulation models. For modeling the air interface, statistic assumptions or traces from the radio simulations can be taken to model the most relevant air interface characteristics and the physical layer effects, e.g. propagation delay, *Bit Error Rate* (BER) over the air interface. Inter-cell interferences and power control are not important for the Iub dimensioning as long as typical error rate and delay values at the air interface are modeled for individual users as well as per cell. The processing delays at the Node B and the UE in the UTRAN are also modeled as additional constant delay components of the end-to-end round trip time. With regard to the UMTS core network, its main impact on the Iub dimensioning is the extra processing delays that need to be added to the end-to-end delay of a user flow, e.g. TCP RTT. Thus, a constant core network delay value is modeled in the simulation model. Similar consideration is done for modeling the external network, e.g. for modeling the Internet an Internet round trip time is added additionally to the end-to-end round trip time.

This framework is also applied to the case when several Node Bs are connected to the RNC in the UTRAN. In this case, for each Node B the related UE entities, the air interface and the Iub interface need to be modeled, and the RNC needs to identify the traffic from different Node Bs and control the resources of each Node B separately.

The modeling of different UMTS networks share the same framework, but the detailed implementations of the UTRAN network or the air interface can be modified. In this thesis, the UMTS Rel99 simulation model is taken as the basic simulation model which is built according to this simulation model framework. Then for each evolved UMTS network, its specific new functions and features can be added to the basic Rel99 simulation model, and the required changes on the protocol structure and the transport network need to be implemented by modifying the corresponding layers in the Rel99 simulation model.

5.3 Simulation Environment

In this thesis, the simulation models were developed with the OPNET Modeler (version 10.0) [Opnet03]. OPNET is one of the most widely used network simulators and has been designed to support the modeling and simulation of communication networks and distributed systems. OPNET provides a comprehensive development environment to analyze both behavior and performance of modeled systems by performing *discrete event simulations*. OPNET has a three-layer modeling hierarchy. The highest layer referred to as the network model allows definition of system topologies. The second layer referred to as the node model allows the definition of node architectures (data flow within a node). The third layer is the process model that specifies the logic or control flow among components in the form of a *Finite State Machine* (FSM).

OPNET provides a set of built-in model libraries providing a variety of communication networks and technologies, e.g. UMTS, ATM or WLAN. In this thesis, the author did not choose the OPNET UMTS module and its library. There are three major reasons. Firstly the OPNET built-in UMTS model is a complex UMTS simulation model including an exhaustive modeling of the radio interface and the core network, which is however not the main focus for the objective of the simulation models in this thesis. Secondly, the OPNET UMTS model only supports packet switched traffic, and does not support circuit switched traffic. But traditional voice traffic is a main traffic type to be considered in the UMTS network. At last, there is no modeling for HSDPA and HSUPA in the built-in OPNET UMTS model family. Therefore, in this thesis the UMTS simulation models were developed separately with own implemented UMTS functions and protocols based on a number of OPNET built-in models, e.g. ATM, IP, and Ethernet models. The following sections will give a detailed introduction of the developed UMTS simulation models in OPNET.

5.4 Simulation Model of UMTS Rel99

This section introduces the modeling of UMTS Rel99 with ATM-based UTRAN. The implementation of the Rel99 simulation model is according to the simulation model framework presented in section 5.2, taking considerations of the overall requirements on the simulation model which are pointed out in section 5.1.

The structure of Rel99 simulation model is shown in Figure 5.2, with the same structure as the simulation model framework presented in Figure 5.1. The required protocol stacks for the

UEs, corresponding nodes, and the UTRAN Iub interface (user plane) as shown in Figure 5.2 are fully implemented in the simulation model. The protocol stack of UTRAN Iub user plane is according to the 3GPP Rel99 specifications, which is introduced in section 3.3.1. For the UEs and the corresponding nodes, the application layer and the corresponding transport layer protocols like TCP or UDP are modeled. In addition, the UEs include the RLC and MAC entities to communicate with the UTRAN. The air interface is modeled between the UE and the Node B, and the core network was between the corresponding node entity and the RNC.

In UMTS Rel99, ATM is defined as the transport technology for the UTRAN transport network. Due to this, the UMTS Rel99 simulation model is implemented based on the OPNET ATM workstation and ATM server model (see Appendix A.8). Their node models are shown in Figure 5.3. Both models have the same node structure, except that the ATM workstation model is working as the client and the ATM server model is used as the server. Both the OPNET ATM workstation and server model consist of the application module, the transport protocol modules (TCP, UDP and IP), the AAL and the ATM module, and the ATM physical layer. These modules represent the essential elements for constructing an ATM-based UMTS network. Based on them, the UMTS related functions and protocol layers are implemented.

Figure 5.3 shows the implementation of the UMTS Rel99 network model according to the designed structure given in Figure 5.2. In this structure, the modified ATM workstation model is used to model groups of UEs and the Node B including the air interface in between, while the modified ATM server model is used to model the RNC, the core network, and the corresponding nodes in the external network. The UMTS radio protocol layers, i.e. RLC, MAC and FP, are implemented above the AAL/ATM.

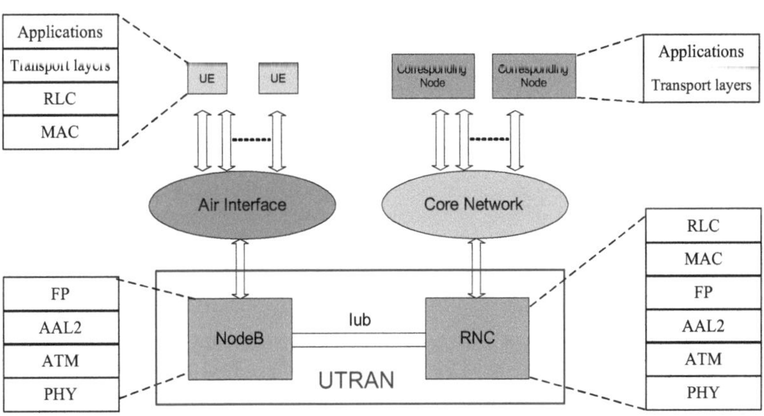

Figure 5.2: *UMTS Rel99 Simulation Model Structure*

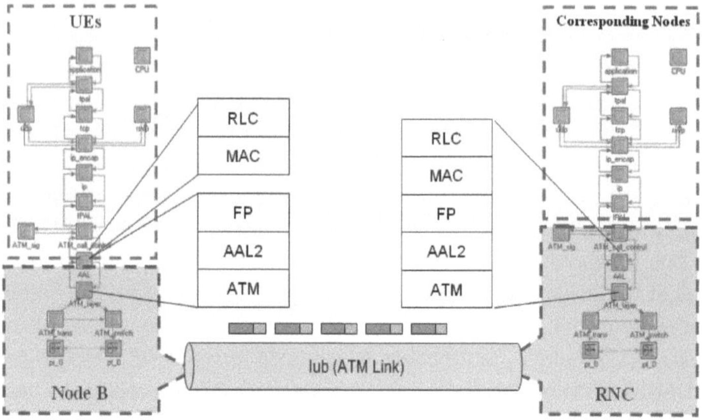

Figure 5.3: *Modeling of UMTS Rel99 (ATM-based UTRAN) in OPNET*

As shown in Figure 5.3, the UE part comprises all the modules above the ATM protocol stack including the implemented RLC and MAC protocol layer functions. The Node B part includes the FP protocol layer, the underlying AAL/ATM layer, and the ATM physical layer. Between the Node B and UE functionalities there is a simplified air interface model. On the other side, the RNC part consists of the physical layer, AAL/ATM layer, and in addition all UMTS radio protocols (i.e. RLC, MAC and FP protocols). Above the RNC part, a simplified core network and the corresponding nodes of the UEs are modeled.

A brief introduction of the developed Rel99 simulation model has been published in [WL+06]. The following introduces in detail the implemented UMTS Rel99 channels, network components, related radio and transport network layer protocols, and specific transport network resource management and traffic control functions.

General Modeling

In the Rel99 simulation model, it is assumed that there is sufficient radio resource (i.e. capacity, power, code) at the air interface, as the resources in the Iub transport network is the major concern in this work. That means the user performance is mainly determined by the transport network of the Iub interface. In this model, each Node B serves up to three cells, which is independent of each other. Each cell can establish its own channels and users. They all share the same Iub link. As only the aggregated traffic of all three cells at the Iub is important, there is no identification of traffic for each cell. Moreover, handover function is not specifically modeled. Because the dimensioning of the Iub interface only takes into account the total amount of the traffic generated by the whole cell. For the Iub dimensioning purpose, it is not necessary to distinguish whether the traffic is for the handover users or not.

Modeling of Channels

Section 3.3.2 introduces the logical and transport channels in UMTS defined by the 3GPP. In this simulation model, only two types of logical channels and three types of transport channels are modeled. The modeled logical channels are Dedicated Traffic Channels (**DTCH**) and Dedicated Control Channels (**DCCH**). Each user connection has a pair of DTCH and DCCH where the DTCH is the traffic channel for transmitting the user data and DCCH is the control channel for sending the corresponding control information for this user connection. According to 3GPP 34.108 [3GP04e], a variety of RABs can be assigned to the DTCH while the DCCH only uses *Signaling Radio Bearer* (SRB) of 3.4 kbps. Both logical channels are normally mapped to **DCH** transport channel. But if the user is transmitting the data via a shared common transport channel, they will be mapped to Forward Access Channel (**FACH**) on the downlink and Random Access Channel (**RACH**) on the uplink. Each UMTS cell models one pair of RACH and FACH, the implemented data rate is 32 kbps for each. While for each DCH, a specific RAB rate is assigned depending on the user requirement and the network resource management functions. For these three transport channels, the corresponding *Transmission Time Interval* (TTI) and the *Transport Block Set* (TBS) are modeled. When coming to the ATM-based transport network, each DCH is mapped onto a separate AAL2 connection, while there is one individual AAL2 connection for RACH and FACH respectively.

UE Modeling

As explained in the overall simulation model framework in section 5.2, the UEs are not modeled as individual nodes in the simulation model but instead each UE connection is represented by an application session and identified by its session identifier. The UE model includes the application layer, the transport protocol, and the involved UMTS radio access protocols. The UE supports a variety of applications and services in this model, where for each application a certain transport protocol can be specified, e.g. TCP or UDP. The OPNET ATM workstation model used only allows applications running over IP using either TCP or UDP protocol, without any circuit-switched path. Therefore, the OPNET ATM workstation model was extended within this work to add another transport path to enable circuit-switched type of applications like voice telephony. With this extension, the simulation model can perform simulations for various traffic mix scenarios: either pure packet-switched traffic which is mainly based on IP, or pure circuit-switched traffic, or both mixed together. Another important feature about the UE modeling is that the UEs can be generated statistically in the simulations. Each UE can be modeled to be active for a limited duration. Depending on different applications, different statistical distributions can be used to define the arrival process of the UEs and their durations. Moreover, the application traffic for a certain user is generated according to the traffic models (a detailed introduction is given in section 5.8). At last, the UMTS radio protocols of a UE, i.e. RLC and MAC as well as the corresponding channels were modeled. Their implementations are explained below.

Air Interface Modeling

In the Rel99 simulation model, the modeling of the air interface is simplified. Two effects are modeled: processing delay over the air interface and the channel condition of each UE. The processing delay is modeled as a constant delay. The user channel condition only models the experienced bit error rate (BER), which is binomially distributed with a mean of 5% per user. With this assumption, there is no differentiation of good and bad users, as all have the same loss probability in the air interface. When BER is greater than zero, there is frame loss over the air interface. This will cause RLC retransmissions, which result in the increase of the end-to-end application delay.

Modeling of Radio Network Layer (RNL) Protocols

The RNL include RLC, MAC and FP protocol layers. Their specifications have been introduced in section 3.3.1. The modeling of the RLC considers both transparent mode (TM) and acknowledged mode (AM). The RLC AM is only used for the packet-switched traffic. For simplification of the simulation model, the basic RLC functions, i.e. segmentation and reassembly, padding, in-sequence delivery of the upper layer PDUs, and Automatic Repeat Request (ARQ) were modeled, but other functions such as flow control, window mechanisms, etc., were not implemented in the model. The modeling of the ARQ function was simplified. Instead of implementing the complex procedure including polling, generating status reports, managing the retransmission buffer, etc., the retransmissions of the RLC PDUs were modeled by delaying the PDUs with certain retransmission delay (either taking empirical values or from a statistical distribution). For the MAC layer, the transport channels like DCH, RACH and FACH were modeled where the user data is mapped onto them. For each transport channel, the forming of the Transport Block Set (TBS) and the corresponding Transmission Time Interval (TTI) were implemented depending on the assigned data rate. The FP layer conveys the data to the Iub transport network layer. In the simulation model, the FP PDU delay, i.e. the delay of transmitting one FP PDU from the RNC to the Node B, is measured at the Node B as an important transport network QoS metric. It is defined that whenever the measured FP PDU delays exceed a predefined delay threshold, the corresponding FP PDUs are discarded. In addition, the protocol overheads for all these layers are implemented according to the 3GPP specifications.

Modeling of the Iub Transport Network

The modeling of the Iub transport network is based on the OPNET provided AAL and ATM modules. They include the detailed modeling of the AAL and ATM protocols according to the standardizations. The ATM module models establishing the ATM VP or VCs, providing various ATM QoS categories and defining the QoS parameters, ATM switching and routing, cell encapsulation, etc. For the Iub user plane, the AAL2 protocol is used. The AAL2 module contains the modeling of segmentation and reassembly, AAL2 multiplexing, etc. The protocol details are introduced in section 3.3.3.2 and 3.3.3.3.

In this work, the AAL2 module was extended to map the UMTS transport channels onto the AAL2 connections. For each transport channel, one AAL2 connection was modeled with a unique connection identifier. Moreover, an AAL2 shaping function was implemented which limits the rate of sending the AAL2 packets to the ATM layer according to the given PCR rate on the VP/VC. At the AAL2 layer, a QoS scheme was implemented to prioritize different QoS classes. In the current simulation model, only two QoS classes were considered: real time (RT) and non real time (NRT). For each QoS category, an individual AAL2 buffer was created. The traffic belonging to the same QoS class is sent to the corresponding buffer, depending on the applications and services of the user connections. And between the two AAL2 buffers, a strict priority packet scheduler was implemented: the RT traffic is given higher priority over the NRT traffic. The implemented QoS structure at the AAL2 layer is illustrated in Figure 5.4.

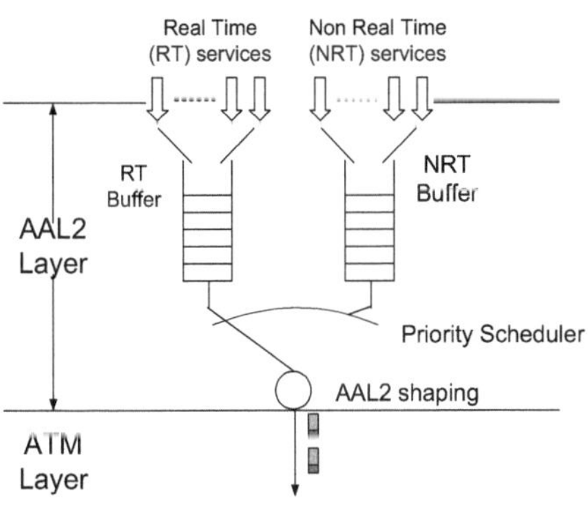

Figure 5.4: *Implemented AAL2 QoS Structure*

The ATM physical layer model is also provided in the OPNET ATM model family. A variety of ATM physical links like E1, STM1 links can be chosen to configure the Iub links offering differing link capacities. Moreover, the simulation model provides the possibility of combining a number of ATM links into a virtual link by using *Inverse Multiplexing for ATM* (IMA) technique. This function is useful if the Iub link capacity needs to be extended, e.g. over a bundle of E1 cables. In addition, in the simulation model 300kbps from the total ATM link bandwidth is reserved for the Iub control plane, i.e. for transporting the signaling traffic such as NBAP.

The simulation model allows establishing multiple Node Bs connected to the RNC. With the use of ATM switches or routers, various network topologies can be configured at the Iub interface. The possible network topologies are introduced in section 4.2.2.

Modeling of the Resource Management Function

In section 3.3.4.3, three important resource management functions are introduced. They are *Connection Admission Control* (CAC), *Bit Rate Adaptation* (BRA) and *Channel Type Switching* (CTS). They were implemented in the simulation model according to the concepts described in section 3.3.4.3.

For the CAC algorithm, the offered transport resource is based on the assigned bandwidth of the VP or VC, i.e. the *Peak Cell Rate* (PCR). Each AAL2 connection corresponding to each UMTS DCH is required to reserve a certain amount of ATM bandwidth for the data transmission on the Iub interface. This is so-called CAC guaranteed bit rate in section 3.3.4.3. In the simulation model, the CAC guaranteed bit rate is a configurable parameter for setting up the simulation scenarios. For each RAB type, there is an individual setting of the CAC guaranteed bit rate. This means, various RAB types with different peak data rates require different bandwidth reservations on the Iub links. In this thesis, the CAC guaranteed bit rate per RAB type is configured with the measured average user throughput for a certain traffic model on the ATM link layer including all the transport network layer overheads. The average throughput is measured during the data transmission period within the duration of the AAL2 connection, as shown in Figure 5.5. The obtained average throughput depends on the peak data rate of the RAB as well as the applied traffic model. Table 5.1 gives an example of measured average link throughputs for various RABs. In this example, an ftp service of a mean file size of 25kbyte is given with different file size distributions. It can be seen that for the same mean file size, the measured average link throughputs (per user connection) using different file size distributions are similar with small deviations, and the value is larger for a higher RAB rate. If the applied traffic model changes, the measured average throughputs can be also different. It is suggested that before performing any network dimensioning, for a given traffic model, an average link throughput of a user connection should be measured for each RAB type to configure its corresponding CAC guaranteed bit rate. The setting of the CAC guaranteed bit rate can be also based on other optimization algorithms.

The BRA and CTS functions are modeled only for the DCH transferring the packet switched traffic, mainly best effort and interactive traffic types. For modeling the BRA function, four parameters were implemented to decide an upgrade or downgrade of the RAB for a user connection, i.e. upgrade threshold, downgrade threshold, monitoring period and waiting time. Moreover, for upgrading the data rate, the CAC function is required to determine whether there are enough transport resources to support a higher data rate. The modeling of CTS requires defining an inactive timer to make a decision on when to switch off the DCH and move to the common channels, and applying the CAC function to check if there is enough transport resource for setting up a new AAL2 connection in case that the user is active again with an increased traffic demand and thus can be switched back to use DCH. In the Rel99 simulation model, the CAC, BRA and CTS functions are working together to attain a cost-efficient utilization of the resources in the UMTS radio access network.

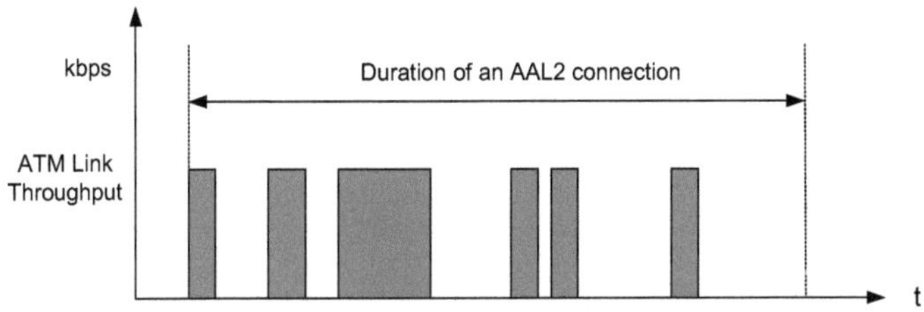

Figure 5.5: *Measuring the CAC guaranteed bit rate*

	RAB 64 kbps	RAB 128 kbps	RAB 384 kbps
Constant distribution	42.5	60.0	76.1
Exponential distribution	43.7	58.6	80.2
Normal distribution	41.1	62.6	78.6

Table 5.1: *Measured average ATM link throughput for different RABs*

QoS Mechanisms in the Transport Network

In the AAL2/ATM transport network in UTRAN, there are two basic QoS methods for differentiating the traffic. The first one is on the AAL2-level with the use of AAL2 prioritizations, as explained in Figure 5.4. The second one is on the ATM level by using different ATM VC/VPs with different QoS categories. For example, the delay sensitive traffic can use CBR VCs while the interactive or best effort traffic can use UBR VCs. Both methods can be used in the simulations.

Summary

The Rel99 Simulation model is a detailed implemented simulation model including the basic UMTS Rel99 functions and protocol stacks with the main focus on the Iub interface. The Iub transport network is fully modeled with the implemented resource management functions, traffic differentiation and traffic control, QoS mechanisms. It is the simulation model for carrying out performance evaluation and dimensioning for Rel99 network, and also as the basic simulation model for building the other evolved UMTS networks.

5.5 Simulation Model of IP-Based UTRAN

One important evolution of the UMTS radio access network is the change of the transport technology from the ATM towards IP. In the framework of this thesis, the performances and dimensioning of the IP-based UTRAN are also an important part. Section 2.3.4.1 explains the motivations and advantages of using IP transport and section 3.4 introduces in detail the protocol structure and main functions of the IP-based UTRAN. This section introduces two implemented IP-based UTRAN simulation models in this thesis: an exact model and a simplified one.

5.5.1 Exact IP-Based UTRAN Simulation Model

This model was implemented as an exact simulation model including the detailed modeling of the IP UTRAN protocol stacks (see section 3.4.1), the main UMTS functions, channels, air interface and the transport network resource management functions. This model was implemented on the basis of the UMTS Rel99 model, by removing the lower AAL/ATM transport layers and replacing them with UDP/IP and Ethernet as the transport layers. Figure 5.6 shows the model architecture for the IP-based UTRAN.

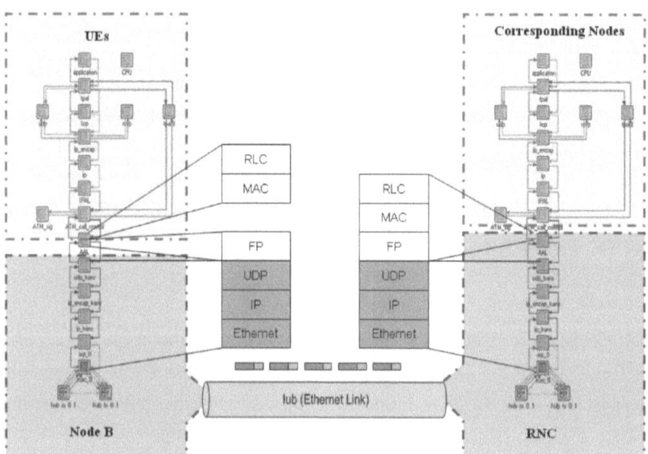

Figure 5.6: *Modeling of IP-based UTRAN in OPNET*

Compared to Figure 5.3, it can be seen that the model architecture is similar to the previous UMTS Rel99 simulation model except for the lower UTRAN transport network layers. In the presented node model, the upper application and transport layer modules are kept for modeling the behavior of UEs and their corresponding nodes. As well the modeling of the UMTS radio network layer protocols (RLC, MAC and FP), the channels, the air interface, and the resource management functions (i.e. CAC, BRA and CTS) are inherited from the developed Rel99 simulation model. However, the modeling of the UTRAN transport network uses the UDP, IP and Ethernet modules to replace the AAL2/ATM modules of the Rel99 simulation model. An overall introduction of this exact IP-based UTRAN simulation model has been published in [LC+07]. More detail description of this simulation model is given in Appendix A.12.

5.5.2 Simplified IP-Based UTRAN Simulation Model

In addition to the exact IP-based UTRAN model, a simplified model was implemented with the application of the IP DiffServ QoS scheme in the IP transport network. The main purpose of this simplified simulation model is to investigate the impact of IP DiffServ QoS on the overall network and end user performances as well as the Iub dimensioning. Therefore, this simplified model implements the detailed DiffServ QoS scheme according to the IETF specification (section 3.4.2.1), and deploys the DiffServ-based QoS structure which is presented in section 3.4.2.3. But on the other hand, the modeling of the UMTS radio protocol stacks, the channels, the air interface and the resource management functions are simplified in order to reduce the complexity of the simulation model to be mainly focused on the IP transport network.

The simplified IP-based UTRAN simulation model was built based on the standard OPNET IP/Ethernet node model, instead of modifying the Rel99 ATM-based UMTS model. In this simplified model, the UE still consists of the detailed models of the application and the corresponding transport protocol layers. But the models of the UMTS radio protocols (RLC, MAC, and FP) are simplified with only taking the overheads into consideration. For each user connection, the RAB rate can be configured where the data rate of the user is limited by the assigned RAB peak rate.

The IP-based transport network is implemented with UDP, IP and Ethernet layers. In the IP layer, the detailed DiffServ QoS function is modeled. The model of the DiffServ is provided in the standard OPNET IP model library. To set up the IP-based UTRAN transport network, the IP Routers are used to connect the Node Bs and the RNC in the simulation scenarios. These IP routers support the IP DiffServ QoS function, *Weighted Random Early Detection* (WRED) discarding function, various queuing managements like *Weighted Fair Queuing* (WFQ) scheduler, strict priority scheduler, etc. In these routers, the IP interface shaping is allowed which is used to allocate the IP bandwidth for the corresponding Iub link.

5.5.3 Summary

For the modeling of the IP-based UTRAN network, two simulation models were built within this thesis. The exact IP-based UTRAN model is a detailed simulation model, which was built based on the Rel99 ATM-based UTRAN model. It inherited the main UMTS functions from the Rel99 simulation model, but replaced the AAL2/ATM protocol stack with the UDP/IP and Ethernet layers, and implemented own specific functions such as LIPE/CIP multiplexing scheme, addressing, QoS schemes, etc. It is a useful simulation model for carrying out the detailed performance analysis and evaluations. The simplified IP-based UTRAN model, instead of providing the detail modeling of the UMTS specific functions, is focused on modeling of the IP transport network with the DiffServ-based QoS structure. As in the framework of this thesis, the DiffServ QoS is an important function to consider for the dimensioning of the IP-based UTRAN, therefore the simplified IP-based UTRAN model is chosen in this thesis to perform the dimensioning for the IP-based Iub interface. The corresponding dimensioning results are presented in Chapter 7 for a single IP-based Iub link and in Chapter 8 for a multi-Iub RAN network scenario.

5.6 Simulation Model of HSPA

The HSPA is the extensions of the UMTS Rel99. In the framework of this thesis, the UTRAN network for the HSPA is based on the ATM. The HSPA simulation model was established in the framework of the HSDPA project by the Communication Networks group in the University of Bremen.

The HSPA simulation model was built separately from the UMTS Rel99 simulation model, but it was developed under the same simulation environment as for the Rel99 simulation model described in section 5.3, and as well it was implemented based on the OPNET ATM workstation and ATM server models adding the HSPA specific functions and protocol layers. The model structure is in accordance with the common simulation model framework presented in section 5.2, including the modeling of groups of HSPA UEs and their corresponding communication nodes in the external networks, the air interface, the Node B, the RNC and the Iub interface. The HSPA simulation model includes the modeling of HSDPA for the downlink and HSUPA for the uplink separately. For each of them, the individual new protocol entities, the air interface, the MAC scheduler, and the related traffic control functions are modeled separately. When performing simulations, HSDPA and HSUPA can run simultaneously or independently. More detailed introduction on the HSPA simulation model is given in Appendix A.10.

5.7 Simulation Model of Combined UMTS Rel99 and HSPA

Usually, the HSPA services are provided on the existing UMTS Rel99 networks. To model such a network with both HSPA and Rel99 traffic, an approach was developed in this thesis to combine the Rel99 simulation model and the HSPA simulation model. The implemented approach is illustrated in Figure 5.7. It models a network with one Node B and one RNC which are connected with each other via a single Iub link. Here the modeled Node B and the RNC support both HSPA and Rel99 services. In this approach, there are two traffic paths: one for HSPA path and the other for Rel99 path. The HSPA path is realized by using the existing HSPA simulation model, which is introduced in section 5.6. The Rel99 path is however using the traces generated from the separate Rel99 simulations for both uplink and downlink, instead of using the developed Rel99 simulation model directly in the scenario. The main reason for not using the implemented Rel99 simulation model is due to that the Rel99 and HSPA simulation models were developed separately in different NSN research projects and they cannot run in the same simulation scenario. However using the trace data for the Rel99 traffic does not cause significant inaccuracy of the results, because in most of cases the Rel99 traffic has a higher priority to transmit than HSPA traffic and thus the characteristics of the Rel99 traffic is not influenced by the generated HSPA traffic. One benefit of using this approach is its simulation efficiency. Because the Rel99 path uses only the traces generated from the separate simulations, it reduces the complexity of the combined simulation model as well as decrease the overall simulation run time.

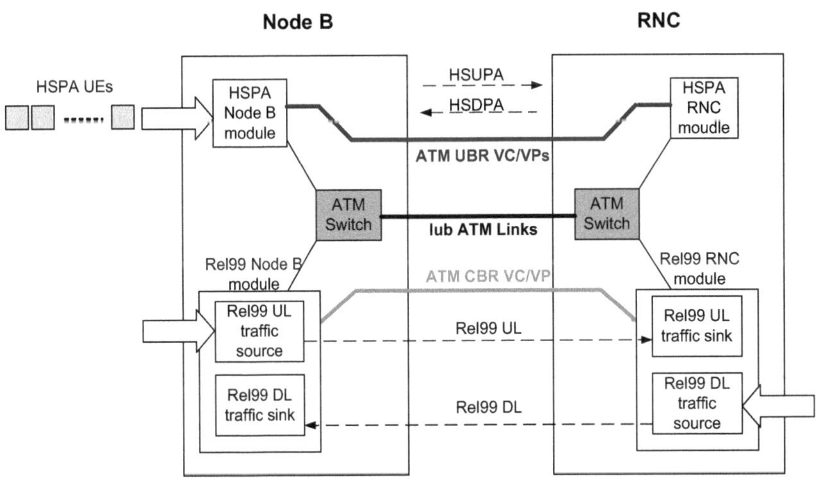

Figure 5.7: *Modeling of combined UMTS Rel99 and HSPA*

As shown in Figure 5.7, the model of the Node B is composed of the HSPA Node B module from the existing HSPA simulation model, and the Rel99 Node B module including an uplink (UL) traffic source and Rel99 downlink (DL) traffic sink. The model of the RNC is composed of the HSPA RNC module from the existing HSPA simulation model, and the Rel99 RNC module including a DL traffic source and UL traffic sink. The HSPA traffic and Rel99 traffic are carried by different ATM VC or VPs. For example, the HSDPA and HSUPA traffic can use either two individual UBR VCs (one for HSUPA and the other is for HSDPA) or a single UBR VC, while the Rel99 traffic is using a separate CBR VC. By defining different ATM QoS categories, and giving different configurations and priorities on the individual VCs for different traffic paths, differentiated QoS support can be attained and the utilization of the transport resources can be optimized. This will be discussed in Appendix A.16.

5.8 Traffic Models

Knowing the traffic demand that a UMTS network has to satisfy is an essential point when coming to the dimensioning of its resources. For this purpose, traffic models are needed in order to predict the user load in the network. Traffic models are used to characterize various service types with certain parameters characterizations. This section gives a survey of various traffic models for voice, web browsing and ftp applications, and emphasizes the traffic models which are used in this thesis.

5.8.1 Voice Traffic Model

Voice traffic is characterized by its symmetric nature, where the inherent QoS allows a very small delay [ETSI98]. For voice traffic, the traffic model can be described as an ON-OFF model with active and silent periods, illustrated in Figure 5.8.

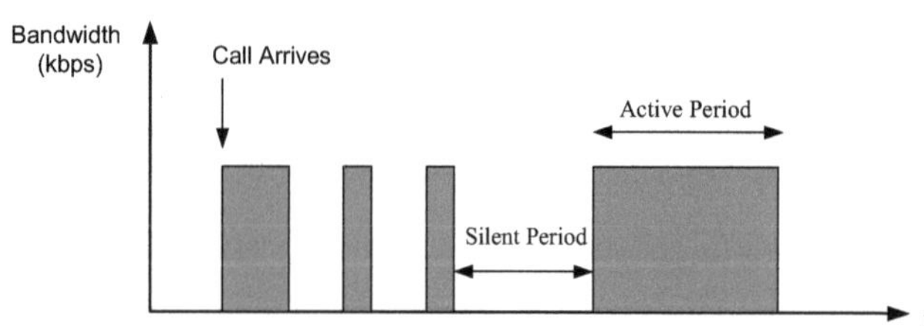

Figure 5.8: *Voice traffic model*

In this thesis, the voice traffic model defined in the MOMENTUM project [TWP+02] is used for simulations and performance analysis. In this voice traffic model, both active (ON) and silent periods (OFF) are exponentially distributed. Mean value for active and silence periods are equal to 3 seconds and independent on the up and downlink. In addition, the call duration is also exponentially distributed. The exponential probability density function is calculated as below:

$$f(t)=\frac{1}{\tau}e^{-\frac{t}{\tau}},t\geq 0 \tag{5.1}$$

Where τ is the average voice call duration, and $\mu = 1/\tau$, is the service rate. In this voice traffic model, an average call duration of 120 seconds is defined. Assuming a large number of independent voice calls, according to Palm's theorem (Theorem 6.1) the call arrival process can be modeled as a Poisson process, which also implies that the call interarrival times (the time between two consecutive calls) are exponentially distributed.

In addition to model the user behavior during a call conversation, a codec model is also required to characterize the properties of the applied speech codec. In UMTS, the *Adaptive Multi-Rate* (AMR) speech codec is adopted as the standard speech codec by 3GPP [3GP99a]. More information of AMR is given in Appendix A.6. Table 5.2 summarizes the parameters for this voice traffic model with AMR 12.2 voice codec.

Voice Traffic Model	
Parameter	Distribution and value
Voice Codec	Adaptive Multi Rate (AMR) 12.2kbps
Silence period	Exponential distribution, mean = 3 seconds
Speech period	Exponential distribution, mean = 3 seconds
Session duration	Exponential distribution, mean = 120 seconds

Table 5.2: *Traffic model for voice [TWP+02]*

In addition, there are also a number of alternative voice traffic models. A. Schieder [Sch03] applied G.723.1 speech codec and a voice traffic model from Brady [Bra69] for modeling VoIP applications. It is also an ON-OFF model where activity and silent periods are generated by an exponential distribution. But the mean value for active period is set to be 1.35 seconds and for silent period is 1.65 seconds. And in [VRF99] a model for voice is described for mobile communications, which includes not only the ON-OFF behavior but also the effect of the voice encoder, compression device and air interface characteristics in mobile communications. It uses a four-state model, where each state defines the generated packet size, the probabilities that a packet is of certain size, and the burst duration with the Weibull distribution.

In the framework of this thesis, the air interface and UMTS protocol layers are modeled separately. Therefore, a voice traffic model only describing the voice user behavior is required. The traffic model from the MOMENTUM project defines a longer mean duration for both active and silent periods than the Brady model and it generates a higher traffic. For the UMTS network, it is a more suitable voice traffic model due to its symmetric property and higher traffic demand. Therefore it is chosen to model the voice traffic within this thesis.

5.8.2 Web Browsing Traffic Model

A general web browsing traffic model is defined in [ETSI98, TWP+02] as illustrated in Figure 5.9. It is defined on the *session* level. The session is unidirectional, i.e. in downlink. A web session contains one or several *packet calls*. Each packet call corresponds to the downloading of a web page. During a packet call several packets may be generated. After the whole web page is entirely downloaded, the user takes a certain amount of time for studying the information before making the request for the next web page. This time interval is called *reading time*.

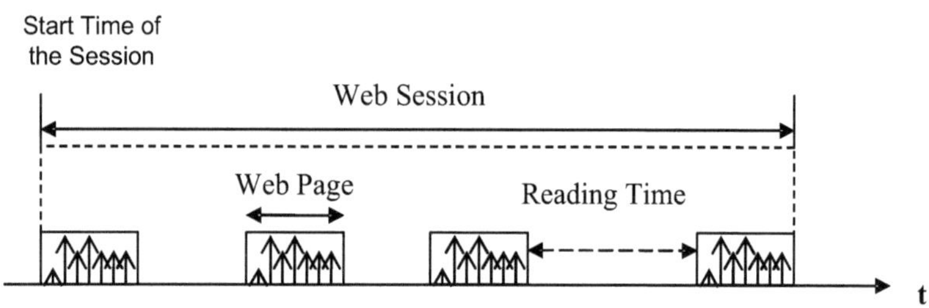

Figure 5.9: *General web traffic model*

The following parameters are defined to describe the characteristics of the web applications [ETSI98, TWP+02]:

- **Session arrival process**
  It defines how often the service sessions are arriving and set up in the network.
- **Number of packet calls (web pages) per session**
  Usually, a web session contains one or several packet calls where each packet call models a single web page.
- **Reading time between packet calls**
  Reading time starts when the last packet of the packet call is completely received by the user and it ends when the user makes a request for the next packet call. The mean reading time is denoted as μ_{IAT}.
- **Number of packets in a packet call**
  According to various networks, different statistical distributions can be used to generate the number of packets. The mean number of packets in a packet call is μ_p.

- **Inter arrival time between two consecutive packets inside a packet call**
  The interarrival time models the available transmission bandwidth. This assumption is based on the idea of having a flow control (like TCP mechanisms) which adapts the flow of packets to the available bandwidth. According to different networks, a variety of statistical distributions can be used.
- **Packet size**
  The packet size is a random variable with the mean packet size μ_{PS}.

The following table summarizes the web traffic models defined by the *European Telecommunications Standards Institute (*ETSI) [ETSI01], *Mobile Wireless Internet Forum* (MWIF) [MWIF01] and the MOMENTUM project [TWP+02].

Model Parameters	**ETSI**	**MWIF**	**Momentum Project**
Packet call Inter arrival Time (Reading Time)	Geometric distribution μ_{IAT} = 5 seconds	Exponential distribution μ_{IAT} = 12 seconds	Exponential distribution μ_{IAT} = 412 seconds
Packet call size	Pareto Distribution Parameters: α=1.1, k=4.5 Kbytes, m=2 Mbytes μ = 25 Kbytes	(Number of packets) x (Mean Packet sizes in a Packet call) μ= 25 x 480bytes = 12 Kbytes	(Number of packets) x (Mean Packet sizes in a Packet call) μ = 25 x 480 bytes = 12 Kbytes
Packet sizes with in a Packet call	Constant (MTU size) 1500 bytes	Pareto distribution parameters: α=1.1, k=81.5 bytes, m=66.666 Kbytes μ_{PS} = 480 bytes	Pareto distribution parameters: α=1.1, k=81.5 bytes, m=66.666 Kbytes μ_{PS} = 480 bytes
Number of packets in a packet call	Based on packet call size and the packet MTU	Exponential distribution μ_p = 25 packets	Geometric distribution μ_p = 25 packets

Table 5.3: *Traffic models for web application*

In the above table, all the three traffic models apply the *truncated Pareto distribution* (Pareto with cut-off) for modeling the packet or packet call size. The probability density function of such a truncated Pareto distribution is given by:

$$f(x)=\begin{cases}\left(\frac{\alpha k^{\alpha}}{x^{\alpha+1}}\right), & k>0,\alpha>0,k\le x<m\\ \beta, & x\ge m\end{cases} \tag{5.2}$$

Where k is the location parameter and α is the shape parameter, and m denotes the maximum value. Here β is the probability for $x \ge m$, calculated as follows:

$$\beta=\int_{m}^{\infty} f(x)dx=\left(\frac{k}{m}\right)^{\alpha},\quad \alpha>1 \tag{5.3}$$

Then the mean packet size μ_{PS} can be calculated by:

$$\mu_{PS}=\int_{-\infty}^{\infty} xf(x)dx=\int_{k}^{m} x\frac{\alpha k^{\alpha}}{x^{\alpha+1}}dx+m\left(\frac{k}{m}\right)^{\alpha}=\frac{\alpha k-m\left(\frac{k}{m}\right)^{\alpha}}{\alpha-1} \tag{5.4}$$

Comparing these three traffic models, it is found that the web traffic model defined by ETSI generates the highest traffic load at the user level, thus it is more suitable to model the web traffic in the nowadays UMTS networks and especially for the HSPA networks which are dedicated for transmitting the high speed packet-switched data. Thus the ETSI web traffic model is used in this thesis for generating the web traffic.

5.8.3 File Transfer Traffic Model

File transfer (ftp) application has similar statistical behavior like web browsing, but the main difference is that one large objects of web browsing is considered as one file transfer. Since one session only contains one file transfer, there is no reading time. One user can have several ftp downloading sessions. The following figure shows the general traffic model for the ftp service. It consists of ON and OFF period where each ON period represents one file transfer.

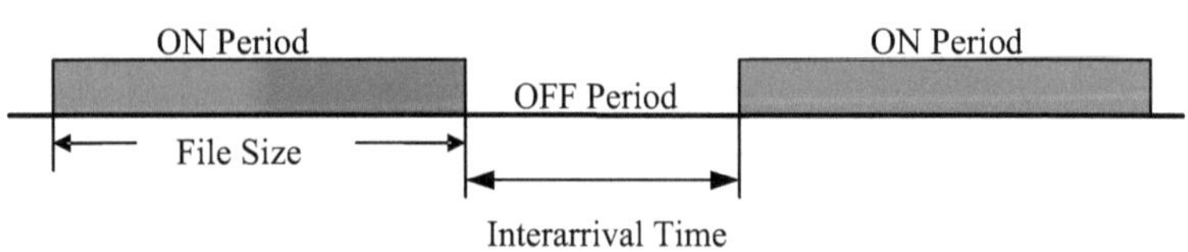

Figure 5.10: *File transfer (ftp) traffic model*

The file transfer traffic model in the MOMENTUM project [TWP+02] defines a Lognormal distributed file size distribution. The probability density function of Lognormal is given below [KLL01]:

$$f(x)=\frac{1}{\sqrt{2\pi}\sigma' x}e^{-\frac{1}{2}\left(\frac{\ln(x)-\mu'}{\sigma'}\right)^2}, x>0 \tag{5.5}$$

where σ' is the standard deviation of the natural logarithms and μ' is the mean of the natural logarithms. In the MOMENTUM file transfer traffic model, μ' is set to 30580 bytes and σ' is 3.6674. However, in this thesis this file transfer traffic model is not used. Instead, for validating the analytical dimensioning models in Chapter 6, a constant file size distribution is used.

6 Dimensioning for Single ATM-Based Iub

This chapter presents the methodologies for dimensioning a single Iub link in the UMTS Rel99 of an ATM-based UTRAN, with both simulation and analytical approaches. Novel analytical models are proposed in this chapter, which allow the dimensioning of a single Iub link for various traffic types and different QoS requirements. The traffic considered in this thesis is classified essentially into two main categories: circuit-switched traffic and elastic traffic (see Chapter 1.4). For these two traffic classes, section 6.2 and section 6.3 present two individual analytical dimensioning models at flow or connection level for their respective user-relevant QoS. Section 6.4 proposes analytical models for dimensioning the Iub link at network level to guarantee the relevant transport network QoS. The validity of the proposed analytical models is demonstrated by means of extensive simulations. Section 6.5 summarizes the main investigations of dimensioning for the ATM-based HSPA networks.

6.1 Overview and Objectives

In this chapter, the dimensioning process considers a single link, as shown in Figure 1.2 (a). It can be applied to the situations that the Node B with a direct link to the RNC or a logical Iub interface with an identifiable bottleneck. In the latter situation, other links might exist on the Iub interface but their influence on the user and network performance is considerably less than the bottleneck and therefore they can be neglected. Overall, the objective is to minimize the bandwidth costs while still meeting a desired degree of QoS. Given a certain traffic demand per traffic class, the amount of aggregated traffic carried on the Iub link need to be estimated. From this, the minimum required link bandwidth should be derived for a specific QoS target.

Every dimensioning procedure requires a model, which provides a relationship between the amount of offered load, the link bandwidth, and the achieved QoS, taking into account the characteristics of the workload and the transmission process. For the circuit-switched traffic, it is assumed that it is generated by real time services with strict QoS requirements, which need to be guaranteed by the network with a guaranteed bandwidth. The resultant analytical dimensioning model needs to consider traffic characterization at connection level and the influence of admission control function. In case of elastic traffic the packet transmission process is mainly affected by the rate-sharing features of TCP. The proposed network dimensioning methodology is based on the processor sharing model. This thesis extends the processor sharing model in order to take into account specifics of the UMTS features and resource management functions, and also consider the case of mixing with circuit-switched type of traffic. To dimension the Iub link for satisfying the network-relevant QoS, e.g. packet delay or packet loss ratio on the Iub interface, queuing models are proposed to estimate the network performances. In this thesis, two arrival process models are presented to capture the characteristics of the aggregated traffic at the packet level on the Iub link. By solving the closed form of delay distributions of analytical queuing models, relevant network performances can be estimated and thus can be used for dimensioning process.

6.2 Dimensioning for Circuit-Switched Traffic

For dimensioning process in the framework of this thesis, the considered relevant user QoS criterion for the circuit-switched type of traffic is call blocking probability as a result of *Connection Admission Control* (CAC). Each connection requires a certain bandwidth in order to achieve the desired quality of service. TNL CAC, as introduced in section 3.3.4.3, is used to decide whether there are sufficient free resources on the transport link (i.e. Iub link) to allow a new connection. A connection can only be accepted if adequate resources are available to establish the connection with the agreed QoS while the quality of service of the existing connections in the network must not be decreased by the new connection, i.e. the available bandwidth on the link needs to be larger than or equal to the requested bandwidth of the new connection. In this way, the Iub link can only carry a maximum number of simultaneous connections (trunks) each occupying a certain bandwidth which is reserved for those connections. This section discusses the feasible analytical models to dimension the Iub link capacity to satisfy a required call blocking ratio of the circuit-switched type of traffic.

6.2.1 Erlang-B Formula

The **Erlang-B formula** has been published by the Danish mathematician A. K. Erlang in 1917 [EHA48]. It is considered to be the classical model to evaluate the performance of a loss system which has finite capacity and limited number of channels (servers, trunks, etc.).

The Erlang-B formula can be used to estimate the probability of a call being blocked and lost in a circuit-switched network. It is used under the following assumptions:

1. The network comprises of n identical channels. A call is accepted for service if any channel is idle. Every call needs only one channel.
2. If all channels are busy, a call is blocked and lost. It is called the Erlang loss model.
3. Furthermore it is assumed that the call arrive process is a Poisson process with rate λ and the service times are exponentially distributed with intensity μ (mean service time = $1/\mu$). Therefore, the Erlang loss model is denoted as M/M/n loss system.

In Erlang's loss model with Poisson arrival process, the offered traffic, denoted by A, is equivalent to the average number of call attempts per mean holding time as given in (6.1).

The unit of the offered traffic A is **Erlang**, which is a dimensionless unit.

$$A = \frac{\lambda}{\mu} \tag{6.1}$$

State transition diagram

Figure 6.1 shows the state transition diagram in order to describe the above loss system with a limited number of channels (n), Poisson arrival process (λ), and the exponential service times (μ).

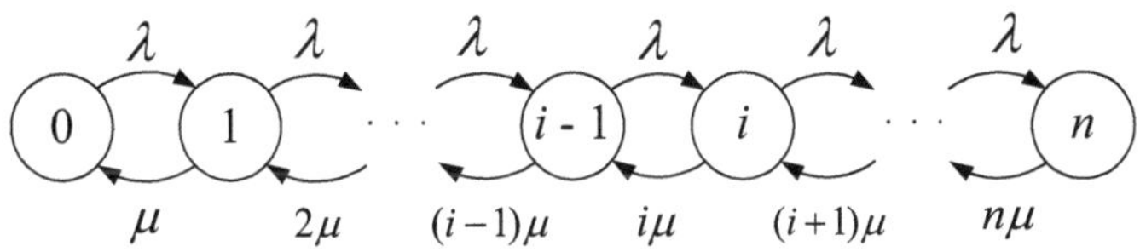

Figure 6.1: *State transition diagram for a system with a limited number of channels (n), Poisson arrival process (λ), and exponential service times (μ).*

The state of the system, depicted as i (i= 0, 1, 2, ... n), is defined as the number of busy channels. As the number of channels is limited to n, the number of states becomes $n+1$. In Figure 6.1, all states of the system are shown as circles, and the rates by which the traffic process changes form one state to another state are shown upon the arcs of arrows between the states. The transitions of the states only occur between the neighboring states. Assuming that the system is in *statistical equilibrium* [Ive04], the system resides in state i with the probability $p(i)$ at a random point of time. When the process is in state i, in case of state transition, it jumps to state $(i+1)$ with λ times per time unit and to state $(i-1)$ with $i \cdot \mu$ times per time unit.

State probabilities

State probability $p(i)$ defines the probability of observing the system with i number of busy channels at a random point of time. Knowing the state probabilities is essential to evaluate the system performances such as carried traffic, lost traffic, and utilization of the channels.

The state probabilities can be obtained by calculating the one-dimensional Markov chain shown in Figure 6.1. Although, for the state probabilities calculation, exponential service time distribution is assumed, it has been proven that the result is also valid for generally (arbitrarily) distributed service times (see [Kle75]), i.e. M/G/n loss system. The calculation of state probabilities requires to set up the equations describing the states of the system under the assumption of statistical equilibrium. There are two ways to set up the equations:

a. Global balance: in statistical equilibrium, the transition rate out of a state equals the transition rate into this state.

$$\begin{aligned} &\lambda \cdot p(0) = \mu \cdot p(1) && (i=0) \\ &\lambda \cdot p(i-1) + (i+1)\cdot \mu \cdot p(i+1) = (\lambda + i\mu)\cdot p(i) && (i>0) \end{aligned} \qquad (6.2)$$

b. Local balance: a fictitious cut is put between the states in (*i-1*) and *i*, then in statistical equilibrium the transition rate from state (*i-1*) to *i* is equal to the transition rate from state *i* to (*i-1*).

$$\lambda \cdot p(i-1) = i\mu \cdot p(i) \qquad (i = 1,2,\ldots n) \qquad (6.3)$$

As the system is always in one of the states, the normalization condition is applied:

$$\sum_{i=0}^{n} p(i) = 1, \qquad (p(i) \geq 0) \qquad (6.4)$$

It is noted that the equation (6.2) involves three state probabilities, whereas (6.3) involves only two. Therefore it is easier to solve the local balance equations. The detailed steps to solve the state probabilities for the one-dimensional state transition diagram (Figure 6.1) using local balance equations can be found in [Ive04]. The state probability formula is given in equation (6.5). It is also called as Erlang's first formula.

$$p(i) = \frac{\frac{A^i}{i!}}{\sum_{i=0}^{n} \frac{A^i}{i!}}, \qquad (0 \leq i \leq n) \qquad (6.5)$$

It can be noted that if n is infinite, i.e. $n = \infty$, the above Erlang loss system becomes a simple Poisson distribution, where $p(i) = \frac{A^i}{i!} \cdot e^{-A}$ (i=0, 1, 2…). An important property of the Erlang loss model (with Poisson arrivals) is its insensitivity property stating that the state probabilities (i.e. the equilibrium distribution of the number of connections present) are insensitive to the form of the service time (or holding time) distribution and requires only the mean service time [Coh79, Ive04].

Erlang-B formula

Knowing the state probabilities, the blocking probability p_b that a new call arriving at the circuit network is rejected can be calculated. According to the PASTA-theorem (*Poisson Arrivals See Time Averages*) [Wol82], the blocking probability p_b is equal to the probability that the system is in state n (i.e. all n channels are busy). Thus p_b can be derived from (6.5) with $i = n$.

$$p_b = p(n) = \frac{\frac{A^n}{n!}}{\sum_{i=0}^{n} \frac{A^i}{i!}} \tag{6.6}$$

This is so called Erlang-B formula (1917, [EHA48]). It is also called as Erlang first formula. It is assumed that the offered traffic A (in Erlang) are offered to n channels. Figure 6.2 plots blocking probability p_b as a function of the offered traffic A (in Erlang) for different n for the Erlang loss model calculated with Erlang-B formula (6.6).

As mentioned above, the Erlang loss model has the insensitive property. This property is also possessed by the Erlang-B formula. That means that the blocking probability p_b is insensitive to the form of the holding time distribution but depends only on the first moment of the holding time. This property will be validated via simulations in section 6.2.4. Another property of the Erlang-B formula is that it can be mathematically generalized to a continuous number of channels, i.e. non-integral number of channels. This is useful for the dimensioning to find out the required number of channels and in turn the bandwidth resources for a given offered traffic and desired blocking probability in the system.

Moreover, the fundamental assumption for the validity of Erlang-B formula is the Poisson arrival process. According to Palm's theorem [Ive04] this assumption can be taken if the traffic is generated by a large number of independent sources.

Theorem 6.1 *Palm's theorem: by superposition of many independent point processes the resulting total process will locally be a Poisson process.*

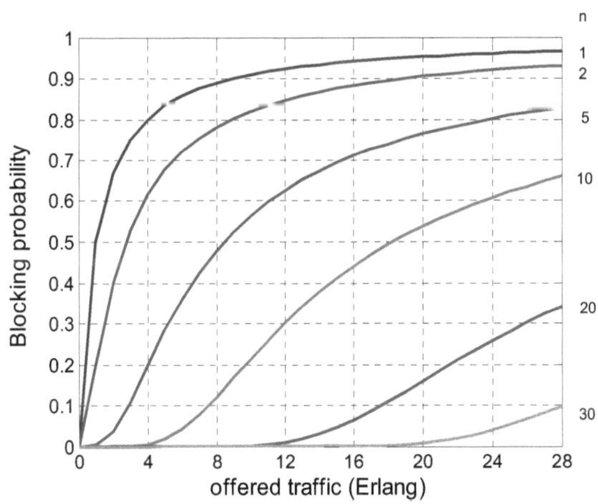

Figure 6.2: *Blocking probability p_b as a function of the offered traffic A in Erlang for various values of the number of channels n with equation* (6.6)

In normal telephone systems, it is usually the case that the call requests are assumed to be a Poisson arrival process. Thus, the Erlang-B formula was widely applied in telephone systems of both fixed and mobile networks. Due to its insensitivity property as mentioned above, it can be generally applied from voice telephony to other circuit-switched traffic which reserves a certain bandwidth.

Traffic characteristics of Erlang-B formula

With the Erlang-B formula (6.6), the important traffic characteristics of the loss system and dimensioning can be obtained. The most important performance measures for loss systems are [Ive04]:

- *Carried traffic* Y: the total traffic carried in the system. It is equal to the probability of channels being utilized, as shown in equation (6.7). A is the offered traffic, so the carried traffic is less than A.

$$\begin{aligned} Y &= \sum_{i=1}^{n} i \cdot p(i) = \sum_{i=1}^{n} \frac{\lambda}{\mu} \cdot p(i-1) = A \cdot \sum_{i=1}^{n} p(i-1) = A \cdot \{1 - p(n)\}, \\ Y &= A \cdot (1 - p_b), \end{aligned} \tag{6.7}$$

- *Lost traffic* A_l: the lost or rejected traffic due to the blocking.

$$A_l = A - Y = A \cdot p_b$$

- *Traffic congestion* C: according to the PASTA-theorem, the blocking probability p_b is also equal to traffic congestion C [Ive04].

$$C = \frac{A - Y}{A} = p_b$$

Channel utilization

Furthermore, from the dimensioning point of view, it is also important to measure the *channel utilization a.* The utilization a depicts the average utilization of a channels, i.e. how often this channel is being used for transmitting traffic.

$$a = \frac{Y}{n} = \frac{A \cdot (1 - p_b)}{n} \tag{6.8}$$

This function is shown in Figure 6.3 over different number of channels for given call blocking probabilities B. It can be observed that the utilization increases with larger number of channels, and also with a higher blocking probability value.

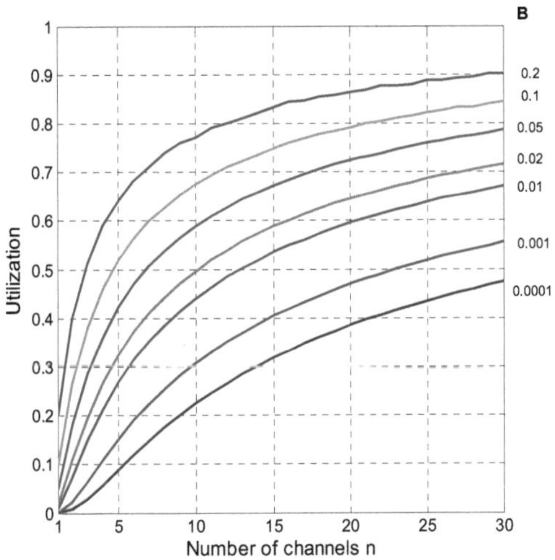

Figure 6.3: *The average utilization per channel a equation* (6.8) *as a function of the number of channels for given values of the blocking*

In this thesis, equation (6.9) is proposed to estimate the link utilization b based on equation (6.8). Assuming that the reserved bandwidth for each connection is r_m and each call generates an actual mean traffic θ (usually $\theta \le r_m$). If the reserved bandwidth for each call is optimally configured without overestimation or underestimation of the bandwidth usage of each call (i.e. $\theta = r_m$), then the link can be maximum utilized without violating the QoS. In this case, the link utilization can be calculated with equation (6.8). But if θ is less than the reserved bandwidth, the link utilization is calculated as a product of the utilization of channels with equation (6.8) and the average traffic carried on each channel (i.e. the amount of actual mean traffic generated by each call θ in a percentage of the reserved channel bandwidth r_m), as given in equation (6.9).

$$b = a \cdot \frac{\theta}{r_m} \tag{6.9}$$

6.2.2 Multi-Dimensional Erlang-B Formula

The Erlang-B formula deals only with a single traffic stream in the system where the arrival process is a Poisson process with arrival rate λ and the service times are exponentially distributed with intensity μ. But for the multiple traffic type systems, especially in today's wireless or mobile networks like UMTS, there are different classes of services with various traffic demands and QoS requirements. Each traffic class corresponds to a traffic stream, thus several traffic streams are offered to the same channel group. When the system is extended to receive a group of independent traffic streams each with its respective arrival rate λ_i and service rate μ_i, then the Erlang-B formula is expanded to a general case of a network with multiple traffic streams. The general form of the Erlang-B formula is the *multi-dimensional Erlang-B formula*.

Now consider a group of n bandwidth units (channels, trunks, etc.), which is offered to two independent traffic streams with arrival and service rates (λ_1, μ_1) and (λ_2, μ_2). Then, the offered traffic is $A_1 = \lambda_1/\mu_1$ for the stream 1 and $A_2 = \lambda_2/\mu_2$ for the stream 2. Let (i, j) denote the state of the system when there are i calls from stream 1 and j calls from stream 2. Each call is allocated to one and only one channel (assuming all channels are identical). And the following restrictions hold: $0 \le i \le n$, $0 \le j \le n$, and $0 \le i + j \le n$. For this case, the two-dimensional state transition diagram is shown in Figure 6.4, which corresponds to a reversible Markov process.

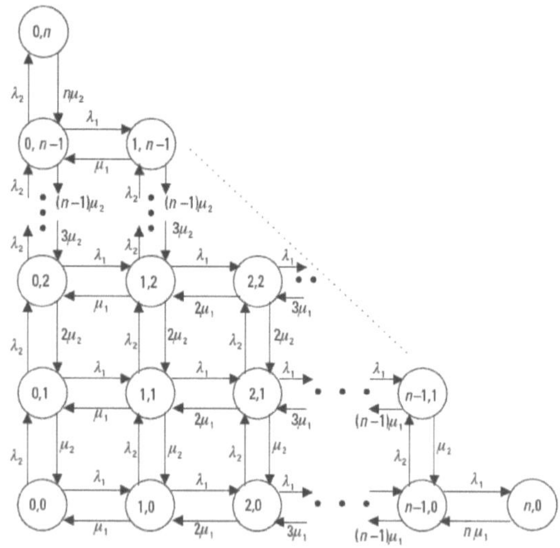

Figure 6.4: *Two-dimensional state transition diagram for a loss system with n channels offered to two traffic streams [Ive04]*

Under the assumption of statistical equilibrium in the system, the state probabilities can be obtained by applying the global balance equations. As the state transition diagram corresponds to a reversible Markov process, the state probabilities can be written in product form [Ive04]. Let $p(i, j)$ denote the probability that the system is in state (i, j), it is expressed in (6.10) in product form [Ive04]:

$$p(i,j) = p(i) \cdot p(j) = K \cdot \frac{A_1^{\,i}}{i!} \cdot \frac{A_2^{\,j}}{j!} \tag{6.10}$$

where $p(i)$ and $p(j)$ are the state probabilities of the one-dimensional Poisson distribution and K is normalization constant to conserve the state probabilities. By the Binomial expansion or by convolving two Poisson distributions, the following aggregated state probabilities can be obtained:

$$p(i+j=x) = K \cdot \frac{(A_1 + A_2)^x}{x!} \tag{6.11}$$

Where K is obtained by normalization in (6.12).

$$K = \frac{1}{\sum_{\upsilon=0}^{n} \frac{(A_1 + A_2)^{\upsilon}}{\upsilon!}} \tag{6.12}$$

As the arrival process is Poisson, due to the PASTA property, the time congestion, the call congestion and traffic congestion are all identical for both traffic streams, and they are equal to the aggregated probability $p(i+j=n)$.

The above two-dimensional system can be generalized to N traffic streams, then the state probabilities are expanded to a general form as given in equation (6.13). This is the general multi-dimensional Erlang-B formula [Ive04].

$$p(i_1, i_2, \ldots\ldots, i_N) = K \cdot \frac{A_1^{i_1}}{i_1!} \cdot \frac{A_2^{i_2}}{i_2!} \cdots\cdots \frac{A_N^{i_N}}{i_N!} \qquad 0 \le i_j \le n, \ \sum_{j=1}^{N} i_j \le n \tag{6.13}$$

In the above, it is assumed that each traffic stream requests the same bandwidth corresponding to one channel and thus the restriction of the number of simultaneous calls for each traffic stream is same. But, in many systems, the bandwidth requested by a call is dependent on the traffic stream (class), i.e. each traffic class may have an individual bandwidth requirement. For example, voice service may require one bandwidth unit per call, and video stream may require two or more. Thus, in this case, the maximum number of simultaneous calls allowed for each traffic stream is service specific. Let i_j denote the number of simultaneous calls for traffic class j and n_j the maximum number of simultaneous calls for traffic class j. They need to satisfy the following restrictions:

$$\begin{aligned} &0 \le i_j \le n_j \le n, \qquad j = 1,2,\ldots\ldots, N \\ &\sum_{j=1}^{N} i_j \le n \qquad and \qquad \sum_{j=1}^{N} n_j > n \end{aligned} \tag{6.14}$$

Due to the above restrictions, the state transition diagram is truncated. Figure 6.5 shows the two-dimensional state transition diagram for the two traffic streams case where the number of maximum calls for traffic stream 1 and 2 is n_1 and n_2 respectively. It is noted that the truncated state transition diagram still is reversible and that the relation in (6.10) still holds, but the normalization constant K is changed. In fact, due to the local balance property, any state can be removed without changing the properties.

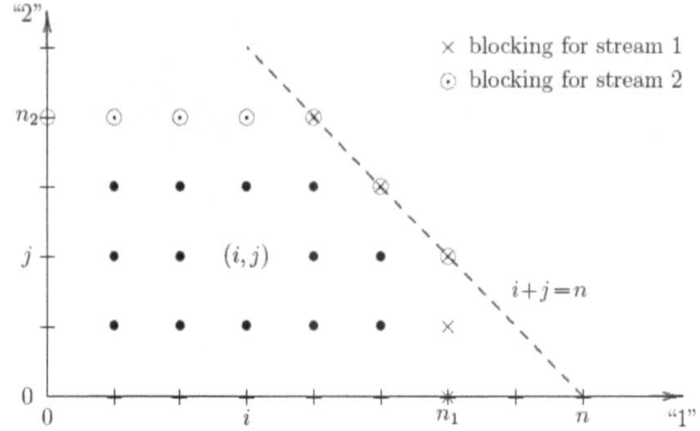

Figure 6.5: *Two-dimensional state transition diagram for a loss system with class limitations: n_1 for the traffic stream 1 and n_2 for the traffic stream 2[Ive04]*

Following gives an example of such a scenario. There are two traffic streams. The resource requirement for stream 1 is 1 bandwidth unit, and for stream 2 it is 3 bandwidth units. The total bandwidth in the system is 10 bandwidth units. Thus, n_1 equals to 10 (=10 units/1 unit) and n_2 equals to 3 (=10 units/3 units). The offered traffic of stream 1 A_1 is 5 Erlang, and the offered traffic of stream 2 A_2 is 1 Erlang. It is assumed that the service rate is equal to 1. Figure 6.6 illustrates the two-dimensional state transition diagram and Figure 6.7 shows the calculated state probabilities $p(i, j)$.

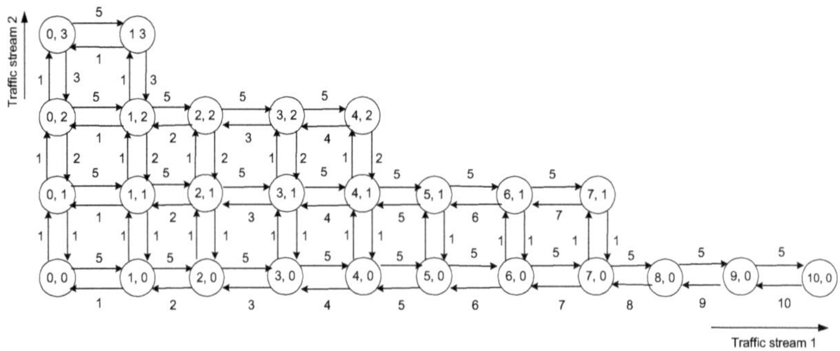

Figure 6.6: *Two-dimensional state transition diagram for a loss system with class limitations: (n_1 =10, λ_1=5, μ_1=1); (n_2 =3, λ_2=1, μ_2=1)*

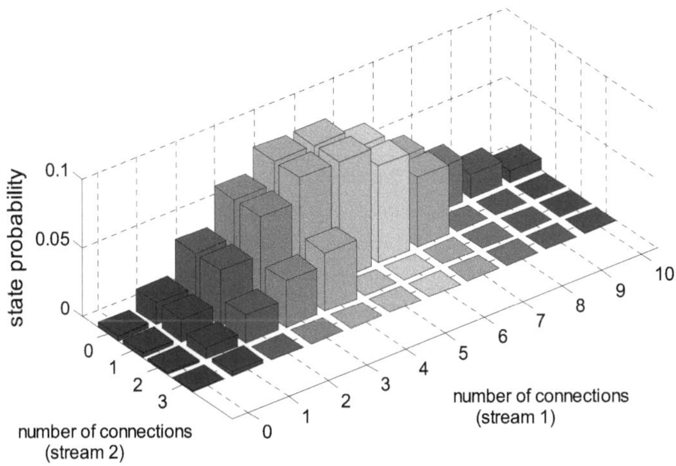

Figure 6.7: *State Probability of two-dimensional Erlang-B formula with class limitations:* *($n1$ =10, $\lambda 1$=5, $\mu 1$=1); ($n2$ =3, $\lambda 2$=1, $\mu 2$=1)*

The exact value of probability of each state $p(i, j)$ is given in Table 6.1.

p(i,j)	i = 0	i = 1	i = 2	i = 3	i = 4	i = 5	i= 6	i= 7	i=8	i =9	i=1 0
j = 3	0.00054	0.0027									
j = 2	0.0016	0.0081	0.020	0.034	0.042						
j = 1	0.0032	0.016	0.040	0.067	0.084	0.084	0.070	0.050			
j = 0	0.0032	0.016	0.040	0.067	0.084	0.084	0.070	0.050	0.031	0.017	0.009

Table 6.1: *State probabilities of two-dimensional Erlang-B formula with class limitations:* *(n_1 =10, λ_1=5, μ_1=1); (n_2 =3, λ_2=1, μ_2=1)*

In this example, the blocking probability for traffic stream 1 is 0.1038 and for traffic stream 2 is 0.3618, respectively, as calculated below.

$$p_b(1) = p(i+j=10) = p(10,0) + p(7,1) + p(4,2) + p(3,1) = 0.1038$$

$$p_b(2) = p(i+j=10) + p(i+j=9) + p(i+j=8) = 0.3618$$

6.2.3 Iub Dimensioning with the Erlang-B Formula

The job of dimensioning is to determine the necessary bandwidth or the required number of channels on the Iub interface to accommodate the given offered traffic with a guaranteed QoS requirement. The following presents the detailed approach to apply the analytical models for the Iub dimensioning. For dimensioning with a fixed blocking probability, as introduced in section 6.2.1 and 6.2.2 the Erlang-B formula (for a single traffic stream) and the multi-dimensional Erlang-B formula (for multiple traffic streams) shall be applied. The required input parameters for dimensioning are the offered traffic per traffic stream, the desired blocking probability as the QoS requirement, and the reserved bandwidth for each call per traffic stream. Table 6.2 gives the detailed parameters that describe a traffic stream.

Traffic Parameters	
r_{peak}	bearer rate in bps (i.e. peak data rate)
BHCA	busy hour call attempts (calls / hour) per user
K	number of users in the system in the busy hour
λ	arrival rate of new calls (calls/second)
T	mean duration per call in seconds, i.e. $T=1/\mu$
activity	activity factor of a call: defined as the ratio of the total duration of ON periods and the total lifetime of the call
r_m	reserved bandwidth for a call in bps, which is used for admission control
N	the maximum number of simultaneous connections allowed by the system
TM	traffic demand in bps

Table 6.2: *Dimensioning with Erlang-B formula (Traffic Parameters)*

It is assumed that the arrival traffic is a Poisson process with arrival rate λ. The arrival rate of the traffic stream λ is derived from equation (6.15). It defines the arrival rate of new calls, i.e. the mean number of new calls every second.

$$\lambda = BHCA \cdot K \tag{6.15}$$

Given the average call duration *T*, the offered traffic *A* in Erlang can be obtained with equation (6.1). With the offered traffic *A*, the created traffic demand of this stream *TM* is calculated as the product of the peak data rate, the activity factor of a call and the offered traffic Erlang, as shown in (6.16):

$$TM = r_{peak} \cdot activity \cdot A \tag{6.16}$$

In case the network has only one single traffic stream, let *C* denote the allocated bandwidth on the Iub interface, then the maximum number of simultaneous connections allowed by the network *N* is calculated below. It is an integer number.

$$N = \left\lfloor \frac{C}{r_m} \right\rfloor \tag{6.17}$$

Thus, with the offered traffic A and maximum allowed number of simultaneous connections N, the blocking probability under the link bandwidth C can be calculated with the Erlang-B formula (6.6). When the network consists of multiple traffic streams, the model is generalized to the multi-dimensional Erlang-B formula. In the concept of dimensioning, the link bandwidth C should to be numerically derived with the Erlang-B formula such that a desired blocking probability (e.g. 1%) is achieved.

6.2.4 Validation of Erlang Model

This section validates the Erlang-B formula (equation (6.6)) for dimensioning of streaming traffic for a guaranteed blocking probability, in the following called CAC reject ratio, through simulations. The employed CAC (Call Admission Control) function in the Iub transport network has been explained in section 3.3.4.3. As the Erlang-B formula is applied for the single traffic stream scenario, thus only one stream service is selected for the simulations.

Voice is a typical circuit-switched type of traffic. The voice traffic model is given in Table 5.2. For call admission control, the required bandwidth for each voice connection is set to 12.4 kbps on the Iub interface (at the ATM level). The validation of the Erlang-B formula consists of three parts: firstly the Erlang-B formula (6.6) is used to estimate the CAC reject ratio of the voice traffic given different offered traffic, and furthermore equation (6.8) and (6.9) are applied to calculate the obtained link utilization; secondly, the derivation of the state probabilities given in equation (6.5) is validated, and as well the insensitivity of the Erlang-B formula to the service time distributions is investigated; at last a group of dimensioning results for achieving certain guaranteed CAC reject ratio are presented, and the simulation results are compared against the calculations derived by Erlang-B formula.

Figure 6.8 shows the calculated CAC reject ratio (left chart) and link utilization (right chart) over various offered voice traffic (in kbps) with Erlang-B formula. The calculated results are compared with the simulated results. The CAC reject ratio is derived from equation (6.6) and the link utilization is based on equation (6.8) and (6.9). The offered voice traffic is calculated with equation (6.16). The investigated scenario consists of one Node B over a single Iub link to the RNC, where the available Iub bandwidth is fixed to 1Mbps. It can be seen that with a limited Iub bandwidth, the increased offered voice traffic results in a rising CAC reject ratio, but the link utilization increases to a maximum value and then stays constant. It shall be noticed that when the offered traffic is above 50% of the link bandwidth, the link utilization is slowly reaching the maximum value. This is the effect of CAC, to ensure the carried traffic on the Iub interface is below the agreed grade of service level and prevent from the link overload.

The link utilization is calculated with equation (6.8) and (6.9). Here the maximum link utilization is only around 0.8. This is due to that the configured reserved bandwidth of 12.4kbps for each voice channel is higher than the actual traffic generated by each voice call (in this example, each voice call creates 10.37kbps traffic on the ATM link in average). So in this example, as the reserved bandwidth overestimates the actual traffic load of each call, thus 20% link bandwidth cannot be utilized which leads to a waste of the transport resources. If configuring the reserved bandwidth for each connection equal to the amount of actual mean traffic generated by each call, then the obtained link utilization can reach 100%.

By comparing the simulation results with the calculated values from the Erlang-B mode, it can be seen that for both CAC reject ratio and link utilization the analytical results match with the simulation results. Therefore, the Erlang-B model is shown to be an accurate model for the streaming traffic in a network using CAC, when the blocking probability is considered as the desired QoS requirements.

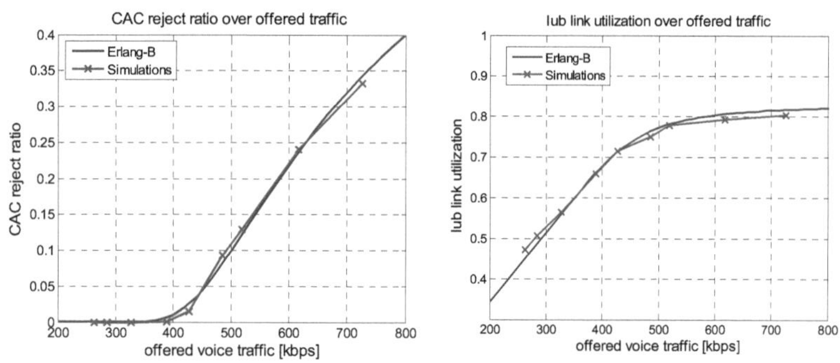

Figure 6.8: *Validation of Erlang-B formula for the voice scenario: CAC reject ratio and link utilization over offered voice traffic*

Figure 6.9 depicts the state probability (i.e. the probability for the number of busy connections) derived with equation (6.5) and validated with simulations. The left figure plots the probability density function of the number of busy connections and the right one plots its cumulative distribution function. In this example, the total bandwidth on Iub link is still 1 Mbps and the offered voice traffic is 0.427 Mbps. As each voice call reserves 12.4 kbps bandwidth, therefore the 1-Mbps Iub link allows maximum 80 simultaneous calls in the system according to equation (6.17). Figure 6.9 shows the probability of the number of busy channels from 1 to 80. It is clearly seen that the calculated probability of each state with equation (6.5) matches well with the probability obtained from the simulations.

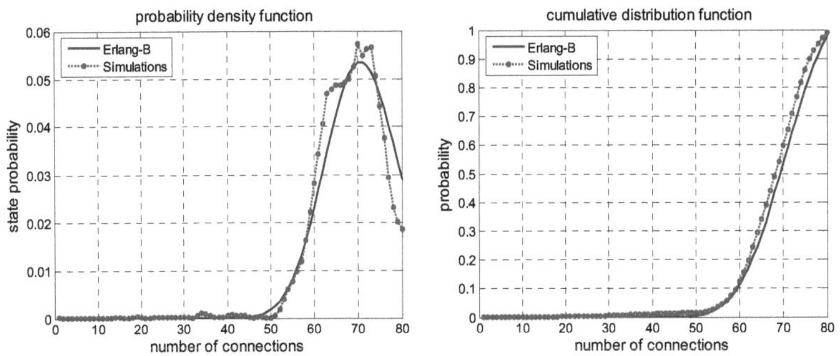

Figure 6.9: *Validation of Erlang-B formula for the voice scenario: state probabilities*

As mentioned in section 6.2.1, an important property of Erlang-B formula is the insensitivity property, which means that the state probabilities are fully insensitive to the form of the holding time distribution but depends only on the first moment of the holding time [Coh79, Ive04]. Here the holding time of a voice call corresponds to a voice session. This property is presented in Figure 6.10. It gives the probabilities of different states, i.e. the number of busy connections, under various holding time distributions with the same mean holding time.

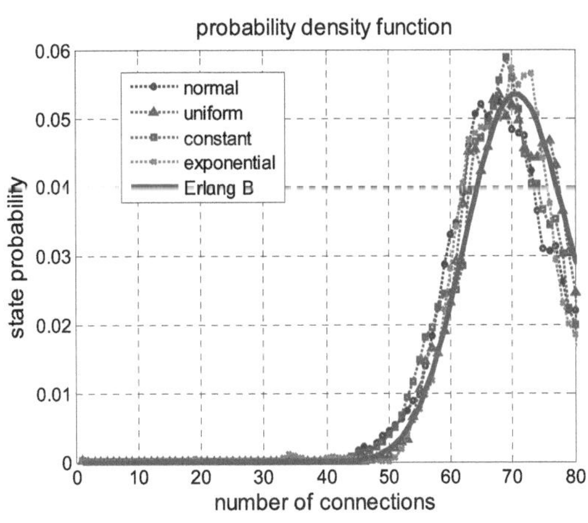

Figure 6.10: *Insensitivity of Erlang-B model (different service time distributions)*

In Figure 6.10, the state probabilities for normal, uniform, constant and exponential holding time distributions are shown, and also validated with the Erlang-B model. This figure demonstrates that the Erlang loss model with the Poisson arrival process and furthermore the Erlang-B formula is not sensitive to the holding time distributions, and thus the probability of the number of busy connections in the system is not influenced by different holding time distributions, but only depends on the mean value. With this property, the Erlang-B formula can be widely applied for a range of circuit-switched traffic with Poisson arrival process but arbitrary holding time distributions.

6.2.5 Dimensioning Results

The following gives the dimensioning with a fixed blocking probability (CAC reject ratio) by applying the Erlang-B formula. For proper operation, a loss system should be dimensioned for a low blocking probability. In practice, the bandwidth is typically designed so that the blocking probability is around 1% to avoid overload.

Table 6.3 shows the required number of channels n in the network for a fixed CAC reject ratio (B= 1%, 5%) for different amount of the offered traffic A in Erlang. The results are derived from the Erlang-B formula (6.6). Because it is configured that each voice connection reserves 12.4 kbps bandwidth, the required number of channels n for a fixed CAC reject ratio can be converted to the required transport bandwidth on the Iub link in kbps. Figure 6.11 and Figure 6.12 present the required Iub bandwidth obtained from both Erlang-B formula and simulations for the fixed CAC reject ratio of 1% and 5% respectively, over different offered voice traffic demands (in kbps). In both figures, the left chart give the required Iub bandwidth in kbps for the guaranteed CAC reject ratio and the right one illustrates the *relative error* that is calculated as the difference in bandwidth of the Erlang-B model and the simulations as a ratio of the simulated values (with equation (4.1)). Moreover, Figure 6.13 presents the Iub link utilization obtained with equation (6.8) and (6.9) and compared with simulations.

Dimensioning of voice traffic with CAC reject ratio B = 1%								
offered traffic A [Erlang]	19.24	47.75	86.81	128.27	166.7	256.82	335.18	461.07
n	29	60	100	144	186	279	357	491
Dimensioning of voice traffic with CAC reject ratio B = 5%								
offered traffic A [Erlang]	19.8	44.49	87.32	130.34	169.61	250.45	332.84	457.2
n	25	49	91	134	173	252	331	447

Table 6.3: *For a fixed CAC reject ratio (1% and 5%), the required number of channels n given the offered traffic A in Erlang*

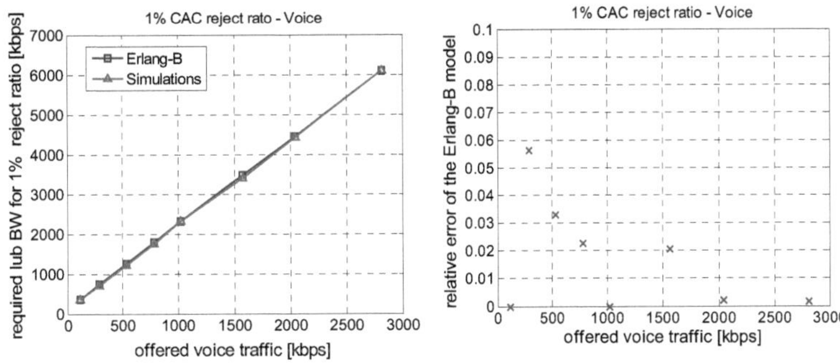

Figure 6.11: *Validation of Erlang-B formula: dimensioning for 1% CAC reject ratio*

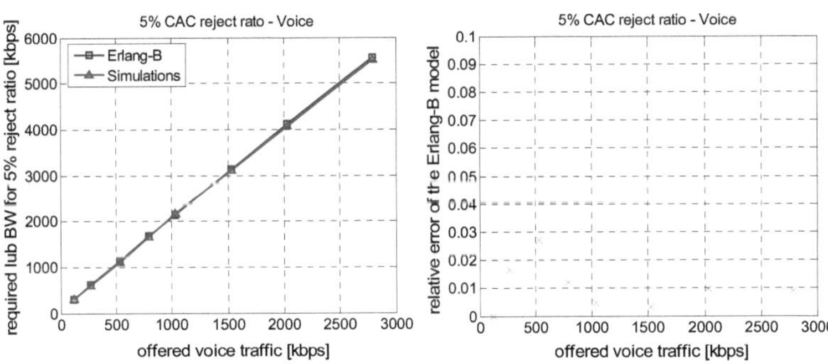

Figure 6.12: *Validation of Erlang-B formula: dimensioning for 5% CAC reject ratio*

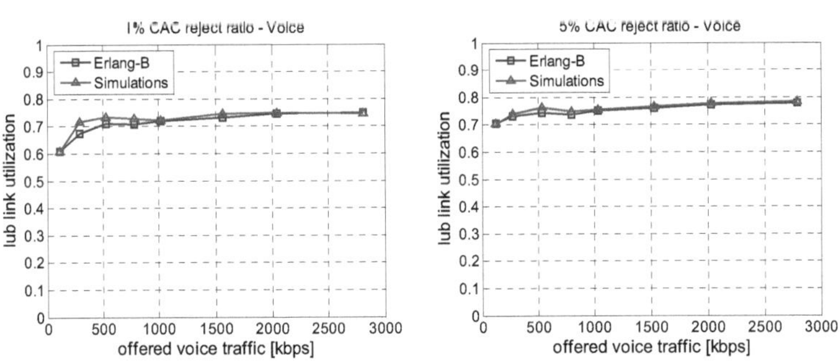

Figure 6.13: *Validation of Erlang-B formula for voice scenario: CAC reject ratio and link utilization over offered voice traffic*

Through comparing the simulation results against the analytical results derived from the Erlang-B model in the above figures, it is observed that the Erlang-B model results in a precise estimation for the dimensioning for a fixed CAC reject ratio. The caused relative error is low, in a range of 0 to 5%. So it can be concluded that the Erlang-B formula is an accurate analytical model for dimensioning streaming traffic with a fixed CAC reject ratio in a single traffic stream scenario. The Erlang-B model is simple to calculate, and furthermore due to its insensitivity to the service time distribution, it can be widely applied to other streaming traffic such as video streaming.

In a multiple traffic class system, where there are different classes of services with various traffic and QoS demands into the network and each traffic class corresponds to a traffic stream, then the multi-dimensional Erlang-B formula can be applied for the dimensioning in this case.

6.2.6 Summary

In section 6.2, dimensioning methodology for circuit-switched traffic was presented. In this thesis, the circuit-switched type of traffic with strict QoS requirements is subject to admission control. In order to guarantee its blocking probability, i.e. connection reject ratio, the Erlang-B model (for single traffic stream) and Multi-dimensional Erlang-B model (for multiple traffic streams) are suggested for dimensioning the Iub link for the circuit-switched traffic. Dimensioning procedure is presented and the characteristics of the Erlang models are demonstrated. The presented results demonstrate the high accuracy of applying Erlang models for the dimensioning of the Iub link for the circuit-switched traffic.

6.3 Dimensioning for Elastic Traffic

This section presents analytical models, which allows the dimensioning of the single Iub link for elastic traffic. As introduced in chapter 1.4, this type of traffic is generated by applications with *non real time* characteristics and is typically carried by TCP. The relevant QoS requirement for elastic traffic flows is *application throughput* (see section 4.1.2.1). Specifically, for a specific object size x_t the overall average transaction time should not go above a certain time threshold T_t. The respective throughput is calculated as x_t / T_t. Time T_t reflects a typical customer's waiting-time tolerance level, which certainly depends on its given maximum data rate and the type of applications. In this thesis, for meeting this QoS target of the elastic traffic, the proposed analytical dimensioning model is based on the *processor sharing model*, which considers the rate-sharing property of TCP. To apply the processor sharing model for dimensioning the Iub link, several assumptions are made and a number of extensions are proposed in this thesis to consider the specific features of UMTS and the related resource and traffic management functions in the UTRAN.

In the following, section 6.3.1 gives an overview of TCP. As for elastic traffic the TCP protocol greatly influences the data transfer process and thus the dimensioning strategy, a good understanding of its fundamental properties is essential. Then the related work and the basic principle of processor sharing are given in section 6.3.2 and 6.3.3. Based on that, the rationale behind of the application of the processor sharing model for the dimensioning of the Iub interface in UMTS is explained. Section 6.3.4 presents the modeling of the UMTS Iub interface as an M/G/R-PS queuing system at flow level. Section 6.3.5 and 6.3.6 introduce the fundamentals of the M/G/R-PS model, which this thesis is based on. Section 6.3.7 presents in detail the main extensions proposed within this thesis for dimensioning the Iub link and section 6.3.9 presents the dimensioning procedure. Afterwards, the main results of employing the proposed M/G/R-PS models for dimensioning the Iub link are presented and they are validated by simulations.

6.3.1 Transmission Control Protocol (TCP)

The *Transmission Control Protocol* (TCP) is the main transport protocol for data transmissions in the Internet. It is a connection-oriented transport protocol, which assures reliable and in-sequence data transfer. It was originally specified in [Pos81]. With the development of wired and wireless networks, many options and features have been proposed for TCP in order to enhance the performance and adapt well to ever changing environments. Hence a variety of versions and implementations of TCP have been developed over the years with main focus on congestion control [Jac88, FH99, Flo01]. Detailed introductions of the TCP protocol can be found in [Ste94]. However, in this thesis the proposed dimensioning models for the elastic traffic only consider the fundamental behaviors of TCP. In the following a brief description of main TCP features and their consequences for the network dimensioning are given.

Sliding Window Mechanism

The *sliding window mechanism* is employed to control the amount of data that the sender is allowed to send to the network without having received an acknowledgement. A *window size* is defined, which constrains the number of packets that can be unacknowledged at any given time. It is a variable that equals to the minimum of the *receiver window size* (notated as *rwnd*) and the *congestion window size* (notated as *cwnd*). Receiver window size is advertised during connection establishment and indicates the maximum amount of data which the receiver buffer is able to hold. The congestion window is influenced by TCP flow control imposed by the sender, i.e. slow start and congestion avoidance (see introduction blow). The TCP sender uses the congestion window to adapt its data rate to the available network bandwidth. Once the sender receives an acknowledgement for the first packet in the window, it "slides" the window along and sends the next packet. The window continues to slide as long as acknowledgements are received. To allow continuous sending without any unnecessary waiting times, the window has to be equal to or larger than the *bandwidth-delay product* of the path between the sender and the receiver. This product represents the "capacity" of the network and is computed by $BW \cdot RTT$. Parameter BW is the available bandwidth in the network, and RTT is the *round trip time*, i.e. the time period between sending a TCP packet and receiving the corresponding acknowledgement. With the sliding window mechanism, TCP can perform efficient transmission by allowing transmitting multiple packets before an acknowledgement; and it can provide flow control considering the receiver buffer size and the available network capacity.

The sliding window mechanism will have impact on the obtained throughput of the TCP flows. In the case of a relatively small receiver window size, the sender may not fully utilize the bandwidth of the network, as the capacity of the receiver buffer limits the rate of the sender to transmit packets. If it is assumed that the receiver has sufficient buffer size to cover the respective bandwidth-delay product of the network, then the resultant throughput will be only dependent on the accessible network bandwidth and the flow control functions at the sender. For an accurate dimensioning, the proposed dimensioning model should be able to capture the effect of the window mechanism.

Slow Start Mechanism

The *slow start mechanism* is a way to avoid network congestion after a new connection is established and the sender starts transmitting data. As the available network bandwidth is unknown to the senders at the beginning, instead of injecting a large amount of data into the network at once, the sender starts transmitting data slowly to explore the network to determine the available capacity. As shown in Figure 6.14, at first the sender starts with only one segment, i.e. the TCP congestion window (*cwnd*) is initialized to one segment. For each received acknowledgement the congestion window is increased by one segment. This leads to an exponential increase of *cwnd* until the receiver window size (*rwnd*) is reached.

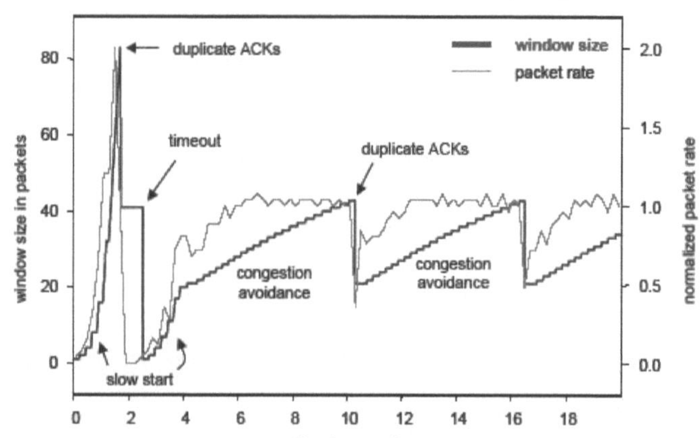

Figure 6.14: *TCP slow start and congestion avoidance [Rie04]*

During the slow start phase, the available bandwidth assigned to the connection cannot be fully utilized. Therefore, the slow start mechanism slows down the transmission process and thus degrades the end-to-end QoS. Especially, this effect becomes more considerable for longer round trip times and small file transfers. Thus, for these cases it is necessary to take the slow start mechanism into account in the dimensioning model.

Congestion Avoidance Algorithm

While the slow start mechanism is used to avoid traffic overload at the beginning of a new connection, the *congestion avoidance algorithm* is designed for congestion situation, which arises during the data transmission phase. The TCP protocol detects congestion and packet losses either through a timeout or through reception of multiple (usually three) duplicate acknowledgements. In both cases, the sender will assume that a packet is lost due to congestion and thus triggers the retransmission of that packet.

Although slow start and congestion avoidance are two independent mechanisms, in practice they are implemented together and make up the congestion control mechanism of TCP. These two algorithms require two parameters, the congestion window *cwnd* and a slow start threshold *ssthresh*. In addition, the advertised receiver window size (*rwnd*) has to be considered. The two algorithms work together as follows:

1. Initially, *cwnd* is set to one segment and *ssthresh* is set to 65535 bytes (maximum possible window size).
2. The TCP output routine follows the sliding window mechanism. The window size at any time is the minimum of *rwnd* and *cwnd.*
3. With every received acknowledgement, *cwnd* is incremented by one segment size (slow start) until *cwnd* reaches *rwnd.*
4. When congestion occurs, the TCP sender reduces *ssthresh* to one-half of the current window size. If the congestion is indicated by a timeout, *cwnd* is reduced to one segment and slow start is invoked. If the congestion is indicated by the reception of multiple (three) duplicate acknowledgements, fast retransmit algorithm is invoked and *cwnd* is set to *ssthresh* plus 3 times the segment size.
5. When new data is acknowledged by the receiver, *cwnd* is increased by the way of either slow start or congestion avoidance. If $cwnd \leq ssthresh$, slow start algorithm is performed; otherwise, congestion avoidance algorithm is in operation. This corresponds to an additive increase of about one segment size per *RTT* period.

The congestion avoidance mechanism leads to the characteristic behavior of additive increase, multiplicative decrease of a TCP source's transmission rate. Whenever TCP traffic traverses a bottleneck, where congestion occurs, the sources start regulating their transmission rates by adjusting the window size. As a consequence, TCP senders do not transmit data with a constant data rate. It varies in order to gain a certain share of the bottleneck capacity. The congestion avoidance mechanism is one of the fundamental causes of traffic "elasticity" as it forces TCP to adapt its transmission rate to the available bandwidth. The congestion avoidance algorithm together with the slow start mechanism provides the basis for the proposed processor sharing dimensioning model.

6.3.2 Related Work

This section gives an overview of dimensioning methods for elastic traffic. Over the past years, there have been extensive researches on promoting processor sharing model for link dimensioning in IP networks, based on the fundamental behavior of bandwidth sharing property of TCP. Roberts proposes the use of processor sharing for network dimensioning in [Rob97]. In [RM98] Massoulié and Roberts suggest applying call admission control for elastic traffic in order to guarantee a minimum achievable throughput for all accepted flows. Lindberger again proves the processor sharing model as an appropriate means of network dimensioning in [Lin99]. He derives the formula to calculate the mean sojourn time of the M/G/R PS model for cases without flow blocking and introduces the "*delay factor*" as a QoS measure for elastic traffic. In addition, he proposes dimensioning for the cases of combined stream and elastic traffic. On the dimensioning for the cases of service integration, Núñez Queija, van den Berg, and Mandjes did further studies in [Qvd+99a, Qvd+99b]. They derive blocking probabilities and mean transfer times for different link sharing strategies between stream and elastic traffic flows. The computations for elastic traffic are based on processor sharing, while for stream traffic a blocking model is used. However, the integrated models are too complex to be applied to dimensioning with a network-wide scope. Beckers et al. [BHK+01] validate the accuracy of the processor sharing model in an ADSL access network. They suggest to consider as well the impact of the round trip times into the model in order to improve the accuracy of estimations. Based on their findings, Anton Riedl also proposes an extension of the processor sharing model to explicitly consider the round trip times and the impact of TCP slow start effect in [RPBP00]. These studies prove that the processor sharing model is an appropriate model for dimensioning the IP networks for elastic traffic.

For elastic traffic, there has been also extensive research on developing detailed models for persistent TCP flows which are based on a very detailed view of TCP (such as in [OKM96, LM97]). They allow thorough performance investigations and as well provide good insight and understanding of TCP's behavior. However, due to their complexity they are not very practical for the purpose of network dimensioning. For network dimensioning a more macroscopic approach is essential. This is why the processor sharing model becomes very popular. It hides the details of the TCP protocol, requires only a few input parameters, and can be solved quickly. Overall, it provides a simple, efficient, yet powerful theoretical framework.

6.3.3 Processor Sharing Model for Elastic Traffic

The *Processor sharing* (PS) model was originally developed to evaluate multi-tasking schemes on computers with a single processor [Kle76]. The computation resource is shared among several jobs by specifying time slots, during which the processing power is exclusively utilized by one of the jobs. Typically, the processor time is assigned in a fair round robin manner. If the time slots are set small enough, the serving of the individual jobs appears to be quasi-simultaneous. The time-sharing process, where the time slots are approaching zero, is denoted as processor sharing [Rie04]. In such systems, it can be assumed that all jobs are theoretically served simultaneously. A lot of work has been published, studying the characteristics and the performance of processor sharing systems (see [Coh79, Ott84, BBJ99]). One performance measure, which is often considered in the context of processor sharing and which is specifically interesting for link dimensioning, is the sojourn time. It denotes the expected time, which a job spends in the system until it is fully served. It is apparent that this time depends on the number of other jobs, which are served during the same time period. Each new job that enters the system reduces the available rate for the existing jobs. A finished job, on the other side, increases it again for the remaining ones.

For the objective of link dimensioning, the processor sharing model is interpreted differently. Here it means that, the bandwidth of a link is shared by a number of elastic flows. What was referred to as a job in the original sense, denotes a single transaction process, i.e. the transmission of an individual file by one TCP connection. Due to the TCP flow control mechanisms as explained in the section 6.3.1, the transmission rate of a file transaction is controlled by TCP to adapt it to the available bandwidth provided by the networks. This dynamic behavior of TCP leads to the 'elastic' property of the traffic flow. If TCP works ideally, i.e. assuming that TCP's feedback and control mechanism is perfect and absolutely fair, the bandwidth of a common bottleneck link would be shared equally among all active TCP flows (connections) traversing that link ([Fan02, RBPP00]). Whenever a new TCP flow starts sending data, all TCP connections, which are already in progress, instantaneously reduce their rate in order to provide a fair share for the new one. When a TCP connection finishes its file transmission, the remaining ones increase their rates to fully utilize the available bandwidth. Therefore, a common link, which is traversed by multiple elastic TCP flows, can be modeled as a processor sharing system [Rie04]. It shall be noticed that processor sharing in this context refers to the flow or connection level. There is no processor sharing at packet level.

The theory of processor sharing on a common link is illustrated in Figure 6.15, which shows two cases with a single shared link. Each TCP connection can transmit data with the peak data rate (r_{peak}). For the capacity C of the shared link, assuming here that $2 \cdot r_{peak} < C < 3 \cdot r_{peak}$. If only two TCP connections are active (Figure 6.15a), each connection can transmit with its peak data rate. However, when a third flow becomes active (Figure 6.15b), the shared link becomes a bottleneck as the link capacity C is now less than the sum of all peak rates. Thus, instead of sending with peak data rate, each flow lowers its transmission rate to *C/3*. In general, if there are *n* elastic flows sharing a common link with capacity *C*, the data rate *r* of each flow equals to $r = \min(C/n,\ r_{peak})$ [Rie04].

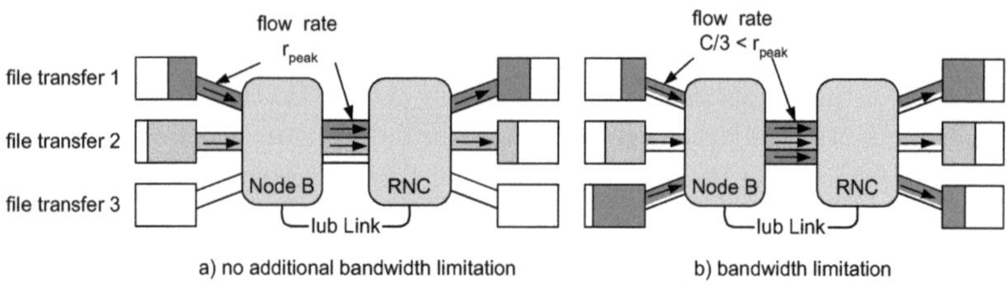

Figure 6.15: *Illustration of rate sharing*

For link dimensioning, the theory of processor sharing can be applied to derive the dependency between the expected transmission time (the sojourn time) and the amount of data, given the peak data rate of the individual connections and the capacity of the shared link. However, it needs to be noticed that the basic processor sharing model only considers the bandwidth sharing property of TCP, which it assumes to be ideal. But in practice TCP flows are not necessarily able to fully utilize their fair share of available bandwidth as a consequence of TCP slow start, packet losses, etc. The basic processor sharing model does not take into account the time spent for TCP connection establishment and termination, and disregards the impact of TCP slow start algorithm and network characteristics such as round trip times, possible packet losses and resultant TCP retransmissions. All these aspects can have considerable impact on TCP's effectiveness to utilize its fair share of available bandwidth. Therefore, to consider these impacts, extensions on the basic processor sharing model have to be made in order to enhance the accuracy of estimations for the expected sojourn time and thus provide a more precise link dimensioning results. In section 6.3.7 a number of proposed extensions of the M/G/R-PS model by the author for the dimensioning of the Iub will be presented in detail.

6.3.4 Application of M/G/R-PS Model for Dimensioning Iub Interface

This section introduces the analytical model for the dimensioning of the Iub link in the UMTS radio access networks. The analytical model is based on the processor sharing model. Figure 6.16 presents the system model of the Iub interface at flow or connection level. The elastic traffic is generated by downloading Internet objects, e.g. file transfers or downloading web objects. Each Node B serves a number of mobile users (UEs), who are transmitting elastic traffic over TCP connections. The TCP connections are established between the UEs and their requesting data servers located in the external networks (e.g. Internet). It is assumed that each elastic traffic flow is related to a single data transfer process and the sojourn time represents the expected transfer time. When considering the *World Wide Web* as the main application, one traffic flow can also represent a number of different web objects downloaded from the same server. Since HTTP/1.1 [FGM+99] allows the download of several web objects within one TCP connection, individual transmission processes can be considered as one transfer process. The size of the transferred data is equal to the sum of the individual object lengths. Measurements of web traffic have shown that the amount of transmitted data shows heavy-tail characteristics [Cha00]. This means that most of the transfer processes are short while there are some extremely long ones. Thus, it can be assumed that the downloaded Internet object sizes are generally distributed, e.g. modeled by a hyper-exponential or Pareto distribution. Furthermore, since each UE sends its Internet request randomly and independently, it can be further assumed that the flow arrival process is modeled as a Poisson process. As according to Palm's theorem (Theorem 6.1), the arrival process can be assumed as a Poisson process if traffic is generated by a large number of independent sources. In summary, the analytical model for dimensioning the single Iub link is built on the above two assumptions that data transfers are according to a Poisson arrival process and the size of the transferred data is generally distributed.

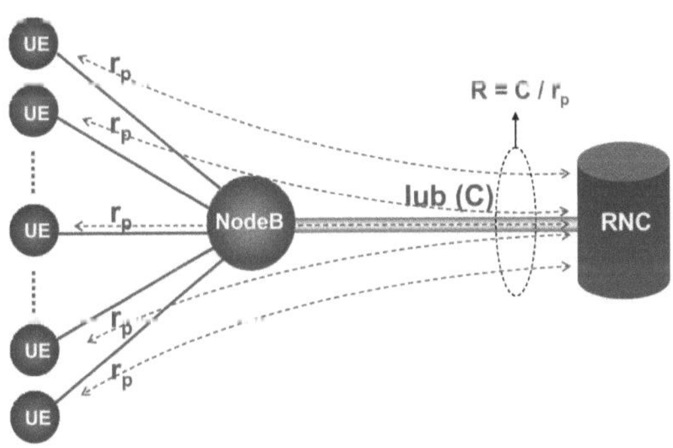

Figure 6.16: *Flow Level Model for Iub Interface*

For data transfer within the UMTS radio access network, each user connection is assigned a certain *Radio Access Bearer* (RAB) (refer to section 3.3.4.1), which defines the peak data rate for this user to transfer data between the UE and the UMTS core network. That means the attainable transfer rate for each individual elastic traffic flow (TCP flow) is restricted to a limited bandwidth (i.e. the maximum data rate defined by the RAB). If all users in the network are assigned to the same RAB type (e.g. RAB 128kbps), i.e. all active flows have the same peak data rate r_{peak}, as illustrated in Figure 6.17, then the Iub link with capacity C with $C > r_{peak}$ can be modeled as a **Processor Sharing** queuing system with $\boldsymbol{R}$ servers where $R = C/r_{peak}$. With the previous assumptions of Poisson arrival process and general service time distribution, the resultant analytical model is **M/G/R-Processor Sharing (M/G/R-PS)** model. According to this model, only up to ***R*** flows can be served simultaneously without rate reduction imposed by the shared Iub link, i.e. each of such flow can fully utilize the capacity of the assigned RAB peak rate. Processor sharing comes only into play when more than *R* flows are active in the system, so that multiple flows start sharing one server capacity, i.e. the peak bandwidth for a RAB. For less than *R* active flows, each one can be served without rate reduction imposed by the shared Iub link. In cases where the individual flows are not subject to any bandwidth restrictions, i.e. a single elastic flow is able to fully utilize the total link capacity *C* in times when no other flow is present in the system (i.e. $r_{peak} \geq C$), then the M/G/R-PS model is reduced to the simple M/G/1-PS model. In general, the M/G/R-PS model can be understood in such a way that several files that need to be transmitted over the common link are broken into little pieces, i.e. individual IP packets of the different traffic flows, and are processed by the link quasi-simultaneously. In this way, large files do not delay small ones too much. And this brings fairness for transferring different files over the same link.

It is important to know that the M/G/R-PS model is insensitive to the service time distribution [Lin99]. Thus, for the Iub dimensioning process, elastic traffic is characterized by the parameters *flow arrival rate* λ_e and *mean size of the transferred data* l_e, or by the product of these two values, the average traffic volume a_e. The QoS of elastic traffic transmission is specified as *application throughput* (transaction speed for transmitting a specific amount of data, as seen in section 4.1.2.1). Furthermore, a generic QoS measure, the *delay factor*, will be introduced in the next section, which can be used to quantify the QoS instead. To apply M/G/R-PS model to derive the average transaction time and its respective throughput or delay factor for the elastic traffic carried on the Iub link, assumptions are made: (1) there are no packets losses and thus no resultant TCP retransmissions; (2) the maximum window size of a TCP connection is large enough to cover the respective bandwidth-delay product of the network; (3) the Iub interface is the main bottleneck between the two ends of a TCP connection (i.e. between the UE and the data server), which would limit the transmission rate of individual flows.

6.3.5 Basic M/G/R-PS Model

Figure 6.17 illustrates the definition of *M/G/R-PS* model. It is a delay system with Poisson arrival process (*M*), general service time distribution (*G*), *R* servers and an infinite number of waiting positions in the system, and each server applies the **P**rocessor **S**haring (*PS*) service discipline. As mentioned above, M/G/R-PS model is applied in cases when the transmission rate of each flow is limited to a peak rate r_{peak}, which is lower than the link capacity *C* (i.e. $r_{peak} < C$). Then the shared link appears as a processor sharing system with $R = C/r_{peak}$ servers.

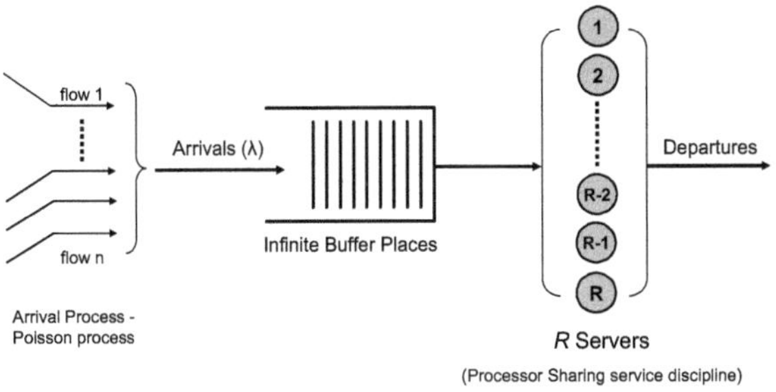

Figure 6.17: *Illustration of M/G/R-PS Model*

For an M/G/R-PS model where $R = C/r_{peak}$, the expected sojourn time (or average transfer time) $E\{T(x)\}$ for an object of size x is given by [Lin99]:

$$E_{M/G/R}\{T(x)\} = \frac{x}{r_{peak}}\left(1 + \frac{E_2(R, R\rho)}{R(1-\rho)}\right) = \frac{x}{r_{peak}} \cdot f_R \qquad (6.18)$$

Here, parameter ρ denotes the mean traffic load (or utilization on the link). It can be derived from the mean flow arrival rate λ_e and mean object size l_e by $\rho = \lambda_e \cdot l_e / C$. E_2 represents Erlang's second formula (Erlang C formula) with R servers and $A = R\rho$ as given in equation (6.19). It is known that the Erlang C formula calculates the delay probability (i.e. the probability that a job has to wait) of *Erlang's delay system* denoted as *M/M/n* [Ive04]. The given equation is only exact, if R is an integer value. However, for fractional R a continuous approximation for the Erlang function is given in [FK78].

$$E_2(R, A) = \frac{\frac{A^R}{R!} \cdot \frac{R}{R-A}}{\sum_{i=0}^{R-1} \frac{A^i}{i!} + \frac{A^R}{R!} \cdot \frac{R}{R-A}} \qquad (6.19)$$

In the equation (6.18), it can be obviously seen that the minimum time for transferring a file of size x is equal to x / r_{peak}. This minimum transfer time is achieved, if at most R flows are sending data. Every additional flow, which becomes active during the transfer process, takes away bandwidth and thus prolongs the transfer. The resultant additional delay is a consequence of the rate reduction imposed by processor sharing, since more than R flows are active. The part for the expected prolongation of the average transfer time in equation (6.18) is called *delay factor* f_R:

$$f_R = 1 + \frac{E_2(R, R\rho)}{R(1-\rho)} \qquad (6.20)$$

The delay factor f_R quantifies the increase of the average transfer time or the reduction of the effective throughput of individual flows as a result of link congestion. It is a quantitative measure of how link congestion affects the transaction times, taking into account the economy of scale effect. It is noted that f_R is always larger than 1.

From the average transfer time given in equation (6.18), the consequent application throughput γ ($\gamma = x / E\{T(x)\}$) can be derived:

$$\gamma = \frac{x}{E_{M/G/R}\{T(x)\}} = \frac{r_{peak}}{\left(1+\frac{E_2(R,R\rho)}{R(1-\rho)}\right)} = \frac{r_{peak}}{f_R} \tag{6.21}$$

For the special case of R=1(i.e. M/G/1-PS), $f_R = \frac{1}{(1-\rho),}$ and $\gamma = C\cdot(1-\rho)$.

Figure 6.18 shows the delay factor f_R and normalized throughput ($\frac{\gamma}{C} = \frac{1}{R\cdot f_R}$) as a function of the offered load (or link utilization) ρ with different R values. It can be seen that for R>1 with small ρ the delay factor f_R is close to 1, and thus the normalized throughput is close to $1/R$. This means that all connections receive an average rate, which is almost equal to their peak rate r_{peak}. As the load increases (higher utilization), the delay factor f_R increases dramatically, and the throughput drops accordingly. Moreover, from the graphs it can be observed that the delay factor reflects the economy of scale characteristics. Links with larger capacities (i.e. higher R) can be higher utilized than smaller ones (i.e. lower R) while still achieving the same delay factor. Furthermore, the sensitivity of the delay factor regarding traffic variation increases more drastically for smaller link capacities (i.e. lower R) when the utilization approaches 1. Therefore, one has to be careful when dimensioning links with lower capacities. In addition, as M/G/R-PS model is insensitive to the service time distribution, the results derived from M/G/R-PS model, i.e. the average sojourn time by equation (6.18) and average throughput by equation (6.21) are independent of object size distributions.

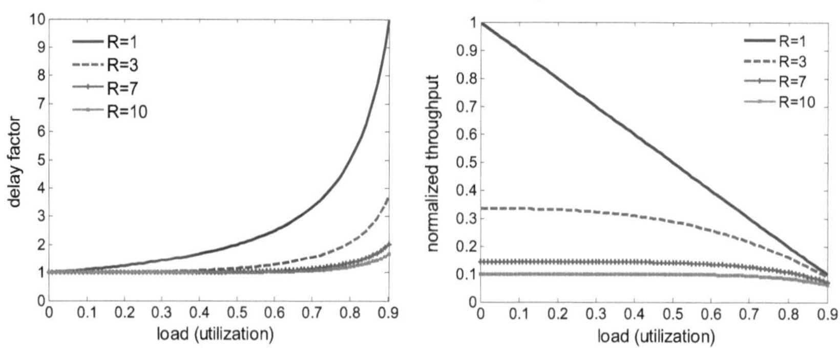

Figure 6.18: *Delay factor and normalized throughput vs. load M/G/R PS Model*

6.3.6 Extended M/G/R-PS Model

The basic M/G/R-PS model assumes ideal capacity sharing among active flows. But in practice the TCP flows are not always able to utilize their fair share of the available bandwidth. TCP's effectiveness of capacity sharing is determined by the TCP flow control mechanism (see section 6.3.1). The TCP slow start mechanism leads to the increase of the transfer time as a result of not completely utilizing the available bandwidth at the beginning of each transmission process. For small transactions and longer round trip times, the impact of the TCP slow start becomes more significant. Because TCP connections spend a considerable amount of their overall transmission time in the first slow start phase. Short transactions might not even get through the slow start phase at all. As a consequence, the achieved throughput will be lower than the theoretically computed one derived from the basic M/G/R-PS model. For network dimensioning, it means that the calculated capacities would not be sufficient to guarantee the desired QoS. To solve this problem, A. Riedl and T. Bauschert propose extending the basic M/G/R-PS model in [RPBP00], which considers the impact of the slow start algorithm. The following gives a detailed introduction of this extended M/G/R-PS model.

Figure 6.19 shows that the data transfer over TCP can be in principle divided into two parts: the first part consists of necessary *RTT*s which are mostly spent waiting for acknowledgements as a result of the slow start. *RTT* is the time period between sending out a datagram and receiving the corresponding acknowledgment (ACK). In the second part, the source starts sending the rest of the data with the available bandwidth, i.e. during the second part the TCP flow can fully utilize the available capacity given by the network.

In order to derive an approximation for the expected transaction time, it is assumed that the total amount of data x is transferred in TCP packets with maximum TCP packet size *MSS* (*Maximum Segment Size*), moreover there are no packet losses during the TCP slow start phase and the maximum TCP window size is large enough to cover the bandwidth-delay product of the network. Let $C_{available}$ denote the designated bandwidth share, which a sender is supposed to utilize in average once its window is large enough. For a given link capacity C, peak rate r_{peak}, and link utilization ρ, this bandwidth per flow can be determined by $C_{available} = r_{peak} / f_R$, where f_R is the delay factor of the M/G/R-PS system. With $C_{available}$ and a given TCP round trip time *RTT*, the time, after which a sender is able to utilize its fair share, can be calculated. This is the case when the congestion window is large enough to cover the bandwidth-delay product, i.e. when it reaches w^* given in equation (6.22).

$$w^* = \left\lceil \frac{C_{available} \cdot RTT}{MSS} \right\rceil = \left\lceil \frac{r_p \cdot RTT}{f_R \cdot MSS} \right\rceil \tag{6.22}$$

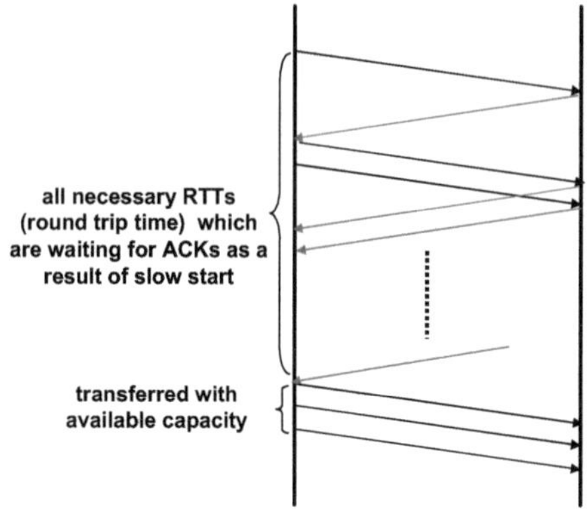

Figure 6.19: *Data transfer in TCP with slow start*

Due to the TCP slow start, the TCP congestion window size increases exponentially (assuming no packet losses in the slow start) with approximately 2^i after i round trip times (i = 0, 1, 2 ...). Thus, the amount of data sent up to the time when the sender can start utilizing its bandwidth share, i.e. the window size reaches or exceeds w^*, is denoted as $x_{slow-start}$ calculated by equation (6.23):

$$x_{slow-start} = \left(2^{n^*} - 1\right) MSS \text{ with } n^* = \left\lceil \log_2\left(\left\lceil \frac{r_p RTT}{f_R MSS} \right\rceil\right) \right\rceil \tag{6.23}$$

In equation (6.23), n^* represents the time step in terms of RTT, at which the sender window is for the first time equal to or larger than w^*.

The approximation of the expected transfer delay $E\{T(x)\}$ for an object of size x (including overhead) is given in equation (6.24) by [RPBP00]:

$$E\{T(x)\} = \begin{cases} \left\lfloor \log_2\left(\left\lceil \frac{x}{MSS} \right\rceil\right) \right\rfloor RTT + E_{M/G/R}\{T(x - x_{start})\} & x < x_{slow-start} \\ n^* RTT + E_{M/G/R}\{T(x - x_{slow-start})\} & x \geq x_{slow-start} \end{cases} \tag{6.24}$$

As seen from (6.24), the computation of the expected time to transfer all data is divided into two parts according to Figure 6.19: the first part gives the sum of all necessary *RTT*s which are mostly spent waiting for acknowledgements as a consequence of the slow start mechanism; the second term considers the time of transmitting the rest of the data with the available share capacity. It shall be noticed that the approximation for the expected transfer time is dependent on the given object sizes.

If the object size x is larger than $x_{slow\text{-}start}$, the transfer will take n^* time steps to reach the state of start utilizing the available bandwidth and further transfer the rest of the data, i.e. $(x - x_{slow-start})$, with the shared available capacity.

If the object size x is smaller than $x_{slow\text{-}start}$, the sender can never reach the state of fair capacity sharing, instead all packets are sent out during the slow start phase. Then the transfer takes $\left\lfloor \log_2\left(\left\lceil \frac{x}{MSS} \right\rceil\right) \right\rfloor$ round trip times waiting for the acknowledgements, during which the amount of data being transferred is denoted as x_{start}:

$$x_{start} = \left(2^{\left\lfloor \log_2\left(\left\lceil \frac{x}{MSS} \right\rceil\right) \right\rfloor} - 1 \right) MSS \tag{6.25}$$

Thus $(x - x_{start})$ gives the amount of remaining data according to the last acknowledgement, which is sent using the available capacity.

In equation (6.24), $E_{M/G/R}\{T(x)\}$ is the sojourn time using the basic M/G/R-PS model according to (6.18). By comparing (6.24) with (6.18), it turns out that the linear relationship between object size x and the expected transfer time disappears when TCP flow control is taken into account. Thus object size distributions with a substantial probability of files mainly driven by slow start phase perceive higher average delays than derived from consideration of mere object size and delay factor. It has to be noted that the system utilization remains the same, i.e. the impact is restricted to the performance experienced by individual flows.

6.3.7 Suggested Extensions on M/G/R-PS Model for Iub Dimensioning

For the objective of Iub dimensioning for elastic traffic, based on the basic and extended M/G/R-PS models introduced above in section 6.3.5 and 6.3.6, several extensions are suggested in this thesis on the M/G/R-PS model to consider the specifics of UMTS including related resource and traffic management functions such as CAC and BRA (see Chapter 3). The following sections (section 6.3.7.1-6.3.7.5) present in detail these suggested extensions, which are developed in the framework of this thesis to apply for Iub dimensioning under different UMTS network and traffic configurations.

Section 6.3.7.1 proposes several adjustments to the above introduced extended M/G/R-PS model to consider variable round trip times and asymmetric data transfer in the UMTS network, as well as to include the TCP connection setup and release times. This adjusted M/G/R-PS model can be applied to the situation where all UEs are assigned to the same RAB type (i.e. have the same peak data rate) and there is no CAC applied to elastic traffic (i.e. elastic traffic is not subject to CAC). Section 6.3.7.2 presents M/G/R/N-PS model for the cases of applying CAC to elastic traffic in the UTRAN transport network, also only a single RAB type is considered for all UEs. Both M/G/R-PS and M/G/R/N-PS models proposed for the Iub dimensioning have also been presented by the author in [LSGT05]. Section 6.3.7.3 expands the M/G/R-PS model to a general form as presented in [LSGT06], which can be applied for having multiple RAB types in UMTS for different use groups with different RAB rates and QoS requirements. Section 6.3.7.4 presents the applications of the M/G/R-PS model for having variable peak rates per flow with the use of BRA function (which adapts the RAB rate dynamically for each user connection during its data transfer). In cases of multiple RABs and BRA, it is assumed that there is no CAC function applied to elastic traffic. Section 6.3.7.5 discusses the scenarios where elastic traffic is mixed with circuit-switched type of traffic. In this scenario, the circuit-switched traffic is given a higher priority to transfer over elastic traffic in the UTRAN transport network. In section 6.3.7.5 analytical approaches are presented for the mixed traffic scenarios to estimate the average transaction times of elastic traffic, taking into consideration the effect of bandwidth sharing.

6.3.7.1 Extension for case with Single RAB and not Applying CAC

At first, a simple scenario is considered where a single RAB type is provided by the UMTS network to all UEs. That means all user connections have the same peak rate r_{peak} to transfer data. Moreover in this scenario it is assumed that elastic traffic is not subject to the *Connection Admission Control* (CAC) function (see section 3.3.4.3). Thus, all elastic traffic supported by the UMTS radio interface is accepted into the UTRAN transport network, and therefore, the maximum number of simultaneous connections to be transported on the Iub link is not restricted. Thus the Iub link with capacity C can be modeled as an M/G/R-PS model with $R = C/r_{peak}$, as explained in section 6.3.4. By applying the M/G/R-PS model, the average user application QoS in terms of mean transfer delay or mean application throughput can be estimated, which is derived from the expected sojourn time of the M/G/R-PS model.

The extended M/G/R-PS model in section 6.3.6 has considered the impact of the TCP slow start and round trip times on the expected transfer delay, however there are still a few improvements that are required and made in the framework of this thesis to be suitable for the UMTS network: (1) It is found that *RTT* in equation (6.24) is assumed to be a constant value, but often it is a varying value as a function of link congestion. Given a limited Iub link capacity, the experienced *RTT* can go up quickly in the case of high link utilization and thus it has significant impact on the achieved end-to-end delay QoS. Therefore, the increases of *RTT* in case of high link utilizations need to be considered. (2) Secondly most of the IP traffic (e.g. Internet) is asymmetric, i.e. the traffic amount on the downlink is dominant. This leads to that the experienced congestion on the downlink is usually more than on the uplink. Moreover, according to the asymmetric traffic property, the UMTS sets up different RAB rates on the uplink and downlink separately and therefore the transfer delay spent on the uplink and downlink can be also different. Thus, the distinction of uplink and downlink is necessary. It requires that the analytical model shall also consider the impact of the asymmetric traffic characteristic. (3) At last, the expected sojourn time given by equation (6.24) does not include the time spent for TCP connection setup and release phase. For an accurate modeling, the extra delay caused by the TCP connection setup and termination is suggested to be taken into account for the overall application delay or throughput performance.

Varying RTT Extension

It is assumed in this thesis that the increase of round trip times are mainly subject to the congested Iub interface. To capture this impact of link congestion on the resultant *RTT*, an adjustment is suggested in this thesis to replace the constant *RTT* with $RTT_{adjust.}$ Let $\mathrm{RTT}_{\mathrm{adjust}} = \mathrm{RTT} \cdot \mathrm{f}_{\mathrm{R}}$ (RTT is the experienced round trip time during the low link utilization and f_R is the delay factor) for equation (6.24). The expression of the average file transfer delay is improved with equation (6.26):

$$E\{T(x)\} = \begin{cases} \left\lfloor \log_2\left(\left\lceil \frac{x}{MSS} \right\rceil\right) \right\rfloor RTT_{adjust} + E_{M/G/R}\{T(x - x_{start})\} & x < x_{slow-start} \\ n^* RTT_{adjust} + E_{M/G/R}\{T(x - x_{slow-start})\} & x \geq x_{slow-start} \end{cases} \quad (6.26)$$

But it should be noted that the calculation of n^* in equation (6.23) is only dependent on the *RTT* experienced under the low link utilization so that n^* is a function of f_R. That means, when the network is congested ($f_R > 1$), n^* becomes low, i.e. the number of time steps when the available shared bandwidth starts to be fully utilized is less.

Uplink and Downlink Differentiation Extension

In order to differentiate the uplink and downlink transmission, an additional parameter *UL_rtt_ratio* is introduced into the model, which is defined as the time spent on the uplink as a ratio of the complete round trip time. Thus (1- *UL_rtt_ratio*) indicates the percentage of the time spent on the downlink transmission. If *UL_rtt_ratio* < 0.5, it means that the uplink path experiences smaller latency than the downlink. It is applied in equation (6.27).

TCP Connection Setup and Release Extension

Figure 6.20 shows the complete TCP procedure of transferring one file. It contains three phases: TCP connection setup, data transfer and connection close. TCP establishes new connections by carrying out the *three-way hand shaking* procedure [Ste94]. The connection requesting instance (a client) sends a *SYN* segment to the server. The server responds to the request by sending its own *SYN* segment and at the same time acknowledging the *SYN* of the client. To conclude the connection setup, the client has to acknowledge the *SYN* of the server. In total, the three-way handshake procedure involves three packets (each 40 bytes long) in one and a half *RTT*.

In the procedure of connection release, each side sends a *FIN* segment indicating that it does not have any more data to transmit. The other side acknowledges this *FIN* segment by sending an *ACK*. This is a complete *RTT*. As TCP connections are full duplex, each side terminates the connection in one direction (this is called half-close). The data transaction time considers only the connection release time on the direction of data being received. Therefore, there are in total two complete *RTT*s and one additional uplink transmission for complete TCP connection setup and release procedure.

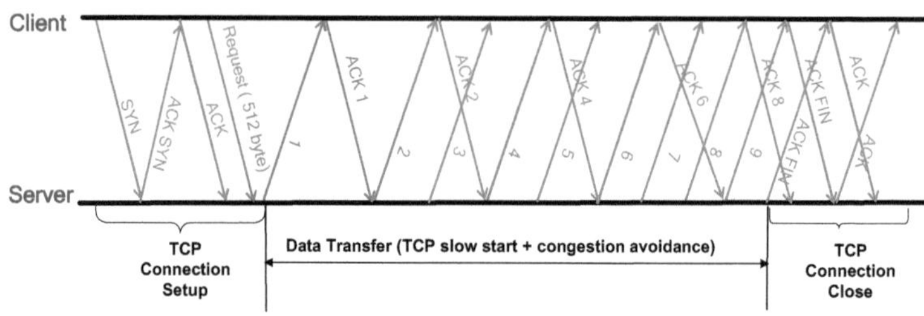

Figure 6.20: *TCP Procedure*

Thus, adding the extra delay for setting up and releasing the TCP connection, the expected file transfer delay *E{T(x)}** is expressed as with equation (6.27). It shall be noticed that the experienced round trip time during the connection setup and release phase is also subject to the link congestion. Thus, in equation (6.27) the varying *RTT* value RTT_{adjust} is used as well for the round trip times.

$$E\{T(x)\}^* = E\{T(x)\} + (2 + UL\_rtt\_ratio)RTT_{adjust} \tag{6.27}$$

6.3.7.2 Extension for case with Single RAB and Applying CAC

This section considers the scenarios when applying CAC to elastic traffic in the UTRAN transport network. The CAC function is often employed in order to guarantee a minimum throughput for each connection over the Iub link, and thus, to avoid instability and low throughput in case of traffic overload (e.g. during busy hours). It is assumed that each elastic flow (i.e. TCP flow) corresponds to one connection on the Iub. In the context of this thesis, the admission control for elastic traffic in the UTRAN is based on limiting the maximum number of simultaneous connections transported over the Iub link. It has been mentioned by Roberts et al. in [MR99] that when a per-flow access control is applied in the system a blocking model needs to be used to estimate the related QoS. Thus for a correct dimensioning model, the previous M/G/R-PS method has to be extended to have the limitation of a maximum number of *N* active connections allowed simultaneously sharing the common resources. For this purpose, the M/G/R/N-PS model is suggested in this thesis by introducing an additional parameter N into the M/G/R-PS model, where N indicates the maximum number of allowed connections.

In the considered CAC scenario, it is still assumed that all UEs have the same RAB type, i.e. with the same peak rate r_{peak}. Additionally, they all request the same *CAC guaranteed bit rate* (see section 3.3.4.3), which is denoted as r_m. It is defined as bandwidth requested by a connection for meeting its QoS, which also implies the minimum bandwidth this connection can get. Given the link capacity C, the link is thus modeled as M/G/R/N-PS model with $R = C/r_{peak}$ and maximum number of allowed flows $N = C/r_m$. The calculation for the M/G/R/N-PS model can be derived directly from the M/G/R/∞-PS model, only that the state space corresponding to the number of flows or connections is now restricted to a total of number of N. The state probability of the M/G/R/N-PS model is calculated in equation (6.28).

$$p(j) = \begin{cases} \dfrac{(1-\rho)\dfrac{R!}{j!}(R\rho)^{j-R}E_2(R,R\rho)}{1-E_2(R,R\rho)\rho^{N-R}\rho} & (j < R) \\[2ex] \dfrac{E_2(R,R\rho)\rho^{j-R}(1-\rho)}{1-E_2(R,R\rho)\rho^{N-R}\rho} & (N \geq j \geq R) \end{cases} \tag{6.28}$$

In the above equation ρ denotes the offered load (or link utilization). With the state probability, the average number of connections $E\{W\}$ can be calculated. By applying Little's law, the average transfer delay $E\{T\}$ can be obtained by equation (6.29).

$$E\{T\} = \frac{E\{W\}}{\lambda(1-p(N))} \quad \text{with} \quad E\{W\} = \sum_{j=0}^{N} j \cdot p(j) \tag{6.29}$$

For network dimensioning, another important QoS is the *blocking probability* (or called *Connection Reject Ratio* in this thesis). When there are already N connections over the Iub link, any additional new connection requests will be rejected. Then the blocking probability $p(N)$ also means the probability of having N connections in the network. It can be derived from (6.28). When $R = N$, $p(N)$ reduces to Erlang's first formula (i.e. Erlang-B formula).

6.3.7.3 Extension for case with Multiple RABs and not Applying CAC

Typically, UMTS provides a variety of bearers called *Radio Access Bearers* (RABs) (see section 3.3.4.1) to support a range of services and QoS. This section discusses the scenarios where multiple RABs are used at the same time in UMTS for transmitting different elastic traffic classes. It is assumed that each traffic class is transferred by one RAB type and specifies a different application QoS target. In this case, the dimensioning needs to fulfill the QoS requirements of individual traffic classes on an end-to-end basis. Moreover, in this scenario it is assumed that each elastic flow is assigned to a fixed RAB type (i.e. the RAB is not changed during the transfer of that flow), and no CAC function is applied to any elastic traffic class, i.e. the Iub link does not restrict the maximum number of simultaneous connections for elastic traffic.

Figure 6.21 illustrates system model of a scenario with multiple RABs for different QoS user groups at TCP flow level. In the shown example there are three UE groups. Each UE group defines a specific traffic class and is assigned a different RAB type with a peak data rate r_i. The index i indicates the user group with traffic class i. The arrival process of each UE group is assumed to be a Poisson process with a flow arrival rate λ_i, resulting in a mean traffic load ρ_i. In this scenario, due to the basic property of TCP, processor sharing still performs among all active TCP flows, regardless of which RAB type they are using. Because the actual transmission rate of each flow is in fact controlled by TCP as a result of TCP flow control mechanism, which is designed to adapt the data rate to the available network bandwidth.

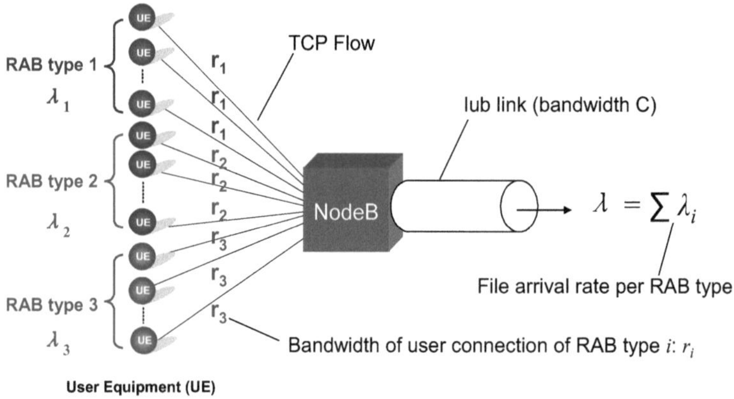

Figure 6.21: *Flow level model for multiple RAB scenarios*

The principle of processor sharing in this case is illustrated in Figure 6.22. Based on this theory, it can be assumed that all active flows are sharing the Iub link capacity quasi equally, and thus the Iub link can be still modeled as a PS system. But since there are multiple peak rates, the original M/G/R-PS model, which is used for the case of one peak data rate, cannot be directly applied. Therefore a general form of the M/G/R-PS model is proposed in the framework of this thesis for the dimensioning of a general multiple RABs scenario.

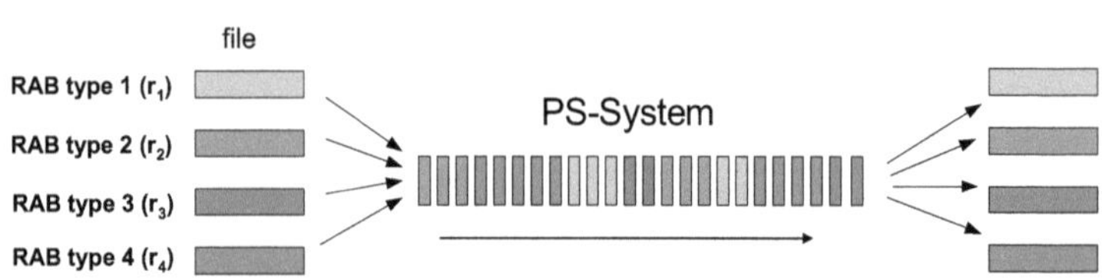

Figure 6.22: *Processor Sharing*

The main idea of this analytical approach is to calculate the application performance for each elastic traffic class individually with the M/G/R-PS model. However, the M/G/R-PS model needs to be extended to consider bearer specific characteristics of each traffic class as well as processor sharing among all traffic flows of different bearers. Based on these considerations, a general form of M/G/R-PS model is proposed in this thesis to apply for individual bearer type or traffic class. The general form is extended from the original M/G/R-PS model, but for each traffic class the derivation of R is dependent on the peak rate of its selected bearer type (RAB type). This is based on the fact that the expected application performance of each traffic class is related to its own peak rate. When the link starts to be congested (i.e. above certain utilization), processor sharing comes into play and thus degrades the overall flow performance. Therefore the individual flow performance is also a function of total link utilization, which is a result of the total traffic from all traffic classes. However, the impact of the link utilization on the application performance of individual traffic class still depends on the bearer type, i.e. delay factor is bearer specific. Nevertheless, in the highly congested states the experienced flow performance of various traffic classes converge since none of them can utilize its peak rate and a real fair sharing is available for each of them.

Based on the above idea, for each traffic class an individual M/G/R-PS model is established with its own R. The individual M/G/R-PS model is then used to calculate the mean transfer delay or mean throughput of elastic traffic flows of individual traffic classes. The following variables are defined in Table 6.4.

Variable	Definitions
i	index of a traffic class
r_i	peak data rate for traffic class i
RTT_i	Round Trip Time for traffic class i
λ_i	flow arrival rate of traffic class i
f_{Ri}	delay factor (QoS target) of traffic class i
$E\{x_i\}$	average file size of traffic class i
ρ_i	offered traffic load for traffic class i, $\rho_i = (E\{x_i\}\ \lambda_i)/C$
R_i	calculated R in M/G/R-PS model for traffic class i
MSS	TCP Maximum segment size
C	link capacity

Table 6.4: *Defined variables for general M/G/R-PS Model*

For traffic class i, the corresponding M/G/R-PS model is based on R_i given by equation (6.30). It is a function of link capacity C and peak rate r_i of this traffic class.

$$R_i = \frac{C}{r_i} \tag{6.30}$$

The total mean traffic load (from all traffic classes) ρ is calculated by equation (6.31).

$$\rho = \sum_{bearers} \rho_i \tag{6.31}$$

With the above R_i and total offered load ρ, the expected sojourn time (or mean transfer delay) $E\{T(x_i)\}$ for traffic class i to transfer an object of size x_i is calculated by generalizing the sojourn delay formula of the basic M/G/R-PS model:

$$E_{M/G/R}\{T(x_i)\} = \frac{x_i}{r_i}\left(1 + \frac{E_2(R_i, R_i\rho)}{R_i(1-\rho)}\right) = \frac{x_i}{r_i} f_{Ri} \tag{6.32}$$

Here delay factor f_{Ri} represents the increase of the average transfer time (or decrease of the average throughput) for traffic class i due to link congestion as a resultant of the total load of all traffic classes. It can be obviously seen that the delay factor depends on the bearer type of the traffic class.

As introduced in section 6.3.6, the basic M/G/R-PS model is extended to consider the impact of TCP slow-start. And further improvements are proposed in section 6.3.7.1 to add the extra delay caused by TCP connection setup and termination, and replace the constant *RTT* with a varying *RTT* which is a function of link congestion. Based on these extensions, the general form of the basic M/G/R-PS model (6.32) is expanded to (6.33). It is a general form of the improved M/G/R-PS model, which is used to approximate the average transfer delay for traffic class i in this multiple traffic classes scenario:

$$E\{T(x_i)\} = \begin{cases} \left\lfloor \log_2\left(\left\lceil \frac{x_i}{MSS} \right\rceil\right) \right\rfloor RTT_{adjust\_i} + E_{M/G/R}\{T(x_i - x_{start})\} & x < x_{slow-start\_i} \\ n^* RTT_{adjust\_i} + E_{M/G/R}\{T(x_i - x_{slow-start\_i})\} & x \geq x_{slow-start\_i} \end{cases} \tag{6.33}$$

Here $RTT_{adjust\_i} = RTT_i \cdot f_{Ri}$, which gives an approximation of the estimated RTT_i as a function of link congestion. It is noticed that $x_{slow-start\_i}$ is also RAB specific. Moreover, by adding the extra delay for setting up and releasing the TCP connection, the expected file transfer delay $E\{T(x_i)\}^*$ is calculated with (6.34).

$$E\{T(x_i)\}^* = E\{T(x_i)\} + (2 + UL\_rtt\_ratio)\, RTT_{adjust\_i} \tag{6.34}$$

6.3.7.4 Extension for case with BRA and not Applying CAC

This section extends the M/G/R-PS model for the scenarios of applying the *Bit Rate Adaptation* (BRA) function for the transport in the UTRAN. As introduced in section 3.3.4.3, BRA is one of the UMTS resource management function dedicated for the packet-switched data traffic. It is used to dynamically change the peak data rate of a user connection during its data transfer process. The data rate adaptation is carried out by performing RAB upgrades to increase data rate and RAB downgrades to decrease data rate. The adaptation of data rate per flow is determined by the activity of the user connection as well as the available network resources. As a result, the peak data rate of elastic traffic flows is not a constant value, and thereby the M/G/R-PS model cannot be directly applied. The main differences between the BRA scenario against the multiple RABs scenario (discussed in the last section) are, that, in the BRA scenario the peak rate per user flow is changing over time and moreover there are no multiple traffic classes with various fixed RAB types and different QoS requirements. The concerned QoS for dimensioning in case of BRA is the average transfer delay or throughput of all traffic flows. In the framework of this thesis, an approximation is suggested for dimensioning the Iub link when BRA function is in use. The main idea of this method is to reuse the single rate M/G/R-PS model, but R is derived from the mean data rate. In the investigated scenarios, it is assumed that there is no CAC function applied to the elastic traffic flows. Therefore the blocking model is not considered in this case.

Assuming that there are in total K RAB types available in UMTS. Let q_j denote the percentage of the RAB type j in use over time. It is calculated as the ratio of the average number of flows assigned to the RAB type j denoted as n_j over the average total number of flows of all RAB types over a long time period. It indicates the average proportion of the RAB type j. Let r_j denote the respective peak rate of RAB type j. Thus the average of the various peak rates r_{peak_avg} is derived from equation (6.35).

$$r_{peak\_avg} = \sum_{j=1}^{K} r_j \cdot q_j \text{ where } q_j = n_j \Big/ \sum_{j=1}^{K} n_j \tag{6.35}$$

An average round trip time RTT_{avg} is calculated in the same way, assuming the round trip time resulting from each RAB type j is RTT_j:

$$RTT_{avg} = \sum_{j=1}^{K} RTT_j \cdot q_j \tag{6.36}$$

With the mean peak rate r_{peak_avg} and the mean around trip time RTT_{avg}, the overall average transfer delay or average throughput can be calculated by the single rate M/G/R-PS model, assuming that all elastic traffic flows have the same average peak rate. Given the link capacity C, the M/G/R-PS model is then described with $R = R_{avg}$ where $R_{avg} = C / r_{peak\_avg}$. Thus, the average transfer delay for an amount of data of size x can be calculated with equation (6.26) and (6.27).

6.3.7.5 Extension for case of Mixing with Circuit-Switched Traffic

Normally, the UMTS network contains both elastic traffic and circuit-switched traffic. This section investigates such a scenario as shown in Figure 6.23. The circuit-switched traffic, which is generated by real-time critical applications with strict QoS requirements, is subject to connection admission control (CAC) function to guarantee its QoS requirement. However, for elastic traffic no CAC is applied. And at the network level, the circuit-switched traffic is given a higher priority for transmission over the elastic traffic, with a strict priority scheduling scheme. In order to determine the application performance of elastic traffic, the consequence of strict priority scheduling and the possible multiplexing gain due to capacity sharing between the two traffic types have to be considered. In the following, an approximation method is suggested to estimate the achievable application QoS for elastic traffic in case of sharing the Iub link with the circuit-switched traffic.

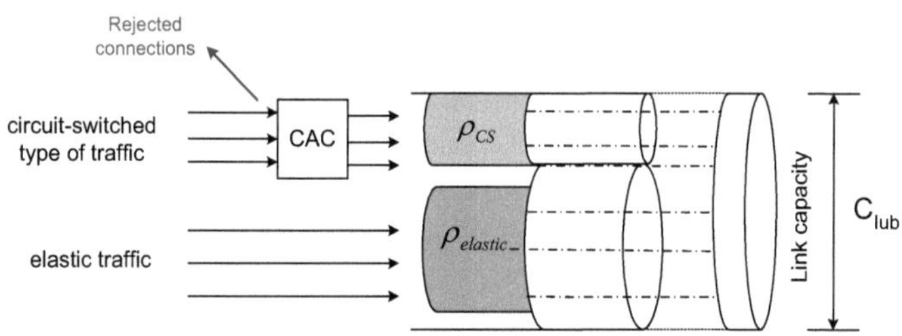

Figure 6.23: *Elastic traffic shares the Iub link with circuit-switched traffic*

Let L_{CS} denote the mean carried load of the circuit-switched traffic after the CAC function, and $L_{elastic}$ denote the mean traffic load of the elastic traffic. Given the total Iub link capacity C_{Iub}, the mean bandwidth $C_{elastic}$ that can be utilized by the elastic traffic in average can be estimated with equation (6.37). Because of the elastic property of TCP, the elastic traffic intends to utilize any available bandwidth on the link. As a result of the priority scheduling, any bandwidth which is not used by the circuit-switched traffic can be taken by the elastic traffic. Thus it can be assumed that in average the elastic traffic can use the bandwidth up to $(C_{Iub} - L_{CS})$. In this way, it is assumed that there is a complete bandwidth sharing between the two traffic types and thus has the possibility of achieving a high utilization.

$$C_{elastic} = C_{Iub} - L_{CS} \tag{6.37}$$

With $C_{elastic}$, a normalized elastic traffic denoted as $\rho_{elastic}$ can be derived by:

$$\rho_{elastic} = L_{elastic} / C_{elastic} \tag{6.38}$$

Assuming that all elastic traffic flows use the same RAB type and thus with the same peak rate r_{peak}, the average application performance of elastic traffic in this case will be still derived from the **M/G/R-PS** model, but **R** is calculated with $R = C_{elastic} / r_{peak}$. Thus the expected delay of transferring an amount of data of size x can be derived from equation (6.26) and (6.27), given the calculated normalized elastic traffic $\rho_{elastic}$ and an estimated *RTT* of elastic traffic which is obtained under the low link utilization.

With this approach, the bandwidth sharing between the two traffic types and their dependency as a consequence of strict priority scheduling are modeled by considering the long time average impact on the elastic flow performances. This will be sufficient for the purpose of dimensioning, which is based on a long time observation. The thorough impact of priority queuing at the network level will be modeled with exact queuing model, which will be described in chapter 6.4.

6.3.8 Validation of Extensions of Processor Sharing Models

This section demonstrates the applicability of the proposed Processor Sharing (PS) models and validates the above proposed extensions for the TCP-based elastic traffic for a desired application QoS. For validations, extensive simulations were carried out and then they were compared with the theoretic expectations from the PS models. The criterion for the application of the PS models is an accurate estimation of the average transfer time over the transferred amount of data. In the following sections validation results are presented. The results demonstrate that the proposed several extensions on the PS models are applicable for the objective of the Iub dimensioning for different network and traffic configurations.

6.3.8.1 Simulation Environment

The simulations were carried out with the developed the ATM-based UMTS Rel99 simulation model, which is introduced in Chapter 5.4. The network scenario consists of a Node B and a RNC with a single ATM link between them. The UEs downloads data from the remote Internet servers (i.e. data is transferred on downlink direction). The data is generated from either ftp or web applications (HTTP version 1.1). In the following, the amount of data per transaction is referred to as file sizes or page sizes. Various file or page sizes and the distributions were investigated (e.g. constant, heavy-tail Pareto, etc.). In all simulation scenarios, flows arrive according to a Poisson process, i.e. the interarrival times are negative exponentially distributed. The used TCP module is Reno with receiver window size of 65535 byte. The configured Iub link is an ATM E1 line (2 Mbps) with a reserved bandwidth of 1500 kbps for the user plane and the remaining bandwidth consumed for signaling and control channels. For all scenarios the simulations run 3600 simulated seconds. The stability and reliability of the measured statistics of average file or page transfer delay from simulations are evaluated by means of *Confidence Intervals*. The evaluation results show that 3600s simulation time is sufficient to reach stable results for all investigated cases in this thesis, as the obtained simulation results are proven to be reliable and stable. Several examples are given in Appendix A.20.

6.3.8.2 Single RAB without Applying CAC

This section validates the extensions proposed in section 6.3.7.1 on the M/G/R-PS model for cases where a single RAB type is served for all elastic flows and no CAC function is applied in the UTRAN transport network. In the following investigated scenarios, all UEs are given the same RAB rate of 128 kbps. For validations, ftp application is used and constant file size distribution is chosen.

This scenario investigates the accuracy of different versions of M/G/R-PS models: the basic PS model in section 6.3.5, the extended PS model in section 6.3.6, and the adjusted PS model proposed in section 6.3.7.1. In this scenario, the application is ftp of a 12 kbyte file size. Figure 6.24 compares the average file transfer delays derived from different processor sharing models against the simulations.

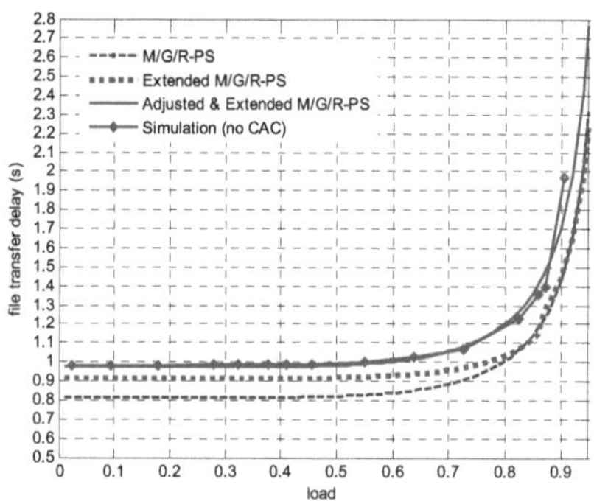

Figure 6.24: *Average file transfer delay over load: ftp scenario, RAB 128 kbps, no CAC, constant file size of 12 kbyte, Iub link = 1 E1 line*

For a 12 kbyte file size, TCP flows remain mainly in the slow start phase. The offset between the basic M/G/R-PS (equation (6.18)) to the extended M/G/R-PS (equation (6.24)) under the low load is due to the underestimation of the delays caused by the slow start during which the available bandwidth cannot be fully utilized. When the load goes higher the shared bandwidth starts to be fully utilized, thus these two curves converge. The adjusted and extended M/G/R-PS model proposed in section 6.3.7.1 with equation (6.26) and (6.27) considers the impact of the slow start as well as the additional delay for setting up and releasing the TCP connection, and moreover the varying round trip time effect. It can be easily seen that the adjusted M/G/R-PS model proposed in the framework of this thesis matches the simulation results best.

The following scenarios validate the accuracy of the M/G/R-PS model for different file sizes and file size distributions. Figure 6.25 presents the transfer delay over different load levels for a small file size of 5kbyte (left chart) and a large file size of 100kbyte (right chart). Both figures demonstrate that the adjusted and extended M/G/R-PS approach can predict accurately the average file transfer delay for different file size ranges. But the basic M/G/R-PS approach underestimates the transaction times: firstly it does not consider the extra delay caused by the TCP slow start and this relative offset is much more significant for smaller file sizes; secondly it does not consider the additional delay for connection setup and release. The extended M/G/R-PS approach improves the accuracy by considering the TCP slow start mechanism but still does not include the effect of varying *RTT* and the additional delays for TCP connection setup and release. From the above results, it can be concluded that the adjusted and extended M/G/R-PS model proposed in section 6.3.7.1 improves the accuracy of the calculation of the mean file transfer delay under different load levels as well as for different file sizes. It gives a precise estimation of the application performance and in turn should enhance the dimensioning results.

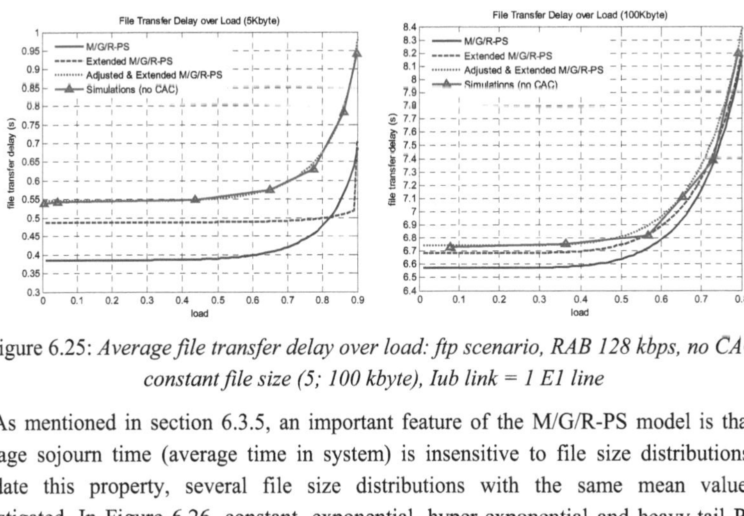

Figure 6.25: *Average file transfer delay over load: ftp scenario, RAB 128 kbps, no CAC, constant file size (5; 100 kbyte), Iub link = 1 E1 line*

As mentioned in section 6.3.5, an important feature of the M/G/R-PS model is that the average sojourn time (average time in system) is insensitive to file size distributions. To validate this property, several file size distributions with the same mean value are investigated. In Figure 6.26, constant, exponential, hyper-exponential and heavy-tail Pareto file size distributions are compared, all with the same mean of 12 kbyte.

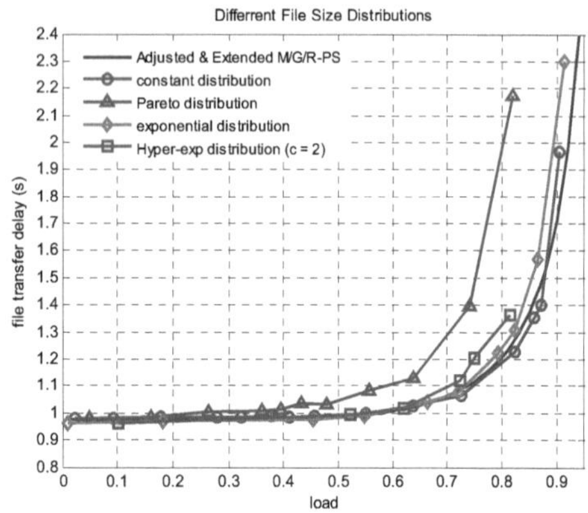

Figure 6.26*: Different file size distributions: ftp scenario, RAB 128 kbps, no CAC, constant file size of12 kbyte, Iub link = 1 E1 line*

In this figure, the parameter c^2 represents the *coefficient of variation* (c is defined as the ratio of the standard deviation to the mean), which is used to estimate the level of burstiness of the traffic. The constant file size distribution represents the deterministic traffic, whose coefficient of variation is 0. For the exponential distribution c equals to 1, it generates pure random traffic. The hyper-exponential distribution produces a bursty traffic with $c^2>1$, in the given scenario $c^2=4$. The Pareto distribution is a heavy tailed distribution, it is configured with $c^2=12$. It can be seen that in the low load situations, the delay from all file size distributions are same. When the load goes higher, the transfer delays get increased in general as a result of the Iub link congestion. Furthermore it is found that constant, exponential and hyper-exponential file size distributions achieve almost similar application performance, which demonstrates the property of independency of the file size distributions to the average sojourn delays. However, Pareto file size distribution experiences significantly longer delays than other distributions when the load is above 50% of the link capacity. This negative effect of underestimating the transfer time for the Pareto distribution is caused by not taking the queuing delays in the buffer of the bottleneck link into account for the experienced *RTT*. This is because that in addition to its transmission time queuing delays over the link also have considerable impact to the round trip times. For a more bursty source traffic, this impact becomes more significant. As with heavy-tail Pareto distribution, the aggregated traffic on the Iub link is much more bursty than other file size distributions, as a result the degradation of the queuing performance under Pareto distribution is more serious. In order to improve the accuracy of estimations for the *RTT* in the case of a highly bursty source traffic, it is proposed by to calculate an estimated round trip time which also includes the queuing delays in the buffer of the bottleneck link. It is assumed that in times where more than *R* flows are active (i.e. the sum of the peak rates exceeds the link capacity C), the buffer fills up quickly and the queue is mostly at its limit. This corresponds to serving B packets of length L_P with capacity C. In times of $n < R$, it is assumed that the queue is approximately empty. Let T_p denote the one-way propagation delay of the link. Thus the average round trip time RTT_{est} can be estimated as suggested by Anton Riedl in [Rie04]:

$$RTT_{est} = P(n \leq R)\cdot(2T_p) + P(n > R)\cdot(2T_p + \frac{B \cdot L_P}{C}) \tag{6.39}$$

The above state probabilities $P(n)$ of having n active jobs in the system are given in [Coh79].

$$p(j) = \begin{cases} (1-\rho)\dfrac{R!}{j!}(R\rho)^{j-R} E_2(R,R\rho) & (j < R) \\ E_2(R,R\rho)\rho^{j-R}(1-\rho) & (j \geq R) \end{cases} \tag{6.40}$$

Figure 6.27 shows the improved results using the above estimated round trip times RTT_{est} for calculating the average transfer delays for the Pareto file size distribution. It is seen obviously that with RTT_{est} , the calculated mean transfer delays with the M/G/R-PS model fit the simulation results much more accurately. Therefore, it is suggested to apply the above approach for the cases for a highly bursty source traffic where the queuing delay on the bottleneck link is significant.

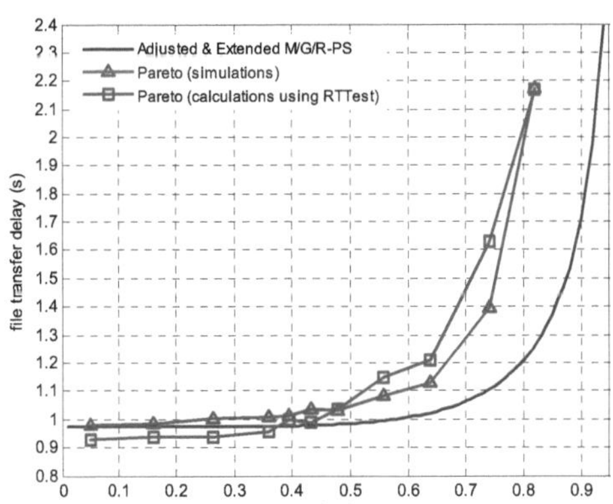

Figure 6.27: *Improvement for Pareto file size distribution with RTT_{est}: ftp scenario, RAB 128 kbps, no CAC, constant file size of 12 kbyte, Iub link = 1 E1 line*

6.3.8.3 Single RAB with CAC

When employing the admission control function (CAC) in the UTRAN, the system can avoid heavy congestions by limiting the number of active simultaneous connections on the link, and thus each user gets a minimum throughput which can guarantee its desired QoS. For this case, the M/G/R/N-PS model shall be applied as suggested in section 6.3.7.2 to estimate both average application delay and connection reject ratio (or called blocking probability). The following scenarios validate the M/G/R/N-PS model for an ftp traffic scenario where the CAC function is applied to control the transport resources of the Iub interface. In the presented scenario, all users choose the same RAB rate of 128 kbps to transfer data and the CAC guaranteed bandwidth per user connection is set to 96 kbps. For the configured Iub link (refer to section 6.3.8.1), the Iub user plane allows a maximum of 15 simultaneous user connections. Figure 6.28 investigates a 12 kbyte file size and Figure 6.29 investigates a 50 kbyte file size, both with constant file size distribution.

Both figures compare the simulation results with the calculations of the M/G/R/N-PS model (here N=15) using formula (6.28) and (6.29) for average file transfer delay (left diagram) and CAC reject ratio or blocking (right diagram). The x axis is the offered traffic load including the rejected traffic, as a percentage of the Iub user plane capacity (1500 kbps). The results verify that for different file sizes the proposed M/G/R/N-PS model gives correct estimations for the average file transfer delay and the call reject ratio. It is also found that the obtained accuracy is higher with a larger the file size. This is because that the presented M/G/R/N-PS model in this thesis has not considered the TCP slow start effect. For 50kbyte file size, the TCP slow start does not cause significant effect. Thus it can be seen that the calculated average delay for the 50 kbyte file size from the M/G/R/N-PS model matches better to the simulation results than the 12 kbyte file size, as well as the call reject ratio.

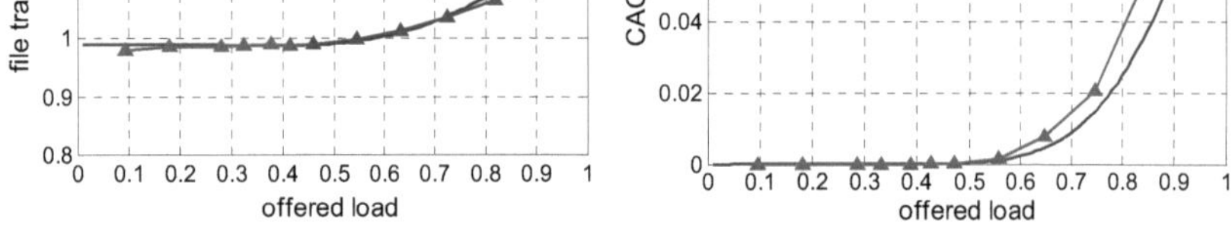

Figure 6.28: *File transfer delay (left), blocking ratio (right): ftp scenario, RAB 128 kbps, CAC reserved rate = 96 kbps, constant file size of 12 kbyte, Iub link =1 E1 line*

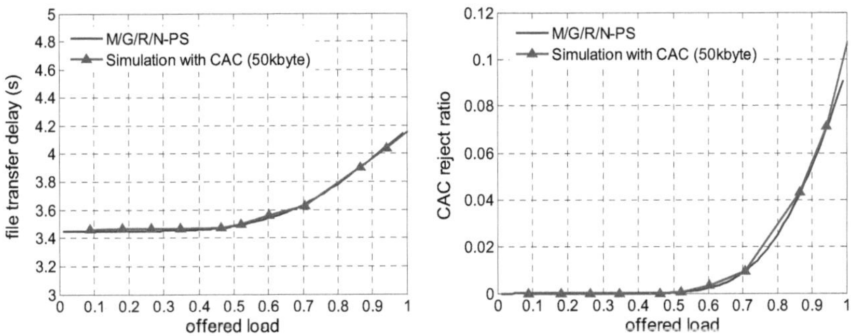

Figure 6.29: *File transfer delay (left), blocking ratio (right): ftp scenario, RAB 128 kbps, CAC reserved rate = 96 kbps, constant file size of 50 kbyte, Iub link =1 E1 line*

The presented example results demonstrate that in case of applying CAC in the transport network for the given traffic, the suggested M/G/R/N-PS model is a correct analytical model to estimate both application performance and connection reject ratio. So that it can be further used for link dimensioning in order to satisfy a desired application QoS or a desired reject ratio.

6.3.8.4 Multiple RABs without Applying CAC

This section demonstrates the general M/G/R-PS model presented in section 6.3.7.3 for multiple RABs scenarios where no admission control is applied in the UTRAN. In the following validation scenarios, constant file size distribution is used.

At first, an ftp scenario is given which mixes three user groups each with a different RAB type: 64 kbps, 128 kbps and 384 kbps, to transfer a 12 kbyte file. The ftp requests are equally distributed among the three user groups (each 33.3%). Figure 6.30 plots the simulated average file transfer delays of individual user groups corresponding to different RAB type and compares them with the analytical results calculated for each RAB type separately using formula (6.31) to (6.34). The load in the graph corresponds to the total file requests of all user groups (calculated with equation (6.31)), which is represented as a normalized total load or link utilization.

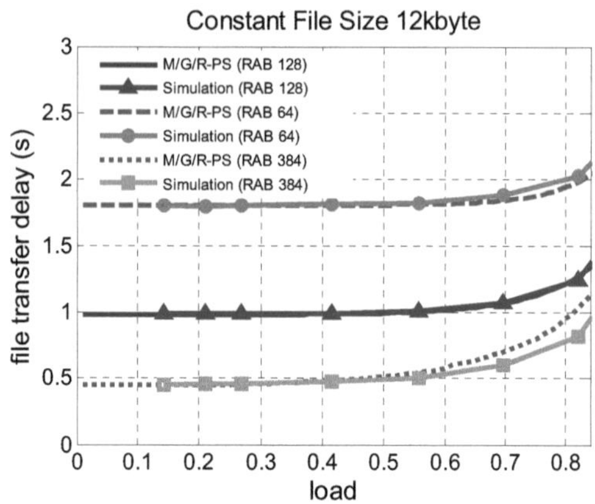

Figure 6.30: *Average application delay over the load for multiple RABs: ftp scenario, no CAC, constant file size (12 kbyte)*

It is seen in Figure 6.30 that the calculated average transfer delay for each individual RAB type is close to the simulation results obtained from the simulations. This means, the general form of the M/G/R-PS model can be applied well to estimate the application performance of each RAB type in a multiple RABs scenario. The results also prove that there is a quasi fair sharing of the available resources among different TCP flows with various RAB types. The load contributed by all RAB types (user groups) is counted for the delay factor of each RAB type, though the parameter R of certain RAB type i in the individual M/G/R-PS model is calculated with equation (6.30). However, it is also obvious to see from the figure that the flows with the low RAB rate achieve longer transaction times than calculated with the M/G/R-PS model, while the high RAB rate flows experience a better delay than calculated. The reason for this unfairness is that the high-peak-rate flows takes advantage of lower *RTT*s and can attain more network resources, especially at the beginning of the transmission phase. This is further proven in Appendix A.14. The following simulation results investigate the dependency of different bearer (or user) distributions on the application performance. Figure 6.31 compares the average transfer delay of each RAB type for transferring a 50 kbyte file (ftp application), under three different bearer distributions:

case 1: 50% RAB 64, 30% RAB 128, and 20% RAB 384
case 2: 33.3% RAB 64, 33.3% RAB 128, 33.3% RAB 384
case 3: 30% RAB 64, 20% RAB 128, and 50% RAB 384

It shows that the performance of each RAB is insensitive to the bearer distributions. This is due to a fair sharing of the available resources is achieved among all ongoing TCP flows present, which is regardless of the peak data of individual flows. Thus the performance of each RAB type is only dependent on its own peak data rate and the total contributed loads from all RAB types. From the dimensioning perspective, the accumulation of the loads of different RABs for dimensioning purposes is essential; otherwise the underestimated multiplexing gain per independent bearer type results in an underutilization of the link (over-achieving the delay requirements).

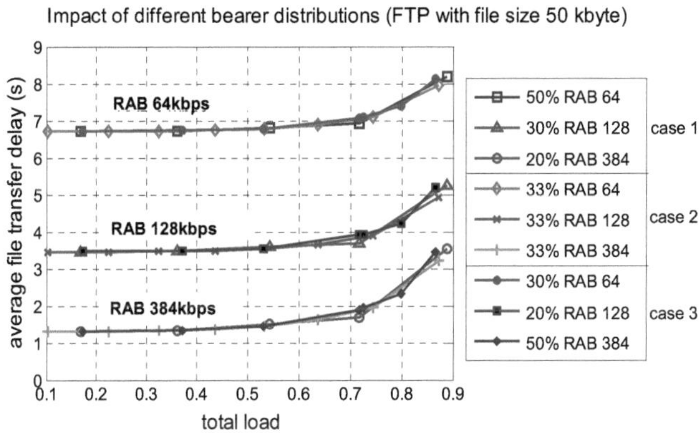

Figure 6.31: *The impact of different distributions of various RAB types: ftp scenario, no CAC, constant file size (50 kbyte)*

The general form of the M/G/R-PS mode presented in section 6.3.7.3 was validated for different bearer distributions and also for web applications. Figure 6.32 shows such a web scenario where the size of a web page is 12 kbyte. For the given two different bearer distributions, the simulated average application delays (i.e. page download delays) over different link utilizations follows approximately the theoretical results derived from the general PS model.

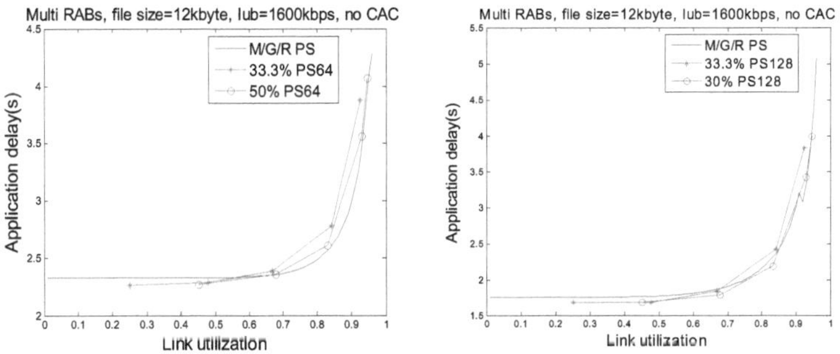

Figure 6.32: *Average application delay over the load for multiple RABs with different RAB distributions: web scenario, no CAC, constant page size (12 kbyte)*

These results demonstrate that the average application delay per RAB type can be derived from the proposal general M/G/R-PS mode, taking the total load of all flows into the link congestion while considering specific RAB peak data rate for calculating the transfer delay per RAB type.

6.3.8.5 BRA without Applying CAC

With the application of *Bit Rate Adaptation* (BRA) function, the peak data rate of elastic traffic flows is changed dynamically during data transfer process. The change of peak rates introduces a problem for deriving R for the M/G/R-PS model. To solve this problem, section 6.3.7.4 proposes to derive R based on the mean peak rate r_{peak_avg}, which is calculated with equation (6.35) given different peak rates and their portions in use over time. In addition, an average around trip time RTT_{avg} is estimated. With r_{peak_avg} and RTT_{avg}, the overall average transfer delay or average throughput can be calculated by the M/G/R-PS model. The following presents a scenario with BRA in use. In the given example, web application is chosen with a constant page size of 12 kbyte.

The validation results are presented in Figure 6.33. The left diagram plots the calculated average page transfer delays over different loads from the proposed approach. The calculations are compared with simulation results. The right diagram gives the relative error of the analytical results. It shows that the obtained relative errors (calculated with equation (4.1)) are below 10% in most cases while at high loads (above 90% link utilization) the relative errors get slightly increased (about 11%).

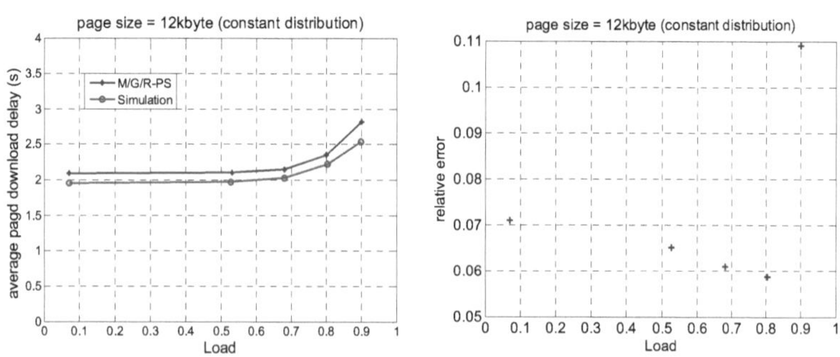

Figure 6.33: *Average application delay over the load for BRA: web scenario, no CAC, page size =12 kbyte, Iub link = 1 E1 line*

The comparison shows that the proposed analytical approach provides an appropriate prediction for average page download delays, though there are some overestimations on the transaction delays in the analytical approach. The overestimation is mainly due to the fact that the applied analytical approach does not consider the exact behavior of HTTP 1.1. With HTTP 1.1, during one web session multiple web pages are downloaded over the same TCP connection. That means, except for the first page which is transferred from the beginning with TCP slow start, the second and subsequent pages are transferred over that established TCP connection, which has already been in the fair-sharing state. As a result, the second and subsequent pages need less time to transfer than the first page, since they skip the slow start phase and thus have no degradation on flow performance caused by the slow start. However, the number of pages to download per TCP connection is a statistical variable. The applied analytical approach does not identify pages which do experience the TCP slow start and which do not. Instead, it assumes that each page is transferred with one separate TCP connection including slow start. Therefore, the analytical approach overestimates the overall average transaction time by taking into account the additional delays for the slow start phase for all transactions. This deviation is more noticeable for short flows, for which the duration of the slow start phase is proportionally long. Speaking in terms of network dimensioning, this would mean that the theoretical calculation overestimates the capacity demands for short flows, which lead to satisfactory QoS and thus on the safe side. For increasing file sizes the theoretical model becomes more accurate. Thus, for larger transactions dimensioning according to the analytical approach can give quite accurate results.

6.3.8.6 Elastic Traffic Mixed with Circuit-Switched Traffic

The following scenario investigates the accuracy of the proposed approximation method for estimating application delays for elastic traffic when mixed with circuit-switched type of traffic, as presented in section 6.3.7.5. In this scenario, there is 40% voice traffic and 60% web traffic (with HTTP 1.1) offered to the Iub interface, where the voice traffic is given higher priority to transport over the web traffic. The voice traffic is generated according to the traffic model defined in Table 5.2. The web traffic is generated with a constant page size of 50 kbyte. It is transmitted with RAB 128 kbps in this example and no CAC is applied to the elastic traffic. In Figure 6.34, the left diagram shows the approximated average transfer delays for a web page over different link utilizations and they are compared with the simulated delay values.

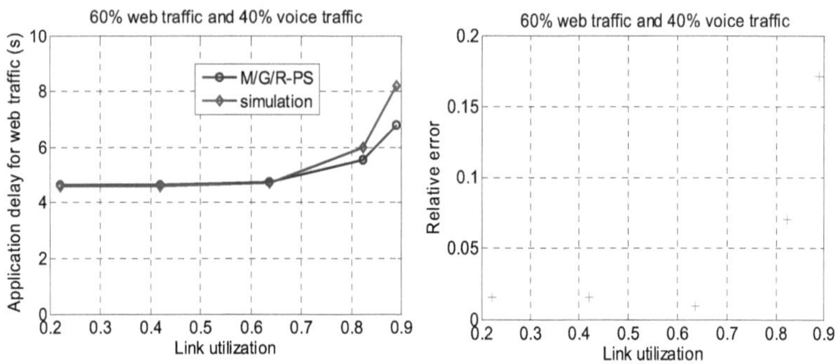

Figure 6.34: Average application delay over the load for web mixed with voice: RAB 128 kbps (for web traffic), no CAC, page size =50 kbyte, Iub link = 1 E1 line

It can be seen that the approximations from the M/G/R-PS model underestimates the delays under the high link utilizations. This is because the elastic traffic starts to experience packet losses at high loads, as a consequence of lower priority of transmissions than the voice traffic. That means, the elastic traffic packets are discarded to bring advantage for the real-time voice traffic. And the discarded packets of elastic traffic will trigger TCP retransmissions, which leads to degradation of the throughput of elastic traffic flows. However, the PS model does not consider packet losses and resultant retransmissions. Thus, the calculated results do not include this degradation on the transfer delays. But, the relative error of the theoretical estimations shown in the right diagram is below 10%, which is acceptable for dimensioning purpose, when the link utilization is within 80%.

6.3.9 Dimensioning Procedures

This section describes in detail the dimensioning procedures, i.e. how to utilize the presented processor sharing models for the purposes of Iub dimensioning. In this chapter, the dimensioning process only considers a single Iub link as illustrated in Figure 6.35(a). The objective of dimensioning is to determine necessary Iub link bandwidth for guaranteeing a desired QoS target with the highest cost-efficiency (i.e. maximum utilization). Figure 6.35(b) shows the input and output parameters of the process sharing models.

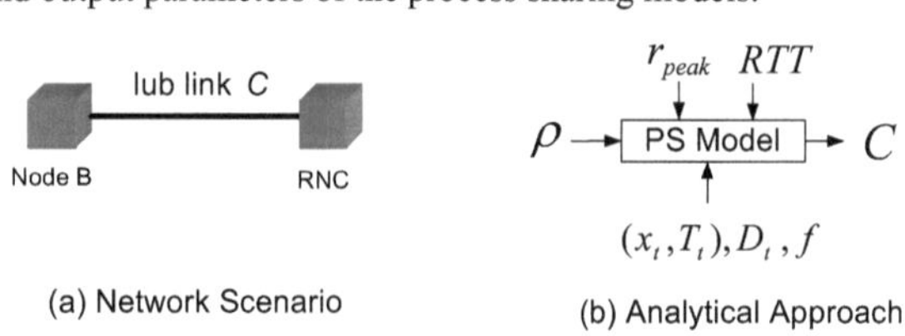

Figure 6.35: *Iub link dimensioning with processor sharing (PS) models*

At first, the amount of elastic traffic ρ, which traverses the Iub link, has to be given. It can be derived from the mean flow arrival rate and mean object size, including the overheads from applications, TCP/IP protocols, as well as from UMTS networks. Furthermore, possible rate limitations confining individual flows (i.e. peak rate of the used RAB type) have to be identified. And in addition, the round trip times experienced under a low Iub link utilization needs to be estimated. At last, an appropriate QoS target much be specified. This can either be a desired average transfer time T_t for an amount of data of size x_t (including all related overheads), or a desired average application throughput D_t which is equal to x_t / T_t, or simply a value for the desired delay factor f.

With these input parameters, the optimum capacity C can be computed by applying the developed M/G/R-PS models. For different UMTS network scenarios, e.g. with a single RAB rate or multiple RABs or applying BRA, with or without admission control CAC etc, the different corresponding extensions need to be applied for estimating the average transfer delay, as presented in section 6.3.7. However, in most cases it is not possible to solve these expected sojourn time equations explicitly for C. Therefore, an *iterative approach* is suggested, which is according to the general dimensioning procedure introduced in Chapter 4.3. That means, the average transfer times are interactively calculated for adapting the capacities C. Starting from a value of ρ, C can be adjusted (incremented or decremented) in small intervals (e.g. 10 kbps) until the desired QoS is achieved, i.e. until throughput $D \geq D_t$ or $T(x_t) \leq T_t$. The obtained optimum capacity C is then the outcome of the dimensioning result for the Iub link for a desired application QoS. The following describes individual dimensioning procedures for different scenarios.

Unlimited Peak Rate

If there is no limitation on the peak data rate on the elastic traffic flows, M/G/1-PS model should be used as R = 1. For negligible round trip times, the optimum capacity C can be calculated explicitly from the basic M/G/1-PS model (equation (6.18)). For a desired throughput D_t, C can be calculated with equation below.

$$C = D_t + \rho \tag{6.41}$$

or in case the QoS requirement is expressed by means of the delay factor f

$$C = \frac{f}{f-1} \cdot \rho \tag{6.42}$$

If the effect of long round trip times should be taken into account, equation (6.24) with R = 1 has to be used. However, it is not possible to solve this equation explicitly for C, thus the *interactive approach* has to be used in this case.

Single RAB without Applying CAC

For a single RAB scenario, if elastic traffic is not controlled by the CAC, equation (6.26) and (6.27) are proposed in section 6.3.7.1 for calculating the average sojourn delay. However, it is also not possible to solve equation (6.26) and (6.27) explicitly for *C*. Therefore, for a desired delay factor f, or transfer delay T_t, or throughput D_t, the optimum Iub link capacity *C* will be derived with the *iterative approach.*

Single RAB with CAC

For a single RAB scenario, when the CAC is applied to elastic traffic, equation (6.28) and (6.29) are suggested to calculate the sojourn times. Same as above, the bandwidth *C* is derived from the *iterative approach.*

Multiple RABs without Applying CAC

In case of multiple RABs scenario, it is assumed that different RAB types are used by different traffic classes. For each traffic class (i.e. each RAB type), there is a different QoS requirement. The goal of dimensioning is to satisfy the QoS requirements of the individual traffic classes on an end-to-end basis. For achieving this goal, the optimum capacity *C* has to meet the desired QoS values of all RAB types. In section 6.3.7.3, a general form of the M/G/R-PS model is presented to calculate the mean sojourn time for each RAB type separately in the multiple RABs scenario. For dimensioning process, for each RAB type *i*, the general form is applied to derive an optimum capacity C_i for meeting its specific QoS requirement with the above *interactive approa*ch. At the end, in order to meet the QoS requirements of all RAB types, the maximum C_i is taken to be the optimum Iub link capacity.

Elastic Traffic mixed with Circuit-switched Traffic

If elastic traffic is transmitted together with the circuit-switched traffic on the Iub link, different approaches can be used for link dimensioning, depending on different policies for integrating elastic and circuit-switched traffic flows.

The analytical model presented in section 6.3.7.5 is applied for the scenarios where both traffic types completely share the network bandwidth, i.e. the elastic traffic can fully exploit the bandwidth that is not used by the circuit-switched traffic as a result of priority scheduling. In this case, the required capacity for elastic traffic $C_{elastic}$ can be derived iteratively from equation (6.26) and (6.27) given a desired application QoS. Then from equation (6.37), the total required capacity for the Iub link C_{Iub} can be obtained with equation (6.43), where L_{CS} denotes the mean traffic load of the circuit-switched traffic carried on the Iub link.

$$C_{Iub} = C_{elastic} + L_{CS} \tag{6.43}$$

In other cases, elastic traffic and circuit-switched traffic can be handled by separated bandwidth portions, as shown in Figure 6.36. One part of the link capacity is exclusively dedicated to circuit-switched traffic which is denoted as C_{CS}, the other part is exclusively dedicated to elastic traffic which is denoted as $C_{elastic}$. Such bandwidth partitioning can be realized by employing specific traffic policy units at the network nodes such as rate limiting function, token buckets, etc. These traffic policy functions assure that individual traffic flows within the portion of each traffic type do not negatively interfere with each other. This allows us to consider elastic traffic and circuit-switched traffic separately and to determine their necessary capacity shares independently with different dimensioning models: for circuit-switched traffic the dimensioning models use Erlang-B formula or Multi-dimensional Erlang-B formula (as presented in section 6.2); whereas for elastic traffic the dimensioning models are based on M/G/R-PS model. In order to obtain the total capacity of the Iub link C_{Iub}, the individual bandwidth shares need to be summed up ($C_{Iub} = C_{elastic} + C_{CS}$).

Comparing these two dimensioning approaches, the latter achieves lower utilization as it does not achieve any multiplexing gain by using bandwidth portioning. In case of complete sharing the bandwidth among the two traffic types, multiplexing gain is attained and thus the achievable utilization is higher. In the framework of this thesis, the elastic traffic completely shares the bandwidth with the circuit-switched traffic, and thus the first dimensioning approach is considered.

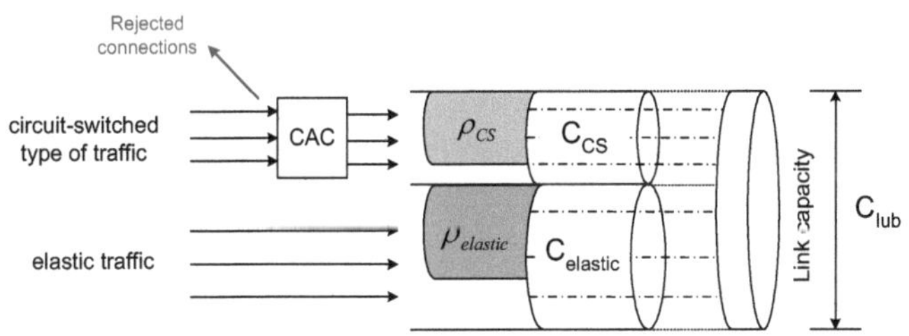

Figure 6.36: *Bandwidth partitioning for elastic traffic & circuit-switched traffic*

6.3.10 Dimensioning Results

This section presents dimensioning results obtained from both simulation and analytical approaches. With the simulation approach, dimensioning is carried out by adapting the Iub link bandwidth iteratively in order to reach a desired QoS in simulations. The dimensioning procedure has been explained with Figure 4.6 in Chapter 4.3. With the analytical approaches, the proposed M/G/R-PS models are applied. The analytical results are derived in accordance with the dimensioning procedures described in the last section. At first, the accuracy of the analytical approaches with the M/G/R-PS models is validated by simulations to dimension a single link to meet a desired delay factor as the QoS target. For analyzing the dimensioning results, different dimensioning metrics are evaluated (definitions of various dimensioning metrics are referred to section 4.3). After validations, the analytical approaches are applied to investigate the dimensioning rules for elastic traffic for a desired application QoS.

Validation of Analytical Approaches for Dimensioning

First of all, a dimensioning example is given, which investigates the applicability of the proposed M/G/R-PS model for dimensioning a single Iub link. In this example, there is only elastic traffic which is generated from the web application with a constant page size of 50 kbyte. All elastic traffic flows are transmitting with the same peak rate of 128 kbps (RAB 128 kbps) and no CAC is applied to elastic traffic. The QoS target is to achieve a delay factor of f =1.25. Figure 6.37 shows the required Iub link capacity (in kbps) over different traffic loads in terms of link throughput (in kbps). The theoretical results are derived with the dimensioning procedure for the case of single RAB without applying CAC, i.e. with the extended M/G/R-PS model presented in section 6.3.7.1. In Figure 6.37, the left diagram compares the calculated Iub bandwidth in kbps and the simulation ones; while the right diagram gives the relative error of the calculated bandwidth from the M/G/R-PS model. It can clearly be seen that the derived required Iub bandwidth from the proposed PS model matches sufficiently well with the simulated results, with relative errors less than 5%. Figure 6.38 presents the resultant normalized capacity (left diagram) and the link throughput (right diagram). From Figure 6.38, a multiplexing gain can be observed for larger traffic loads. As the normalized capacity, i.e. overdimensioning factor, approaches one for increasing traffic load. As a result, the link utilization increases for larger traffic aggregations. This also means that for a higher traffic load, the achieved cost-efficiency of dimensioning is higher as a result of the increased multiplexing gain.

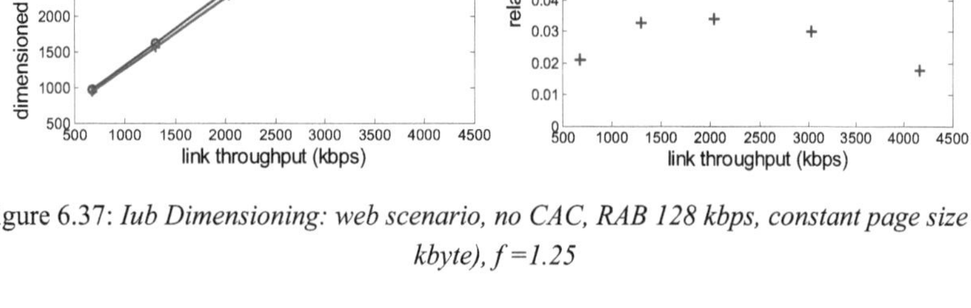

Figure 6.37: *Iub Dimensioning: web scenario, no CAC, RAB 128 kbps, constant page size (50 kbyte), f =1.25*

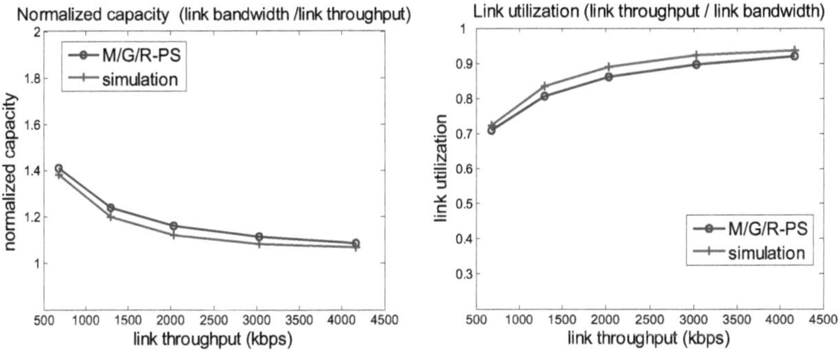

Figure 6.38: *Normalized capacity and link utilization*

Similar validations were also done for other scenarios. An example of dimensioning for the mixed traffic scenario is given in Appendix A.17. From all investigations, it is found that the calculated link bandwidths derived from the presented PS models generally match well with the simulations. In addition, the accuracy of different extensions on the M/G/R-PS model have already been validated in section 6.3.8 for estimating application QoS of elastic traffic under different scenarios. Therefore, it can be concluded that the proposed PS models can provide sufficiently accurate estimation for the required link bandwidth for elastic traffic to meet a desired application QoS.

Investigation of the Dimensioning Rules

After demonstrating the analytical approach for dimensioning the Iub link, the analytical dimensioning approach can be used to investigate the dimensioning rules. In the following, several examples are presented to demonstrate the applicability of the analytical approaches for deriving dimensioning rules. Figure 6.39 illustrates the normalized capacity as a function of the offered traffic load (link throughput) with different QoS targets (three different target delay factors f = 1.1, 1.3 and 1.6). Figure 6.40 presents the normalized capacity over traffic load with various peak rates (RAB types). Figure 6.41 shows the results for different traffic mixes (elastic web traffic mixed with circuit-switched voice traffic in different percentages). The voice traffic is transmitted with a higher priority over the Iub link than the web traffic, and the Iub link is completely shared between web traffic and voice traffic.

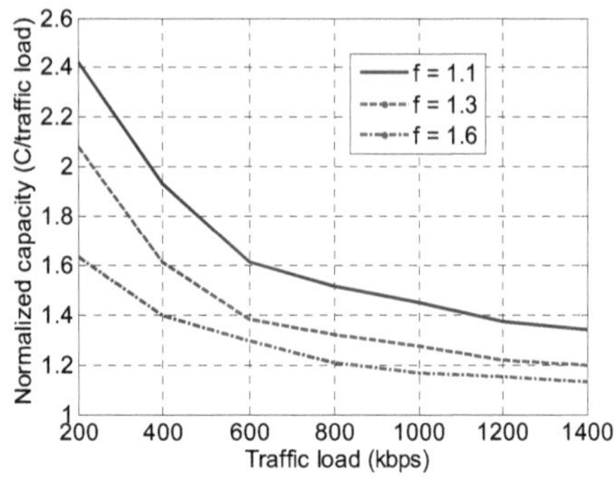

Figure 6.39: *Capacity vs. traffic load (with different delay factors): web scenario, constant page size of 12kbyte, RAB 128 kbps, no CAC*

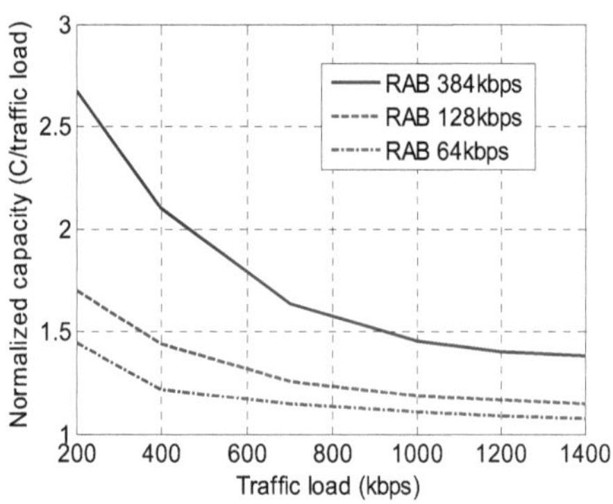

Figure 6.40: *Capacity vs. traffic load (with different peak rates): web scenario, constant page size of 12kbyte, target delay factor $f = 1.5$ for all RABs, no CAC*

It can be observed from these figures that an improved multiplexing gain is achieved for larger traffic load as the overdimensioning factor approaches one for increasing traffic load. The graph in Figure 6.39 depicts the dependency of the required link capacity on the delay factor: the higher the desired delay factor, the lower is the QoS requirement, and thus the less bandwidth is required. The influence of the peak rate on the link capacity is demonstrated in Figure 6.40. Larger peak rates require larger bandwidth values. In Figure 6.41, the impact of percentage of mixed voice traffic on the required link capacity is shown. The more portion of circuit-switched voice traffic, the higher gain can be achieved and thus less demand of the total link bandwidth.

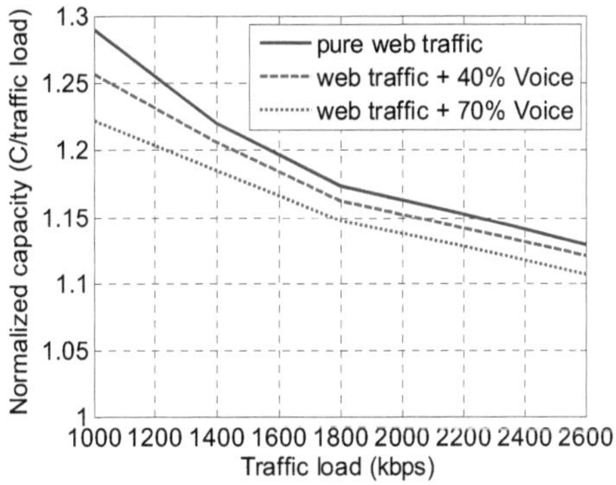

Figure 6.41: *Capacity vs. traffic load (different traffic mixes of web and voice traffic) for web traffic: RAB 128 kbps, no CAC, page size =50 kbyte, delay factor f = 1.25*

6.3.11 Summary

In Chapter 6.3, the dimensioning process for elastic traffic was addressed. Analytical dimensioning models are presented, which are based on the theory of processor sharing. A number of extensions on the M/G/R-PS model are suggested in this thesis for dimensioning the UMTS Iub interface in different scenarios, which consider the UMTS specific features and resource management functions. The proposed analytical approaches are demonstrated by simulations. Overall, the simulation results comply with the theoretical expectations sufficiently well. This suggests that the extensions on the M/G/R-PS model which are proposed in this thesis can be applied to dimension the Iub link for elastic traffic for different scenarios to guarantee a desired application QoS.

6.4 Dimensioning for Transport Network Performance

As discussed in section 6.3.10, the transport network performance is also a critical QoS requirement for dimensioning the Iub interface. The most important performance aspects in the UTRAN transport network are the packet delay or packet loss ratio (packet losses are actually caused by excessive long packet delays) over the Iub interface. This chapter will present analytical models for dimensioning of the Iub link in order to guarantee the Iub transport delay performance, referred to as TNL performance in this thesis. The proposed dimensioning approaches will be based on analytical queuing models on the packet level to estimate the Iub delay. In this thesis, two analytical models are proposed: the *Markov-modulated Poisson process* (**MMPP**) and the *Batch Markovian Arrival Process* (**BMAP**).

The structure of this chapter is organized as follows. Section 6.4.1 presents the Iub system model on the packet level. Section 6.4.2 analyzes the characteristics of the aggregated traffic on the Iub. Section 6.4.3 presents the analytical dimensioning models based on MMPP, and section 6.4.4 presents the dimensioning method with BMAP. Both analytical models are used for building queuing models to analytically approximate the aggregated traffic on the Iub and estimate the resultant Iub transport delays. In both sections, it is explained in detail how to apply these queuing models in the dimensioning framework, the required procedure for estimating the necessary Iub bandwidth for satisfying certain Iub delay performance and finally validate them through simulations.

6.4.1 System Models for the Iub Interface

This section is intended to analyze the system models of the Iub interface on the packet level, in order to set up appropriate analytical queuing models for estimating the overall Iub transport delays and in turn to dimension the Iub link properly. To understand the system models, it is necessary to analyze the delay over the Iub interface, including its components and their relations to the given Iub link capacity. With the detailed Iub delay analysis, the generic system models can be proposed for setting up queuing models for calculating the Iub transport delay.

6.4.1.1 Delay Analysis on the Iub

In UMTS, each Node B serves a number of mobile users in the cell with each of the users generating different traffic streams related to different applications. The traffic from all users served by the same base station is aggregated and then transported over the Iub link. As seen in Figure 6.42, each user traffic stream (voice, video or data) is carried by a *Dedicated Transport Channel* (DCH) between RNC and Node B. The packet flows for each DCH from the upper layer down to the *Frame Protocol* (FP) layer has been explained in section 3.3.3.1. The DCH traffic stream is handed over from the FP layer to the *Transport Network Layer* (TNL) in the form of FP PDUs. At the Iub transport network layer, FP PDUs from each DCH are sent to a common AAL2 buffer every TTI. Thus, the FP PDUs streams of all DCHs are aggregated at the AAL2 queue.

In the UTRAN the Iub delay is in fact the FP PDU delay, i.e. the delay of a radio frame across the Iub. From the system point of view, the FP PDU delay comprises all delays accumulated on the Iub interface (see Figure 6.42): (i) the segmentation and reassembly delay; (ii) the queuing delay in the AAL2 buffer; (iii) the queuing delay in the ATM buffer; and (iv) the transmission delay or processing delay over the Iub link.

(i) The segmentation and reassembly delay includes the delay of segmenting FP PDUs into AAL2 Packets and AAL2 packets to ATM cells, and the corresponding reassembly for the opposite task. This delay is usually neglected due to a very low value.

(ii) As shown in Figure 6.42, the DCH traffic from each user is aggregated at the AAL2 layer, the arrival traffic of the AAL2 queue can be seen as the superposition of the arriving FP PDU streams of all users. In order to avoid overload on the ATM link, usually the aggregated AAL2 traffic is limited by the available Peak Cell Rate (PCR) of the ATM (section 3.3.3.2), known as AAL2 shaping. With this shaping at the AAL2 layer, the service rate of the AAL2 queue is deterministic.

(iii) For the ATM buffer, the service rate is the link rate, which is also deterministic. Due to traffic shaping on the AAL2 layer, the data rate on the ATM layer is below the configured peak cell rate. Therefore there will be no congestion or overload situation on the ATM link in general. As a result, the queuing delay in the ATM buffer is quite small and stable.

(iv) The transmission delay over the Iub link is rather low due to a high transmission speed. For transmitting an ATM cell on a 2 Mbps E1 line it is 0.212 ms. The node processing delay such as switching is very small and thus can be neglected.

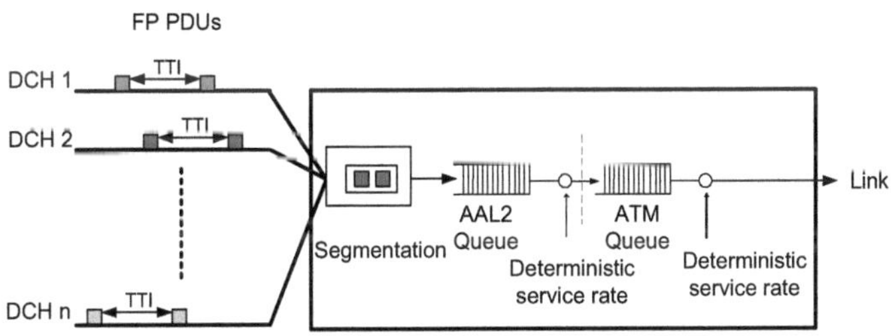

Figure 6.42: *Delay in the UTRAN Iub*

The following gives an example of the UTRAN FP PDU delay, AAL2 queuing delay and ATM cell delay from the simulation. The simulation is set up for one Node B connected with the RNC via a 2 Mbps Iub line, transmitting web traffic with an average web page size of 15 kbyte (Pareto page size distribution). In this example, the FP PDU delay is shown for the downlink, i.e. the delay of transporting FP PDUs from the RNC to the Node B, measured at the Node B side. The AAL2 queuing delay is the queuing delay of the AAL2 buffer at the RNC side. The ATM cell delay consists of (iii) the queuing delay in the ATM buffer and (iv) transmission delay on the Iub link. Table 6.5 gives the values of the mean and the variance of the ATM cell delay, AAL2 queuing delay and FP PDU delay from the simulation.

	Mean (s)	Variance
ATM Cell Delay	0.000521	1.836e-8
AAL2 Queuing Delay	0.090760	0.07549
FP PDU Delay	0.091789	0.07500

Table 6.5: *UTRAN Delay Components*

The results from Table 6.5 show that the ATM cell delay is much smaller than the AAL2 queuing delay and the FP PDU delay, while the AAL2 queuing delay is quite close to the FP PDU delay. The gap between the AAL2 queuing delay and the FP PDU delay is mainly caused by the ATM cell delay and additional segmentation and reassembly delay. Furthermore, the variance of ATM cell delay is rather small, which means that the ATM cell delay is kept quite stable. The results imply that the FP PDU delay is mainly composed by the AAL2 queuing delay. That means, the AAL2 queuing delay can be used to approximate the Iub transport delay.

6.4.1.2 System Models

Based on the above delay analysis on the Iub, it can be concluded that the Iub transport delay (i.e. FP PDU delay) can be approximated by the AAL2 queuing delay. So for the sake of the Iub dimensioning to decide an optimum Iub bandwidth with the minimum transport cost for satisfying Iub transport delay requirements, the estimation of the AAL2 queuing delay is important. If the AAL2 queuing delay can be estimated analytically given certain Iub link bandwidth, then the optimum Iub capacity which meets the defined Iub transport delay requirements can be obtained numerically. In queuing theory, the queuing delay is determined by the queuing models which consist of arrival traffic process, queue management, packet scheduling and service process, etc. So for setting up the analytical dimensioning approach, it is essential to construct appropriate queuing models to analytically calculate the AAL2 queuing delay and hence apply them to dimension the Iub capacity properly to meet required Iub delay constraints. Naturally, the construction of queuing models depends on the structure of the system models.

Generally, the system model of the AAL2 layer may consist of a number of AAL2 queues for serving different traffic types with different priorities. Depending on different QoS structures and packet scheduling on the AAL2 layer, the system models are configured differently. In the scope of this thesis, two system models are considered. In the ATM-based transport network two QoS classes are regarded on the AAL2 packet level: *Real Time* (RT) traffic and *Non Real Time* (NRT) traffic. The applied packet scheduling between the two traffic classes is strict priority, where the RT traffic is given higher priority while NRT traffic uses any bandwidth left by the RT traffic. Thus, according to different traffic scenarios two system models are structured as illustrated in Figure 6.43: (a) Single-service system: there is only one traffic type in the system; (b) Multi-service system: there are both real time traffic and non real time traffic.

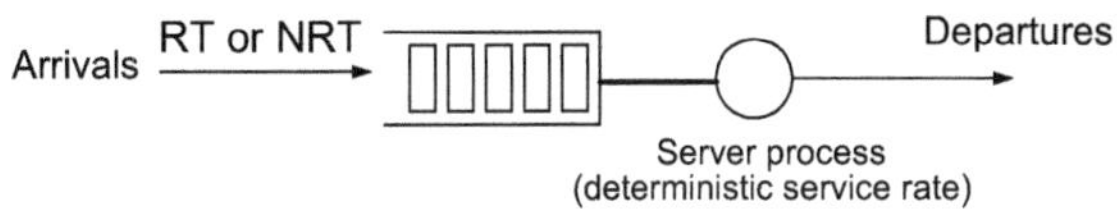

(a) Single-service system

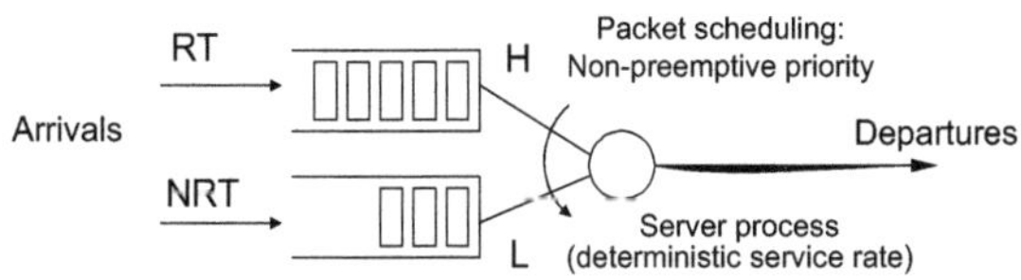

(b) Multi-service system

Figure 6.43: *System Models (AAL2 packet level)*

In case (a), there is only one service type, either RT or NRT traffic. It is used for the scenario where only voice traffic or only packet-switched traffic is transmitted over the Iub. As seen in Figure 6.42, the arrival traffic to the AAL2 buffer is a superposition of the FP PDUs streams of all users. The service rate of the AAL2 queue is deterministic due to the shaping applied at the AAL2 layer. Thus the resulting queuing delay in the AAL2 buffer strongly depends on the characteristics of the arrival traffic and the distributions of the FP PDU sizes. Due to the stochastic arrival process, traffic can be very bursty in nature and the AAL2 queuing becomes significant.

Case (b) is a more realistic scenario as usually the networks will hold multiple services simultaneously. It is applied for the scenario of both RT traffic and NRT traffic being transmitted on the Iub link, for example the conversational voice traffic mixed with the web traffic. In this case, the system model consists of two AAL2 queues each serving different type of traffic, and one server with deterministic service rate, but the RT traffic is taken higher priority over NRT traffic. The low priority queue is mainly subject to the Iub congestion situation, and therefore its queuing delays are the main issue for the Iub dimensioning. Thus the major goal of applying queuing models is to estimate the distribution of the low priority queue, and in turn to allocate the bandwidth accordingly for satisfying a required delay bound for the NRT traffic. Here, the queuing delay of the low priority queue depends not only on the properties of the arrival traffic of the low priority queue, but also the characteristics of the arrival traffic of the high priority queue.

The next section investigates the properties of the arrival traffic in the above system models, in order to choose appropriate arrival process models for building accurate queuing models to calculate the Iub delays.

6.4.2 UMTS Iub Traffic Characteristics

This section analyzes the characteristic of the carried Iub traffic, i.e. the arrival traffic of the above system models, in terms of its variation and burstiness, considering different traffic scenarios.

6.4.2.1 Squared Coefficient of Variation

The *Squared Coefficient of Variation* (COV^2) represents the burstiness of the traffic. The COV^2 of a constant distribution is 0 and the COV^2 of an exponential distribution is 1. If $COV^2 > 1$, the traffic is considered as bursty. Figure 6.44 and Figure 6.45 show the measured COV^2 for the packet interarrival time of voice and packet-switched traffic from the simulations. The results in the two figures can be summarized as follows:

(1) COV^2 for voice traffic varies between 1.2 and 1.8, with or without packet-switched traffic.

(2) COV^2 for packet-switched traffic with BRA varies between 11 and 12.

(3) COV^2 is different for different packet-switched RAB types. COV^2 is also different for different traffic mixes. In general, higher RAB rate has a higher COV^2 value. When mixing with voice, the COV^2 of packet-switched traffic is larger than that without mixing with voice.

Therefore from the above investigations, it can be concluded that regardless of circuit-switched voice or packet-switched traffic, the arrival traffic is considered as bursty traffic as their $COV^2 > 1$.

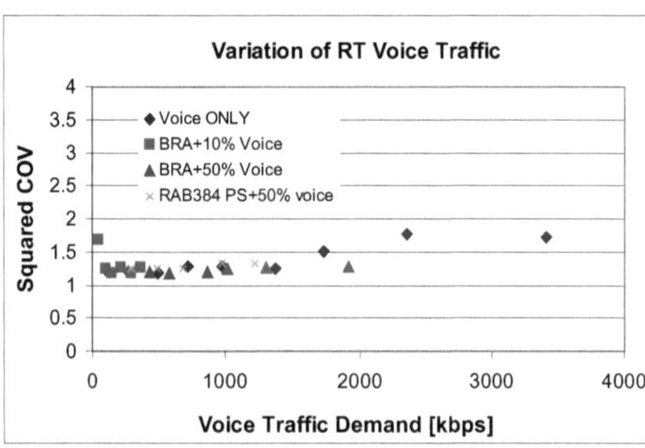

Figure 6.44: *Squared CoV of packet interarrival time for RT voice traffic*

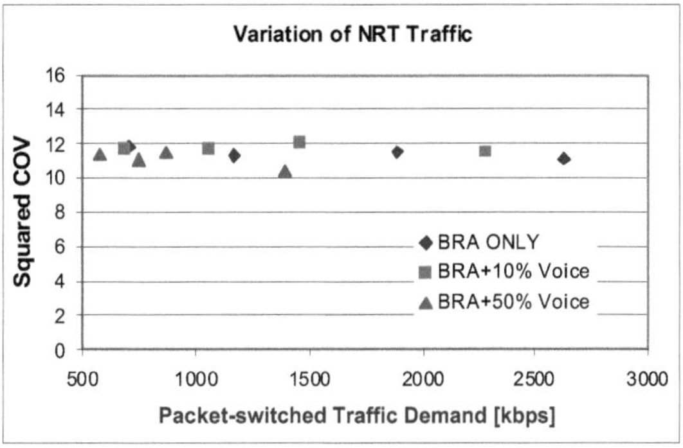

Figure 6.45: *Squared CoV of packet interarrival time for NRT traffic*

6.4.2.2 Self-Similarity of Elastic IP Traffic

In the last decade, many investigations focused on the characterization of measured IP traffic in local and wide area networks. The most important result of these studies is that the IP traffic has fractal-like behavior, which indicates the burstiness, long-range dependence and self-similarity. Self-similarity means, the measured traffic rates are statistically the same in all time scales, i.e. small and large time scale [WPT98]. Usually, the degree of self similarity is mathematically measured in the light of the Hurst parameter H. The detailed definition of self-similarity and the Hurst parameter is given in Appendix A.7. If $0.5<H<1$, the traffic is considered as self-similar. This is the so-called *Hurst effect*. The self similarity of IP traffic can be caused by the heavy tail file size distributions at the application layer, other possible reasons are the TCP congestion control and TCP retransmission scheme for reliable transport.

When transferring the elastic IP traffic in the UMTS network, the IP packets are converted into corresponding shaped FP PDU streams. For each TCP round trip time, one or several TCP packets are sent in a time, which cause the burst of the traffic. It is expected that the aggregated UMTS Iub traffic, i.e. the superposition of FP PDU streams of all users, inherit the burstiness and self-similarity property of the IP traffic. This can be proven by the following example. Figure 6.46 shows the *Cumulative Distribution Function* (CDF) of the FP PDU inter-arrival time in case of carrying pure elastic IP traffic (all users transmit with RAB 128 kbps). It can be seen that the CDF curve has several abrupt steps instead of a smooth curve. Figure 6.47 plots the property of the given aggregated UMTS traffic. The value of the Hurst parameter H is obtained by the skewness of the linear regression. In this figure, its value equals to 0.69 (i.e. H>0.5). This demonstrates that the aggregated UMTS traffic indeed has the self-similarity property.

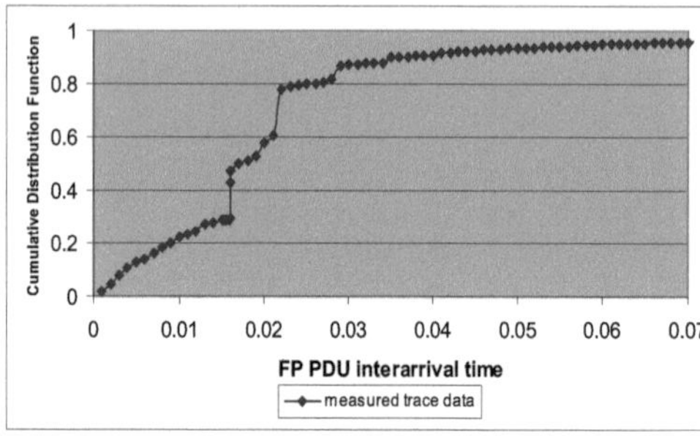

Figure 6.46: *CDF of FP PDU inter-arrival time*

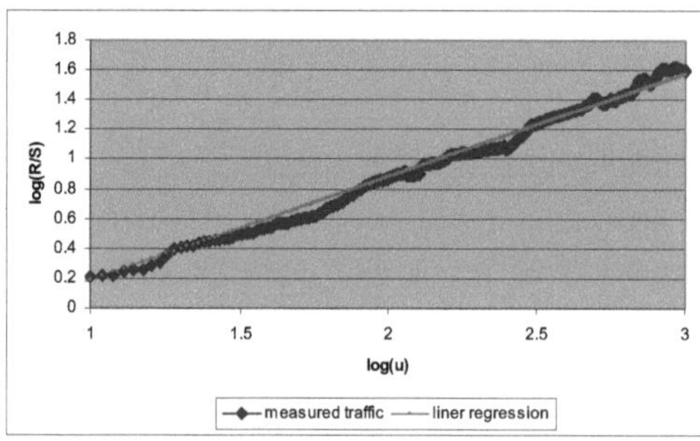

Figure 6.47: *Self-similarity of the aggregated UTRAN traffic (Hurst Parameter)*

In order to capture the important statistical properties of the arrival traffic, analytically tractable models are preferred. However, the characterization of the aggregated traffic across the Iub interface cannot be represented with the simple Poisson traffic model any more, since the Poisson model is typically used for studying the classical telephone networks and it is not able to capture the bursty nature of the UMTS traffic. In this thesis, two arrival process models are proposed: the *Markov Modulated Poisson Process* **(MMPP)** and **the** *Batch Markovian Arrival Process* (**BMAP**) models.

6.4.3 Analytical Dimensioning Models with MMPP

This section presents the dimensioning approach based on MMPP for the Iub dimensioning devoted to the TNL delay performance. The mathematical definition of the MMPP model is introduced in the following section 6.4.3.1. Section 6.4.3.2 and 6.4.3.3 present the corresponding queuing models with the MMPP arrival process and give their analytical calculations on the queuing delay. The closed form of the queuing delay distributions and their approximations are given. Section 6.4.3.4 explains in detail how to apply these queuing models in the dimensioning framework to estimate the necessary Iub bandwidth for satisfying a certain Iub delay performance. In section 6.4.3.5, the proposed analytical dimensioning approach is validated by simulations.

6.4.3.1 Markov Modulated Poisson Process (MMPP)

The Markov-Modulated Poisson Process (MMPP) is an appropriate and practical analytical model, which is able to qualitatively model the time-varying arrival rate and capture some of the important correlations between the interarrival times while still remaining analytically tractable. Therefore, the MMPP model has been extensively used to study communication networks such as modeling ATM links and the superposition of packetized voice and packet data stream. The use of MMPP in queuing theory has been well discussed and described in the literature. A useful reference for the MMPP is [FMH92]. There is also much discussion on various fitting procedures and the accuracy of MMPP models, for example [NSV99, Mei87].

MMPP is a doubly stochastic Poisson process where the intensities of the Poisson processes are defined by the states of an independent Markov chain. Figure 6.48 illustrates a two state MMPP model. Thus, the Markov chain can be said to modulate the Poisson process, hence it is named Markov-modulated Poisson process. This modulation introduces correlations between successive interarrival times in the process. Equivalently, an MMPP process can be constructed by varying the arrival rate of a Poisson process according to an *m*-state irreducible continuous time Markov chain which is independent of the arrival process.

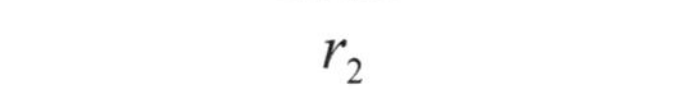

Figure 6.48: *Two State MMPP*

When the Markov chain is in state *i*, arrivals occur according to a Poisson process of arrival rate λ_i. The MMPP is parameterized by the *m*-state continuous-time Markov chain with infinitesimal generator $\boldsymbol{Q}$ [OAS90] which is the generator matrix of the modulating chain, and the arrival rate matrix $\boldsymbol{\Lambda}$ which is the diagonal matrix of arrival rates in each state of that chain. They are defined by

$$Q = \begin{bmatrix} -r_1 & r_{12} & \dots & r_{1m} \\ r_{21} & -r_2 & \dots & r_{2m} \\ . & . & \dots & . \\ r_{m1} & r_{m2} & \dots & -r_m \end{bmatrix} \quad \text{where } r_i = \sum_{\substack{j=1 \\ j \neq i}}^{m} r_{ij}$$

$$\Lambda = \begin{bmatrix} \lambda_1 & 0 & \dots & 0 \\ 0 & \lambda_2 & \dots & 0 \\ . & . & \dots & . \\ 0 & 0 & \dots & \lambda_m \end{bmatrix} \tag{6.44}$$

In the above definition, $r_{ij} \geq 0$ is the rate at which the modulated process transits from phase i to phase j. λ_i is the arrival rate when the modulating chain is in phase i. For the m-state Markov chain, there are m Poisson arrival rates, i.e. $\lambda_1, \lambda_2, \lambda_3, \ldots, \lambda_m$.

A two-state MMPP denoted as MMPP(2) is described as a Markov chain with two states and two different intensities, sometimes referred to as Switched Poisson process (SPP). If the two intensities are equal the model becomes an ordinary Poisson process. The special case when one intensity is zero is called Interrupted Poisson Process (IPP).

6.4.3.2 MMPP(2)/D/1

In the above section 6.4.1.2, two system models on the AAL2 packet level are presented for different traffic scenarios. For case (a), there is only one service type, either RT or NRT traffic. According to its system model, with MMPP as the arrival process model, the resultant queuing model is MMPP/D/1 where the service process is deterministic. Thus the AAL2 delay or the Iub transport delay can be derived from the calculation of the queuing delay of MMPP/D/1 model. The 2-state case has received attention as a simple tractable process which can closely approximate much more complicated processes and predict queuing delays very accurately. Therefore in this thesis, a two-state MMPP (denoted as MMPP(2)) is suggested for modeling the arrival traffic. The following presents the analytical approach for calculating the waiting time distribution for the MMPP(2)/D/1 system.

It is given that the MMPP(2)/D/1 with the infinitesimal generator $\boldsymbol{Q}$ of the underlying Markov chain and with the arrival rate matrix $\boldsymbol{\Lambda}$:

$$Q = \begin{bmatrix} -r_1 & r_1 \\ r_2 & -r_2 \end{bmatrix} \qquad \Lambda = \begin{bmatrix} \lambda_1 & 0 \\ 0 & \lambda_2 \end{bmatrix} \tag{6.45}$$

Let $\boldsymbol{\pi}$ be the steady-state vector of $\boldsymbol{Q}$, and $\boldsymbol{e}$ be the unitary column vector. Assuming the unit service time, the traffic intensity is then given by $\boldsymbol{\rho} = \boldsymbol{\pi} \boldsymbol{\Lambda} \boldsymbol{e}$.

Lucantoni presented the Laplace transform of the virtual waiting time distribution (CDF) for the MMPP/G/1 queue in [Luc91]:

$$\tilde{F}(s) = s(1-\rho)g[sI + Q - \Lambda(1 - H(s))]^{-1}e \tag{6.46}$$

Here $H(s)$ is the Laplace transform of the service time distribution.

For the MMPP/D/1 system, due to the deterministic service time distribution with the unit service time, the Laplace transform of the service time distribution $H(s) = e^{-s}$. So from the equation (6.46), the *complementary cumulative distribution function* (CCDF) of the virtual waiting time of the MMPP/D/1 can be derived. Let $W(x) = \Pr(W>x)$ be the complementary cumulative distribution function of the virtual waiting time W. The Laplace transform of $W(x)$ is given in equation (6.47):

$$\tilde{W}(s) \equiv \int_0^\infty e^{-sx} W(x) ds = \frac{1}{s} - s(1-\rho)g[sI + Q - \Lambda + \Lambda e^{-s}]^{-1}e \tag{6.47}$$

Where $\mathbf{g} = (g_1, 1-g_1)$ is the steady-state probability vector associated with the stochastic matrix $\mathbf{G}$ satisfying

$$G = e^{(Q - \Lambda + \Lambda G)} \tag{6.48}$$

[KSC98] proposed a simple, empirical real-time approximation approach for calculating the complementary distribution of the virtual waiting time for the MMPP(2)/D/1 system. In this approach, $W(x)$ can be asymptotically approximated as a *two-term* exponential function:

$$W_{appr}(x) = a_1 e^{s_1 x} + a_2 e^{s_2 x} \quad (a_1 > 0, a_2 > 0, \quad s_2 < s_1 < 0) \tag{6.49}$$

whose Laplace transform is given by

$$\tilde{W}_{appr}(s) = \frac{a_1}{s - s_1} + \frac{a_2}{s - s_2} \tag{6.50}$$

Note that s_1 and s_2 are defined as asymptotic decay rates, while a_1 and a_2 are the asymptotic constants. The estimation of the asymptotic decay rate s and asymptotic constant a can be obtained by matching the poles, the residue of the dominant pole, and the mean waiting time between the Laplace transform of the exact distribution and the approximation distribution.

The detailed computation procedure for obtaining the asymptotic approximation of the complementary distribution of the virtual waiting time $W_{appr}(x)$ in equation (6.49) is explained as follows.

(1) Compute the asymptotic decay rate s_1 and s_2:

s_1 and s_2 are chosen as the negative poles of $\tilde{W}(s)$, i.e. the roots of the equation from the denominator of $\tilde{W}(s)$:

$$\det\left[sI + Q - \Lambda + \Lambda e^{-s}\right] = 0 \tag{6.51}$$

Equivalently,

$$\begin{aligned}&\lambda_1\lambda_2 - e^{s}(2\lambda_1\lambda_2 + \lambda_2 r_1 + \lambda_1 r_2 - \lambda_1 s - \lambda_2 s)\\&+e^{2s}\left\{\lambda_1\lambda_2 + \lambda_2 r_1 + \lambda_1 r_2 - (\lambda_2 + \lambda_1 + r_1 + r_2)s + s^2\right\} = 0\end{aligned} \tag{6.52}$$

Among the negative roots of (6.52), the parameters s_1 and s_2 are chosen from the first two close to 0 negative values ($s_2 < s_1 < 0$).

(2) Compute the vector *g*:

The vector ***g*** is given by equation (6.53), derived from the stochastic matrix **G**:

$$g = (g_1, g_2) = \left(\frac{G_2}{G_1 + G_2}, \frac{G_1}{G_1 + G_2}\right) \text{ where } G = \begin{bmatrix} 1-G_1 & G_1 \\ G_2 & 1-G_2 \end{bmatrix} \tag{6.53}$$

The calculation of ***g*** is introduced in [FMH92] and [Luc91]. The detailed computation steps for **G** are:

(a) Start with $G_1 = 0$,

(b) Compute cyclically

$$\begin{aligned}G_1 + G_2 &= 1 - H(r_1 + r_2 + \lambda_1 G_1 + \lambda_2 G_2) = 1 - e^{-(r_1 + r_2 + \lambda_1 G_1 + \lambda_2 G_2)}\\G_1(r_2 + \lambda_2 G_2) &= G_2(r_1 + \lambda_1 G_1)\end{aligned} \tag{6.54}$$

until G_1 and G_2 become stable, i.e. G_1 and G_2 shall converge. Otherwise exchange indices and start the procedure again.

(c) Then, ***g*** can be directly computed by equation (6.53).

(3) Compute the asymptotic constants a_1 and a_2:

It is proposed here to approximate the asymptotic constants a_1 and a_2 by matching the mean waiting time. So we need to calculate the mean waiting time, denoted as ϖ, which is calculated from [Luc91, eq. (47)]:

$$\varpi = \frac{1}{2(1-\rho)}\left[3\rho - 2\{(1-\rho)g + \pi\Lambda\}(Q + e\pi)^{-1}\Lambda e\right] \tag{6.55}$$

Then we match the above ϖ to the mean waiting time of the approximated complementary distribution of the virtual waiting time, $W_{appr}(x)$ in equation (6.49). Thus, the asymptotic constant a_1 and a_2 need to satisfy:

$$\begin{aligned}\varpi &= -\frac{a_2 s_1 + a_1 s_2}{s_1 s_2}\\a_1 + a_2 &= 1\end{aligned} \tag{6.56}$$

Then a_1 and a_2 can be calculated from (6.56). Here the sum of a_1 and a_2 should be equal to 1 (normalization of $W(x)$ in equation (6.49)).

(4) Calculate $W_{appr}(x)$:

With the asymptotic decay rate s_1 and s_2 and the asymptotic constant a_1 and a_2, $W_{appr}(x)$ given in equation (6.49) can be obtained. Furthermore, the waiting time distribution $F_{appr}(x)$ (CDF) can also be derived from $W_{appr}(x)$, i.e. $F_{appr}(x) = 1 - W_{appr}(x)$.

6.4.3.3 MMPP(2)/D/1 – Non-Preemptive Priority

This section deals with the system model case (b), where both RT traffic and NRT traffic are transmitted on the Iub link, but the RT traffic is given priority over the NRT traffic. To estimate the distribution of the low priority queue, the considered queuing model is MMPP/D/1 with non-preemptive priority scheduling.

An analytical approach for calculating the queuing delay distribution in an MMPP/G/1 non-preemptive queue is presented in [HZ05]. In this approach, the high priority traffic is modeled as a Poisson arrival process with rate λ_h while the low priority traffic is modeled as an MMPP arrival process, which is characterized by a phase transition matrix $\mathbf{R}_l$ and arrival rate matrix $\mathbf{\Lambda}_l$. The high and low priority packets are assumed to have the same service time distribution with a mean service rate of μ.

Figure 6.49 illustrates the possible delays a packet may experience at the AAL2 queue. First define a number of random variables corresponding to a tagged packet which arrives at the system in phase j.

- B denotes the full service time
- U denotes the partial service time
- V denotes the residual service time
- A denotes the number of packets queuing at the beginning of the service time during which the tagged packet arrives;
- X_j denotes the number of packets that arrived during the partial service time U.
- Q_j denotes the queuing time for the tagged packet.
- W_j denotes the total time the tagged packet spends in the system. It is the sum of the queuing time and the service time B of a packet.

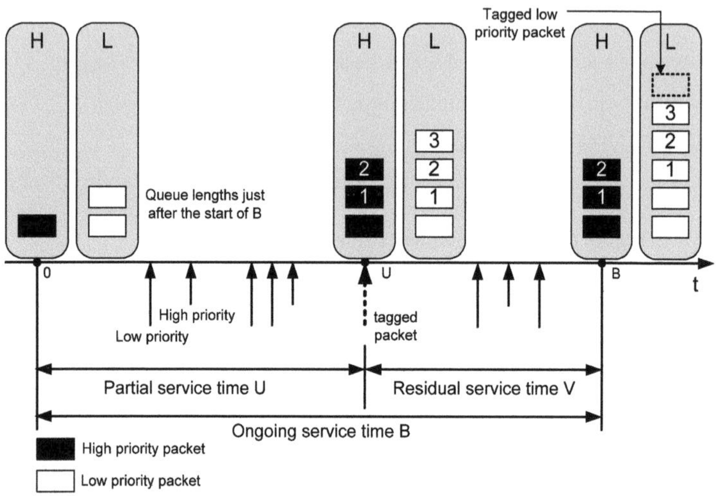

Figure 6.49: *Basic idea to study the delay distribution of a priority queue [HZ05]*

Firstly consider a high priority tagged packet. As shown in Figure 6.49, this packet sees the server busy with some packets queuing (the queue could also be empty). Since it is a high priority packet, it can only see the high priority packets which are queuing ahead of it, if there are any. Low priority packets, even any that came earlier, are transparent to the tagged packet. In this case, the queuing delay of this tagged packet consists of three parts:

(1) the first part is the time for the server to serve all the high priority packets that were already in the queue at the time the on-going service began;

(2) the second part comes from serving those high priority packets that enter the system during the partial service time U;

(3) the third part is the residual period V.

For a low priority packet, its queuing delay is the sum of four parts, as shown in Figure 6.49.

(1) the first part is the sum of the high-priority busy periods generated by all the packets (A) in the queue at the time the on-going service began;

(2) the second part comes from the sum of the busy periods generated by all the packets (X) that enter the system during the partial service time U;

(3) the third part arises from the busy periods generated by the high priority packets that arrive during V;

(4) the fourth part is the residual period V.

In this thesis, the dimensioning is according to the performance of the low priority queue, as it is found that in all investigated cases the high priority traffic can overachieve its QoS due to a higher priority. Thus, the following analytical calculations are focused on estimating the queuing delay distribution of the low priority traffic. In [HZ05], the *Laplace-Stieltjes transform* (LST) of the delay distribution of a low priority packet in an MMPP/G/1 non-preemptive queue is given as follows:

$$E\left[e^{-sW}\right]=\begin{cases}\vec{\phi}_{hl}B^*(s)(s+(1-M^*(s))\lambda_h)[sI+R_l+(M^*(s)-1)\Lambda_l]^{-1}\vec{e}^T(s>0)\\ \vec{\pi}\vec{e}^T\,(s=0)\end{cases} \tag{6.57}$$

The mean sojourn time for the low priority packet is [HZ05]:

$$\overline{W}=\mu^{-1}-\frac{1}{1-\rho_l-\rho_h}\begin{Bmatrix}(\vec{\phi}_{hl}(1-\lambda_h M^{*(1)}(0))-\vec{\pi}(I+\Lambda_l M^{*(1)}(0))(R_l+\vec{e}^T\vec{\pi})^{-1}(\Lambda_l/\mu)\vec{e}^T\\ -0.5M^{*(2)}(1-\rho_h)\vec{\pi}\Lambda_l\vec{e}^T-0.5M^{*(2)}(0)(1-\rho_h)(1-\rho)\end{Bmatrix} \tag{6.58}$$

$M^*(s)$ is the LST of the high priority busy period distribution, which is calculated as

$$M^*(s)=B^*\left[s-(M^*(s)-1)\lambda_h\right] \tag{6.59}$$

However, the cumulative delay distribution function (i.e. CDF) for low priority packets $W(x)$ is difficult to obtain by implementing the inverse Laplace transform for equation (6.57). Therefore it is proposed in this thesis to approximate it with a two-term exponential function:

$$W_{appr}(x)=1-e^{s_1x}-e^{s_2x}\quad(a_1>0,a_2>0) \tag{6.60}$$

The asymptotic decay rate s_1 and s_2 can be derived from the first and second moment of the delay distribution for a low priority packet.

6.4.3.4 Iub Dimensioning with MMPP

As introduced in section 6.4.3.1, a two-state MMPP model is characterized by four parameters: λ_1, λ_2, r_1 and r_2. λ_1 denotes the mean arrival rate of source 1 and λ_2 is the mean arrival rate of source 2. $1/r_1$ is the mean holding period of the source 1 and $1/r_2$ is the mean holding period of the source 2. The setting of these four MMPP parameters is based on certain MMPP fitting methods. The fitting method used in this thesis is the *interarrival time distribution inference procedure* [NSV99]. In this approach, the four MMPP parameters are estimated based on the *Cumulative Distribution Functions* (CDF) of the interarrival time distributions. The interarrival time distributions for the AAL2 packets can be based on a 2-Phase Hyper-exponential distribution, denoted as Hyper (2), which is described by the following three parameters:

a_1 - the mean arrival rate of flow 1;
a_2 - the mean arrival rate of flow 2;
p_1 - the probability of flow 1.

The main reason of using a 2-Phase Hyper-exponential distribution for the interarrival time distribution is because the arrival traffic is usually rather bursty, where the coefficient of variation is larger than 1 as shown in section 6.4.2.1. Hyper-exponential distributions using 2 phases can model this variation by introducing two Poisson arrival processes where one

Poisson process models very small packet interarrival times and the other models the larger packet interarrival times. Moreover, the Hyper(2) model requires the minimum number of parameters, i.e. the mean arrival time and the variance of arrival times. It has a freedom of tuning the parameters given the mean and variance, with various methods.

As in the arrival of the AAL2 CPS packets, there are bulk arrivals. Because in the Rel99 model, the arrival of each FP PDU will be converted into a stream of AAL2 CPS packets and they are put into the AAL2 buffer at the same time. That means, a number of AAL2 CPS packets from the same FP PDU are coming to the AAL2 buffer with "zero" interarrival time. In order to model the bulk arrival of AAL2 CPS packets, a parameter tuning algorithm is proposed in equation (6.61) to calculate the above three Hyper-exponential parameters (i.e. a_1, a_2, and p_1). The calculation is based on the mean time between arrivals μ and the variance of arrivals v. To model the bulk arrival, let $(1/a_1)$ be close to 0 (e.g. 1e-10).

$$\begin{aligned}
a &= \frac{1}{2}\cdot a_1^{\,2}\cdot(v-\mu^2)\\
b &= \frac{1}{2}\cdot a_1^{\,2}\cdot(v+\mu^2)\\
p_1 &= a/(1-2\cdot a_1\cdot\mu+b)\\
p_2 &= 1-p_1
\end{aligned} \tag{6.61}$$

$$a_2 = \frac{(1-p_1)}{(\mu-p_1/a_1)}$$

With this approach, the Hyper-exponential parameters from the observed mean time between arrivals and the variance of arrivals can be obtained. However it is known that the Hyper-exponential distribution does not have any correlation, i.e. there is no correlation between packet arrivals in the Hyper-exponential distribution; so in order to capture the correlations another parameter *б* is needed, which is defined as the decaying rate of the auto-correlation function. The *Autocorrelation Function* (ACF) for a samples series X_i, (i =1, 2, 3….) is defined in equation (6.62), where k is the lag.

$$R(k) = \frac{E[(X_i-\mu)(X_{i+k}-\mu)]}{\sigma^2} \tag{6.62}$$

Usually for *Long Term Dependence* (LRD) traffic, the Hurst parameter H is used to indicate the self-similarity. And the decaying rate of the ACF, i.e. *б*, can be estimated with the Hurst parameter with *б = 2*(1-H)*. Then the Hurst parameter value can be obtained from the measurement.

With the three hyper-exponential distribution parameters, i.e. a_1, a_2, p_1, and an additional parameter *б*, the four parameters of the MMPP model can be derived with the following formulas.

$$\lambda_1 = \frac{1}{2}\left[p_1(1-\sigma)(a_1 - a_2) + \sigma a_1 + a_2 + \sqrt{\xi}\right]$$

$$\lambda_2 = \frac{a_1 a_2 \{\lambda_1 - p_1(a_1 - a_2) - a_2\}}{\lambda_1 a_1 - \lambda_1 p_1 (a_1 - a_2) - a_1 a_2}$$

$$r_1 = \frac{(a_1 - \lambda_1)(a_2 - \lambda_1)}{(\lambda_2 - \lambda_1)} \tag{6.63}$$

$$r_2 = \frac{(\lambda_2 - a_1)(\lambda_1 + r_1 - a_1)}{(a_1 - \lambda_1)}$$

where $\xi = \left[p_1(1-\sigma)(a_1 - a_2) + \sigma a_1 + a_2\right]^2 - 4\sigma \cdot a_1 a_2$

6.4.3.5 Validations

This section validates the proposed analytical dimensioning approaches by comparing the analytical results with the simulation results. The following gives some validation examples. The dimensioned Iub bandwidth from the simulations is obtained from the UMTS simulation model. The simulation scenario is a single Iub link scenario, consisting of one Node B and one RNC connected with each other via the Iub link. In the investigations, the QoS requirements in UTRAN is that the acceptable FP PDU delay considered for voice service is within 10 ms for 99% of voice packet transmissions and below 30 ms for data services for 99% of the packet-switched traffic packet transmissions. Those packets whose experienced FP PDU delays violate the given values would be discarded. That means, a maximum of 1% packet loss ratio is allowed for both voice and packet-switched data services.

For all scenarios the simulations run 3600 simulated seconds. The stability and reliability of the measured statistics of packet loss ratio and queuing delay distribution from simulations are evaluated by *Confidence Intervals* (CI) and *Limited Relative Error* (LRE). A few examples are given for the CI in Appendix A.20 and for LRE in A.21.

Figure 6.50 compares the dimensioning results for the voice only scenario from the simulations using the Rel99 UMTS model (system simulation) and the analytical solutions presented in section 6.4.3.2 (analytical calculation). In addition, simple simulations of the MMPP/D/1 queuing model in OPNET (queuing simulation) are carried out as an additional validation method to validate the accuracy of the calculation results. Figure 6.50 demonstrates that on the one hand the MMPP/D/1 is an accurate model for deriving the required Iub bandwidth for the defined TNL performance, i.e. 1% packet loss ratio, and on the other hand demonstrates that the analytical results are all are close to the simulation results form the Rel99 UMTS system simulations. The relative error of the analytical models is below 10%.

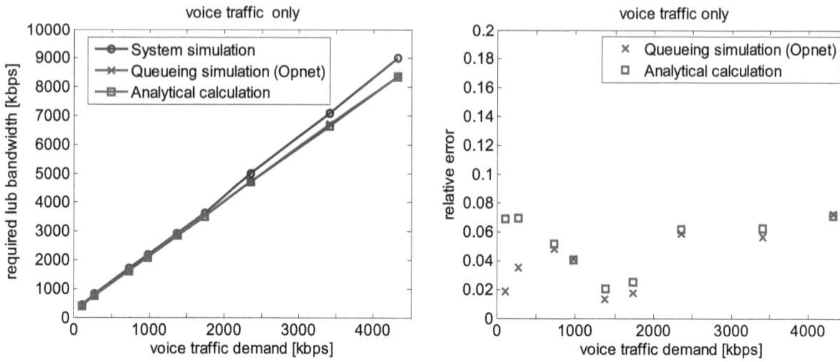

Figure 6.50: *Validation of MMPP/D/1 model (voice scenario)*

Figure 6.51 compares the dimensioning results of the voice only scenario from the OPNET simulation compared with the results derived from different analytical models, i.e. Hyper(2)/D/1 (denoted as (H2/D/1)), M/D/1, and MMPP/D/1. Figure 6.52 shows the relative error of various analytical models.

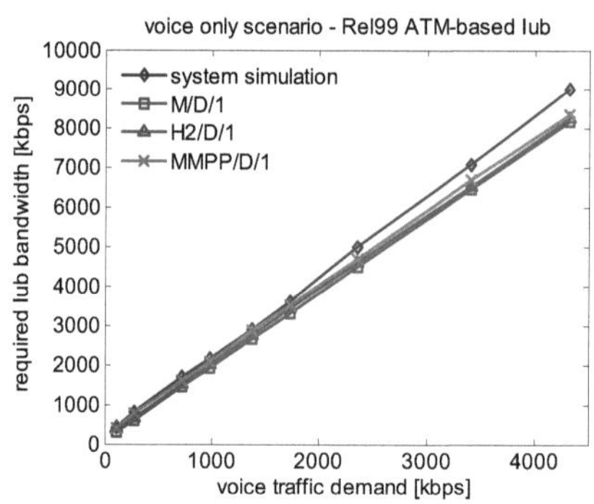

Figure 6.51: *Analytical models vs. simulation for Iub bandwidth (voice scenario)*

As can be seen in Figure 6.51 and Figure 6.52, the M/D/1 model has the lowest accuracy as it assumes Poisson arrivals whose coefficient of variation equals to 1, but the arrival traffic behaves much more bursty with a coefficient of variation larger than 1. In addition, it does not consider the bulk arrival of the packets. The H2/D/1 model improves the accuracy, since it considers the coefficient of variation of the arrival traffic by using the two phases to model this variation by means of introducing two Poisson arrival processes where one Poisson process to model very small packet arrival times and the other to model the larger packet arrival times. Moreover, in order to model the bulk arrival of AAL2 CPS packets, the tuning approach in section 6.4.3.1 is used to calculate the Hyper-exponential parameters. But the H2/D/1 model does not have any correlation between packet arrivals. Among these three analytical models, the MMPP/D/1 model has the highest accuracy whose error is within 10%

range. Because the MMPP/D/1 model is derived from the H2/D/1 model, and additionally includes certain correlation of the arrival traffic. By comparing the dimensioning results from the OPNET simulation with the results derived from different analytical models, it can be concluded that in the voice scenario using the MMPP/D/1 model to derive the required Iub bandwidth for the 1% packet loss ratio is accurate (less than 10% relative error).

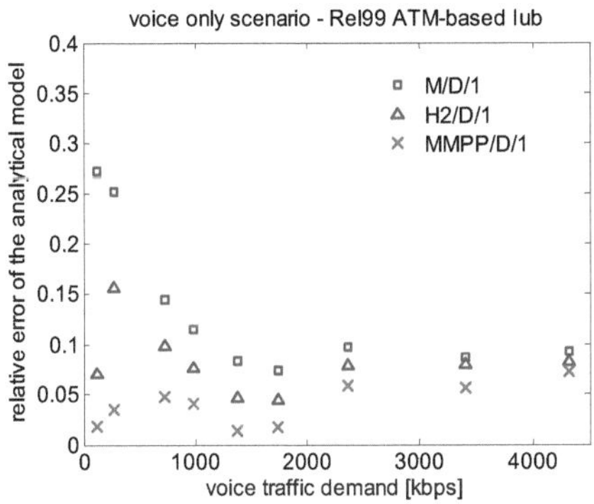

Figure 6.52: *Relative error of the analytical model (voice scenario)*

For the web traffic only scenario, the corresponding queuing model is MMPP/D/1. Figure 6.53 compares the dimensioning results for a web traffic scenario (with BRA function) obtained from simulations with UMTS Rel99 simulation model and the results derived from MMPP/D/1 model. Figure 6.54 shows the error of the MMPP/D/1 model. It is seen from the results that the MMPP/D/1 model can estimate the required Iub bandwidth accurately for the pure web traffic scenario, and the obtained relative error of the analytical model is less than 5%.

For the mixed traffic scenario, the queuing model MMPP/D/1 with non-preemptive priority shall be applied for dimensioning. Figure 6.55 compare the dimensioning results of web traffic (using BRA) mixed with 10% voice obtained from the simulation with the results derived from the MMPP/D/1 priority model (section 6.4.3.3). It is seen that the dimensioned Iub bandwidth derived from the MMPP/D/1 priority model is close to the results obtained from system simulations.

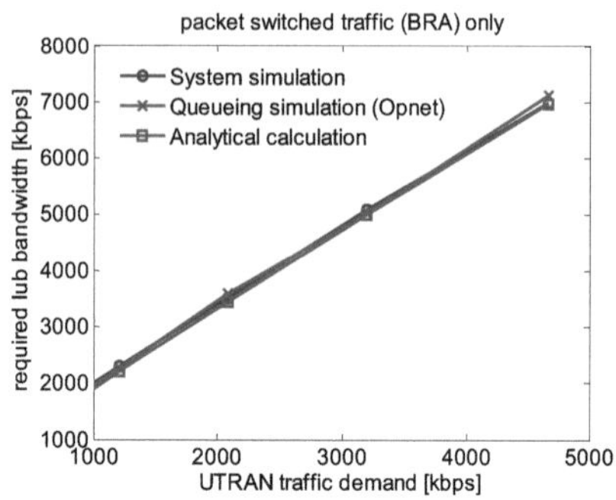

Figure 6.53: *Analytical Models vs. Simulation (web traffic with BRA)*

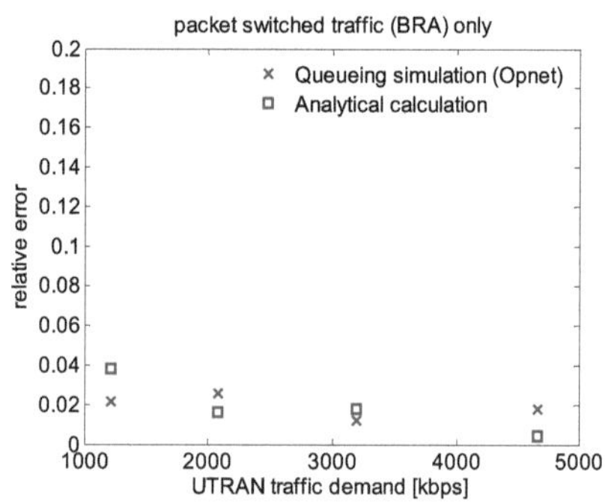

Figure 6.54: *Relative Error of the analytical model (web traffic with BRA)*

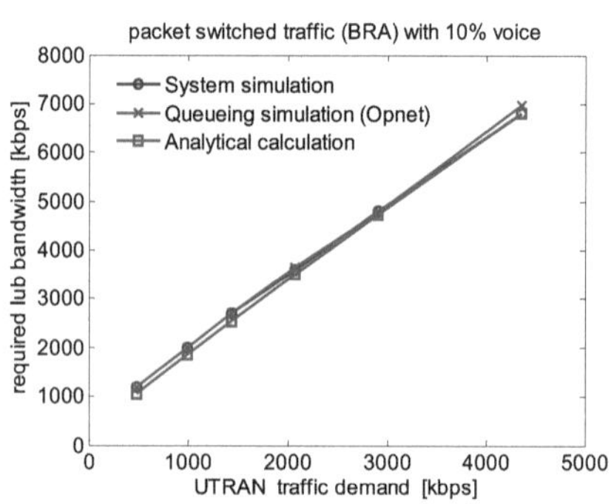

Figure 6.55: *Analytical Model vs. Simulation (90% web with BRA +10% Voice)*

From the above presented validation results, it can be concluded that the proposed analytical models, i.e. the MMPP/D/1 model for pure voice scenario and the packet-switched traffic scenario, and the MMPP/D/1 priority model for the mixed traffic scenario, can achieve an accurate approximation for the dimensioning of the Iub for satisfying the transport network performance, i.e. 1% packet loss ratio as the desired dimensioning target.

6.4.4 Analytical Dimensioning Models with BMAP

This section presents the dimensioning approach based on BMAP for the Iub dimensioning devoted for the TNL delay performance. The following section 6.4.4.1 gives a general introduction of the BMAP model. The section 6.4.4.2 explains the construction of the BMAP model for analyzing the transport network of UMTS , and compares the analytical results with the simulation results. The detailed results on modeling the arrival process and estimation of the transport network delay performance using the BMAP model have been presented by the author in [LLGT06, LLSTG06].

6.4.4.1 Batch Markovian Arrival Process (BMAP)

The *Batch Markovian Arrival Process* (BMAP) is the generalization of Phase-type (PH) distribution [Ris02]. BMAP is represented by an infinite number of matrices $D(m)$ for $m > 0$ allowing for an arbitrarily large batch size. The BMAP can be understood as in Figure 6.56, when the *K-th* packet comes, this packet process starts transitions between transient states. When the next new packet (here is *(K+1)-th*) arrives, the *K-th* packet process leaves the transient state and immediately goes into absorbing state. At the same time, the *(K+1)-th* packet process starts transitions and repeats the same behavior.

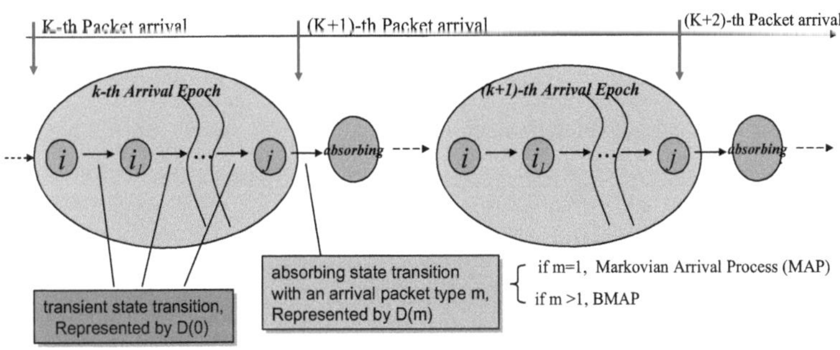

Figure 6.56: *Packet arrival process (BMAP model)*

The BMAP is characterized by a finite and absorbing Markov chain [Luc91]. Considering a two-dimensional Markov process $\{N(t), J(t)\}$ on the state space $\{(i, j); i \geq 0, 1 \leq j \leq n\}$, $N(t)$

counts the number of arrivals in the interval $(0,t]$, and $J(t)$ represents the underlying Markov chain. Transitions from a state (i,j) to a state $(i+m,k)$, $m \geq 1, 1 \leq j,k \leq n$, correspond to batch arrivals of size m, and thus the batch size can depend on j and k. An infinitesimal generator Q has the structure:

$$Q = \begin{Bmatrix} D(0) & D(1) & D(2) & \dots & D(m) \\ 0 & D(0) & D(1) & D(2) & \dots \\ 0 & 0 & D(0) & D(1) & \dots \\ 0 & 0 & 0 & D(0) & \dots \\ \dots & \dots & \dots & \dots & \dots \end{Bmatrix} \tag{6.64}$$

Here, $D(0)$ is the matrix (the dimension equals to the number states of $J(t)$), which has negative diagonal elements and nonnegative off-diagonal elements, representing a state transition without arrivals (batch size = 0). $D(m)(m \geq 1)$ is the matrix (the dimension equals to the number states of $J(t)$), which has nonnegative elements representing a state transition with an arrival batch size m. The basic characteristics of a BMAP are:
(1) the density function:

$$f_m(t) = e^{D(0)t} \cdot D(m) \tag{6.65}$$

(2) the complementary cumulative distribution function:

$$F^c(t) = e^{D(0)t} \tag{6.66}$$

If we define π as the initial probability vector of the underlying Markov chain $\{J(t)\}$, π satisfies

$$\pi D = 0, \pi 1 = 1 \tag{6.67}$$

Here, $D = \sum_{m=0}^{\infty} D(m)$ and the 1 is a column vector of ones.

The inter-arrival time distribution function for the batch size m is [KKSC02]:

$$F(t) = \pi(1 - e^{D(0)t})(-C)^{-1} D(m)1 \tag{6.68}$$

Here, 1 is the unit matrix. The arrival rate of the process is:

$$\rho = \pi \sum_{m=1}^{\infty} mD(m)1 \tag{6.69}$$

The importance of BMAP lies in its ability to be more effective and powerful traffic model than the simple Poisson process or the batch Poisson process, as they can effectively capture dependence and correlation, salient characteristics of the arrival process.

The BMAP model is characterized by parameters *D(0)*, which represents the transition probability between the transient states without arrivals, and *D(m)*, which is the transition probability within an arrival of packet with batch size *m*. The BMAP can capture the packet arrival process and packet length process. In previous investigations, the BMAP model has been applied to analyze the aggregated traffic modeling of IP networks [KLL03].

The BMAP model is chosen as the arrival process model as it considers different lengths of packets and batch arrivals. The BMAP model defines *D(0)* and *D(m)* ($m \geq 1$). *D(m)* represents the packet arrival process. And *m* corresponds to the different lengths of the packet of measured traffic. For example, if there are three different RAB types in the UMTS network, that means there are three different packet lengths (i.e. FP PDU sizes) corresponding to three different RAB types in the UMTS network. Correspondingly the BMAP will consider three different packet lengths, i.e. FP PDU sizes. In this way BMAP can provide a rather accurate model for characterizing the aggregated traffic, even on the self-similarity and burstiness properties of the IP traffic

Normally, the BMAP traffic arrival process is difficult to calculate analytically. Usually, the parameter estimation method is used for estimating the model parameters based on the observations of the packet inter-arrival time and packet size from the measured traffic. There are a few current studies on the parameter estimation for the BMAP model. Ryden has employed the *Expectation Maximization* (EM) algorithm for the MMPP model [Ryd96]. Breuer and Lindemann developed the method of the EM algorithm for the BMAP [Bre02] [LL01]. In this thesis, the EM for BMAP is implemented based on the above two methods from Ryden and Lindemann. In principle, the EM algorithm is divided into E-step (*Expectation*) and M-step (*Maximization*). The E-step is to compute the expectation of likelihood function for the observed data; the M-step is to estimate the parameter by maximizing the expectation of the likelihood function from the E-step. These two steps are repeated until a predefined maximum number of iterations is reached or until a convergence criteria holds. In this thesis, the BMAP results are generated using a software package IP2BMAP [LL03].

Figure 6.57 compares the CDF of the FP PDU inter-arrival time from the measured trace data, from BMAP using the EM algorithm and the Poisson process. It can be obviously seen in this figure that the CDF of the BMAP model accurately matches the CDF of the measured traffic; while the CDF of the Poisson process behave much more smoothly than the measured traffic, which means the Poisson process conceals the bursty property of the traffic. This figure demonstrates the advantages of using the BMAP model compared with Poisson process to analyze the aggregated UMTS traffic.

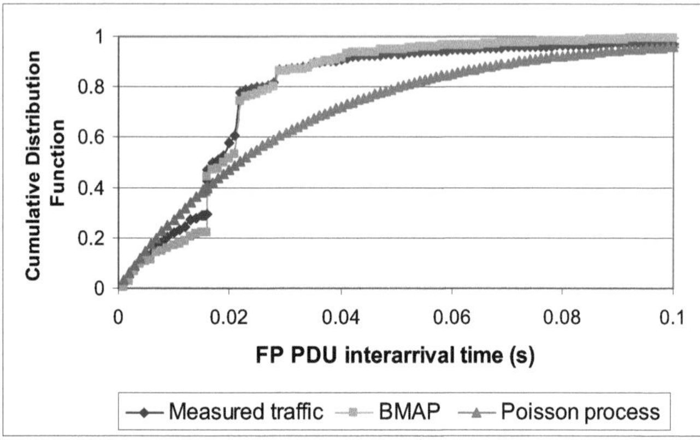

Figure 6.57: *CDF of FP PDU inter-arrival time – single RAB scenario*

Figure 6.58 plots sample paths of the aggregated UMTS traffic stream taken from the simulations (left) and sample paths of the aggregated traffic stream applying the parameterized BMAP model (right). The figure shows that the customized BMAP captures the average transferred data rate in packet per second. The burstiness of the measured FP PDU rate and the analytical FP PDU rate obtained by the BMAP are similar. The mean arrival FP PDU rate of the measured traffic is 703.90 packets/second, while the mean arrival FP PDU rate of BMAP is 722 packets/second. The standard deviation of arrival FP PDU rate of the measured traffic is 7657.9 packets/second and from BMAP is 7520.5 packets/second.

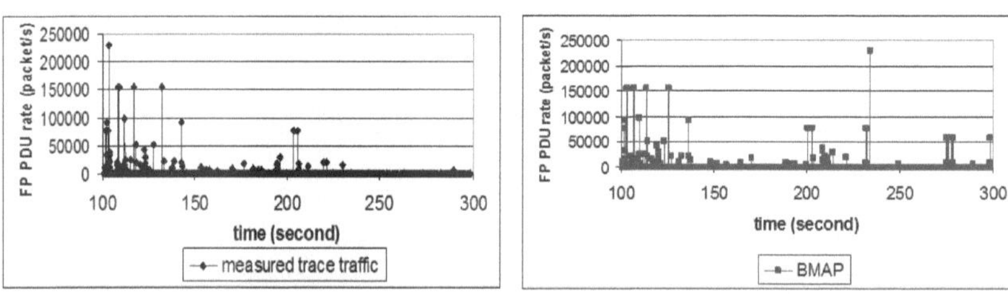

Figure 6.58: *Measured user traffic (left) and customized BMAP (right)*

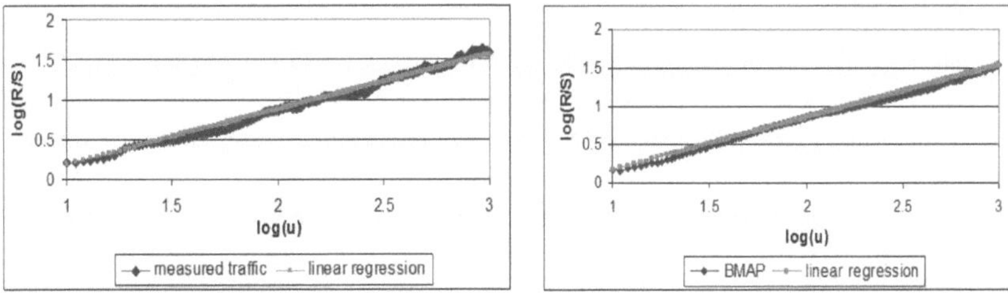

Figure 6.59: *R/S statistic of the measured traffic (left) and customized BMAP (right)*

The observations are further emphasized by the analysis of the traffic properties. Figure 6.59 presents the R/S statistics of the measured user traffic (left) and the customized BMAP (right). The degree of traffic self-similarity can be expressed by the slopes of linear regression plots of the R/S statistics. The value of H for the BMAP model is 0.68, which indicates a significant amount of traffic burstiness, compared with the Hurst parameter of the measured traffic (H = 0.69). This result also proves that when transferring the elastic IP traffic, the carried traffic of UTRAN in forms of aggregated FP PDU streams is self-similar as $H > 0.5$.

Figure 6.60 gives another example of three Packet Switched RAB types being used at the same time, i.e. RAB 64kbps, 128kbps and 384kbps. The application is web browsing with Pareto page size distribution and an average page size of 15kbyte. The HTTP requests are equally distributed among these three RAB types. The Iub link is a 2Mbps E1 line and the *Permanent Virtual Circuit* (PVC) on the ATM layer is set to 1.6 Mbps. Figure 6.60 compares the CDF of the FP PDU inter-arrival time obtained from the measured simulation trace and from BMAP. It is obvious that the CDF of BMAP model matches the CDF of the measured traffic very well. The average FP PDU inter-arrival time measured from the simulation is 5.8ms while from BMAP is around 5.7ms.

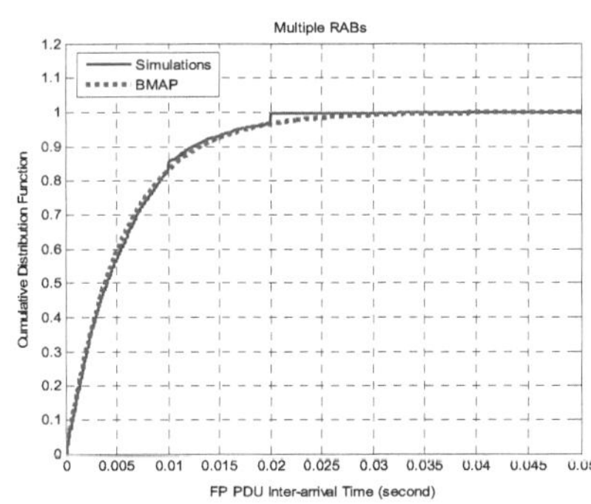

Figure 6.60: *CDF of FP PDU Inter-arrival time – multiple RABs scenario*

6.4.4.2 BMAP/D/1

If the arrival process of the AAL2 queue is modeled using the BMAP and the AAL2 buffer size is assumed to be unlimited, thus the AAL2 queue can be modeled as BMAP/D/1 queuing system. With this queuing system, the queuing performance given for example in queuing delay or queue length can be derived.

However, the BMAP trace arrival process is in fact fairly difficult to apply analytically and its important metrics are often calculated numerically and require parameterization when fitted to the measured trace data. The waiting time distribution in the BMAP/D/1 queue can be found in [Luc93], which is quite complex and involves many separate techniques and theories. In this thesis, rather than solving the analytic equations for the waiting time, the BMAP/D/1 queue was simulated using the OPNET simulator with a traffic source, a queue and a traffic sink, as shown in Figure 6.61. The source traffic of the queue is a trace generated using the BMAP parameterized with real data using IP2BMAP software. In the simulated queue system, there is only one server and the service rate is set to deterministic.

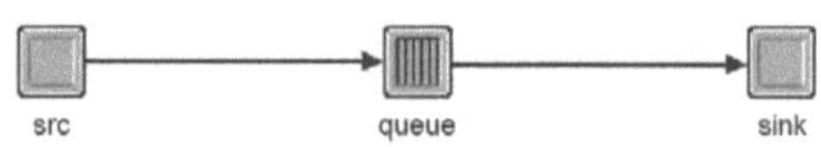

Figure 6.61: *BMAP/D/1 Queue in OPNET*

The following validates the accuracy of the BMAP/D/1 model to estimate the AAL2 queuing delay. In the given example scenario, only RAB 64kbps is used for transmitting the web applications with an average web page size of 25kbyte (Pareto page size distribution). The Iub link is a 2 Mbps E1 line and the bandwidth for the user plane is set to 1.55 Mbps.

Figure 6.62 presents the results from BMAP/D/1 to derive the CDF of AAL2 queue length in bits (left diagram) and queuing delay in seconds (right diagram) compared with simulations. The results show the distribution of the AAL2 queue length and queuing delays obtained from the BMAP/D/1 model are pretty close to the distribution of the measured values from the simulation. From this example, the BMAP/D/1 model is demonstrated to be an accurate analytical model to calculate the queuing delay in the AAL2 buffer.

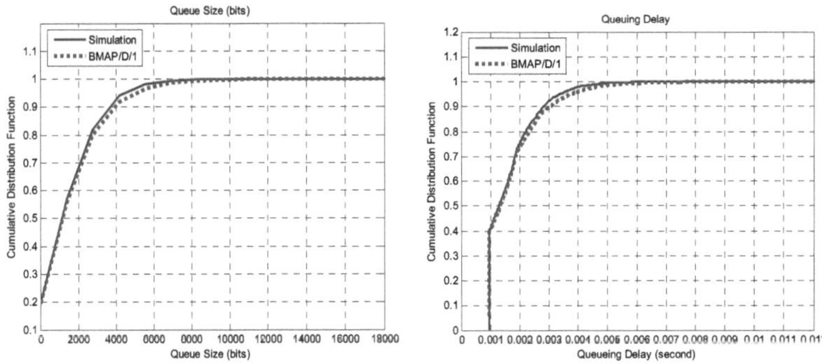

Figure 6.62: *CDF of FP PDU Inter-arrival time (simulation vs. BMAP)*

6.5 Dimensioning for HSPA

The HSPA (HSDPA and HSUPA) traffic is typically characterized by high peak data rate and high burstniess, which leads to higher transport capacity requirements on the Iub interface than the Rel99 traffic. In this thesis, the dimensioning for HSPA traffic is focused on the ATM-based Iub transport and the defined QoS target of dimensioning is the end user application performance, i.e. the application delay or throughput.

The dimensioning for HSPA is presented in detail in Appendix A.15. The dimensioning and investigations are based on extensive simulations using the developed HSPA simulation models (details of the HSPA simulation models are given in section 5.6 and Appendix A.10). In Appendix A.15 (I), a general framework is proposed for dimensioning of the Iub interface for HSPA traffic. This framework determines several important factors that will have impact on the bandwidth dimensioning for the Iub interface, such as the user applications and the traffic models, the available cell capacity and the number of UEs within one cell, the decision of the air interface scheduler, and the required QoS targets. Appendix A.15 (II) presents the detailed dimensioning results using HSUPA as an example, which have been published in [LZW+08]. The presented results consist of the investigations of the influence of different traffic models and the number of users on the obtained performance as well on the Iub dimensioning, and furthermore the impact of QoS targets on the Iub bandwidth dimensioning. From the presented dimensioning results obtained from the simulations, it can be concluded that the dimensioned Iub bandwidth is strongly dependent on the QoS target, and more importantly on the aggregated Iub traffic characteristic which is an outcome of the HSPA air interface scheduling influenced by the number of HSPA users in the cell and the traffic demand of each UE.

Appendix A.16 presents the investigations of applying traffic separation approach to transport HSPA (HSDPA/HSUPA) traffic together with the Rel99 traffic over the same ATM-based Iub interface. The concept of traffic separation in the ATM-based UTRAN is to apply separate ATM Virtual Path (VPs) or Virtual Circuits (VCs) with different ATM QoS categories to transmit different types of traffic with different priorities. Normally Rel99 and HSPA traffic have rather different QoS requirements: Rel99 mainly carries delay sensitive traffic like voice or streaming services; whereas HSPA traffic is primarily interactive and background traffic which is insensitive to the delay, therefore applying traffic separation technique to separate the HSPA and REl99 traffic over different paths with different priorities in the UTRAN transport network enables QoS differentiations for HSPA and Rel99 traffic while at the same time aims to achieve a maximum utilization of the transport bandwidth. Typically, Rel99 traffic is transported with ATM CBR (Constant Bit Rate) service category. It is defined as highest priority traffic class, where bandwidth is reserved up to requested Peak Cell Rate (PCR) with guaranteed cell loss ratio and cell transfer delay. This also implies high transport costs. While the transport of HSDPA and HSUPA traffic uses ATM traffic class UBR (Unspecified Bit Rate) or UBR+. UBR is a best effort service and is the lowest class of service in ATM. It is defined as low priority traffic class, which utilizes all bandwidth unused by the high priority traffic. Therefore it does not provide any guarantees for bandwidth, cell loss ratio and cell transfer delay. This traffic class has much lower transport costs. UBR+ is similar to UBR, but bandwidth is guaranteed up to a minimum rate MDCR (Minimum Desired Cell Rate). With UBR+, the HSPA traffic can be guaranteed up to MDCR. The detailed introduction of the basic concept of traffic separation and different possible traffic separation configurations for transmitting Rel99, HSDPA and HSUPA traffic in the same radio access network is given in Appendix A.16 (II) and (III).

The impact of applying traffic separation is analyzed in Appendix A.16 (IV). The presented results show that using traffic separation technique greatly improves the end user performance as well as the transport network performance, which in turn saves the bandwidth on the Iub link for achieving the same QoS level. Therefore, it brings a more efficient utilization of the transport resources in the UTRAN transport network. By investigating different MDCR settings, it is concluded that MDCR should be chosen as a compromise of the system performance and the Iub link utilization, and also dependent on the QoS target defined by the network operator. Appendix A.16 (V) compares different traffic separation configurations, i.e. 2 VPs or 3 VPs, to transmit HSDPA, HSUPA and Rel99 traffic within the same transport network. The traffic separation approaches are developed by the author in this thesis together with colleagues in the MATURE and HSPA project (see section 5.7). The presented simulation results demonstrate that both 2 VPs and 3 VPs configurations can be applied as the traffic separation solutions for the transport of HSPA and Rel99 traffic in the same radio access network. Through comparing the network and user performances of the two traffic separation configurations, using 3 VPs is slightly better than the 2 VPs due to a better protection for the signaling traffic but at the expenses of paying for one additional VP.

Therefore by taking considerations of both QoS and network costs, both 2 and 3 VPs solutions can be considered for the network operators or service providers to plan a UMTS access network to separately transport Rel99 and HSPA traffic. All these results have been presented in [LSGT08, LZW+09].

The traffic separation approach is a generic approach for transmitting different services (each with different QoS constrains) over different paths in the transport network. As another example, it can be also applied in a Carrier Ethernet-based UTRAN (using Pseudo-Wire Emulation (PWE) to emulate the ATM service over Ethernet in the transport network): a UBR or UBR+ path can be used for transmitting the best effort data service while the delay sensitive real time service like voice telephony is transmitted over a CBR path.

7 Dimensioning for Single IP-Based Iub

After studying the dimensioning of the ATM-based Iub, this chapter is focused on dimensioning the IP-based Iub of a single link. Different than the ATM-based UTRAN, the IP-based UTRAN employs IP as the underlying transport technology and thus has a different protocol stack and transport network. Additionally, it applies IP QoS schemes in the transport network to serve various service classes with different priorities. In the framework of this thesis, the investigated IP-based UTRAN applies the *Differentiated Service* (DiffServ) QoS scheme and a combined *Weighted Fair Queue* (WFQ) and *Strict Priority* (SP) scheduling strategy in the IP transport network, in order to distinguish different service flows and guarantee the desired QoS for various service classes. For dimensioning such an IP-based URRAN, in this chapter analytical models are proposed to determine the required IP bandwidth for the Iub link to satisfy user defined QoS of various service classes. Same as for dimensioning the ATM-based Iub, two fundamental traffic types are studied: elastic traffic and circuit-switched traffic. Section 7.2 presents analytical dimensioning models for circuit-switched traffic, and section 7.3 proposes analytical dimensioning models for elastic traffic. In section 7.4, the performances of ATM and IP based UTRAN and their required Iub link bandwidths are compared.

7.1 Overview and Objectives

In this chapter the dimensioning process for the IP-based UTRAN is focused on guaranteeing *user-relevant QoS* as the main objective of dimensioning. In the framework of this thesis, the considered user QoS for circuit-switched traffic is the connection reject ratio and for elastic traffic it is the end-to-end application throughput or delay (see section 4.1.2.1). For these two QoS targets, the M/G/R-PS models are proposed to dimension for elastic traffic and the Erlang models to dimension for circuit-switched traffic for the ATM-based Iub in Chapter 6. This chapter will discuss the applicability of both analytical models for dimensioning the IP-based Iub, and propose required extensions on both models to consider the IP transport specific protocols, functions, and applied DiffServ QoS scheme.

In addition, investigations are done to compare the IP-based with the ATM-based UTRAN. Section 7.4 presents main conclusions of these comparisons and a few examples. More results can be found in [LC+07] and Appendix A.22.

7.2 Dimensioning for Circuit-Switched Traffic

Same as in the ATM-based Iub, the related QoS criterion for the circuit-switched traffic in the IP-based UTRAN is still the connection reject ratio as a result of *Connection Admission Control* (CAC). In section 6.2.1 and 6.2.2, Erlang-B formula and multi-dimensional Erlang-B formula are proposed to estimate connection reject ratio (i.e. blocking probability) given a certain offered traffic and link capacity, for single traffic stream and multiple traffic streams scenarios respectively. These two models have been studied for the traditional circuit-switched telecommunication network and in Chapter 6.2 they are applied for dimensioning the Iub interface in the ATM-based UTRAN and they have been demonstrated to be accurate dimensioning models for the circuit-switched traffic to satisfy a desired connection reject ratio.

In the IP-based Iub, the applied CAC function is similar as in the ATM-based Iub except that the corresponding transport overheads are different in these two transport networks due to different underlying transport technologies being used. This results in different requirements on the physical link bandwidth. Usually the IP-based Iub has more transport overheads due to larger overheads of UDP/IP and layer 2 overheads (e.g. Ethernet). However, this only have impact on the demand of the bandwidth for each connection in the CAC algorithm. To accommodate the increased transport overhead, each connection needs a higher bandwidth reservation on the IP-based Iub link. As the concept of the CAC function is not changed, the derivation of the call blocking probability (i.e. CAC reject ratio) is still based on the Erlang-B formula (for single traffic stream) and multi-dimensional Erlang-B formula (for multiple traffic streams). The detailed calculation and application of the Erlang models for dimensioning can be referred to Chapter 6.2.

7.3 Dimensioning for Elastic Traffic

This section presents analytical models for dimensioning the transport bandwidth on the single IP-based Iub link transmitting elastic traffic that are carried by the TCP/IP. As introduced in section 3.4.2, the IP-based UTRAN network deploys an IP DiffServ-based QoS structure for performing QoS support and management. This QoS structure is based on DiffServ QoS scheme with an integrated *Weighted Fair Queue* (WFQ) and *Strict Priority* (SP) scheduling. For this IP QoS architecture, an analytical approach is proposed in this section for dimensioning the IP-based Iub link for supporting a wide range of elastic service class each with different QoS requirements. For the elastic traffic, the relevant QoS for dimensioning is the end user application performance with respect to the average per user throughput or the mean delay for transferring a certain amount of data (see section 4.1.2.1).

The objective of bandwidth dimensioning is to guarantee the desired QoS requirements of various services in a cost efficient way in terms of achieving a high transmission efficiency and maximal utilization of the IP transport bandwidth.

7.3.1 Analytical Models for the IP-Based Iub Dimensioning

In the ATM-based Iub, as presented in section 6.3, for the elastic traffic the proposed analytical dimensioning models to achieve the end user application performance are based on the M/G/R-PS model, which models the Iub link traversed by a number of elastic TCP flows as a **Process Sharing** queuing system on the flow level. In the IP-based Iub network, though the underlying transport technology is changed from ATM to IP and as well the employed QoS scheme is adapted from ATM QoS to IP QoS, the M/G/R-PS model is still valid and applicable for modeling the IP-based Iub link. As due to the behavior of TCP protocol, all elastic traffic flows going over the same IP-based Iub link will share the IP bandwidth resources equally, and thus the system is still essentially behaving as a Processor Sharing queue. This section further extends the M/G/R-PS model to dimension the IP-based Iub link for elastic traffic, where the DiffServ-based QoS structure is deployed on the IP transport.

As seen in the IP DiffServ based QoS architecture described in section 3.4.2.3, the elastic traffic is accommodated by Assured Forwarding (AF) or Best Effort (BE) PHBs, whereas the circuit-switched traffic is carried by the Expedited Forwarding (EF) PHB. The elastic traffic is corresponding to *Non Real Time* (NRT) traffic and the circuit-switched traffic corresponds to *Real Time* (RT) traffic. The bandwidth allocations among different AF and BE PHB classes are controlled by the WFQ scheduling. If each AF/BE PHB queue is continuously sending data, then the *k*th queue receives its share of the available bandwidth BW_k as calculated in formula (3.1). That means, a fraction of the total capacity is supposed to be allocated to each PHB. But if one PHB queue does not have data to send (i.e. not utilizing its allocated bandwidth), then in accordance with the concept of WFQ, its spare bandwidth shall be fairly shared among the other queues according to their weights.

The concept of dimensioning for elastic traffic over AF or BE PHBs is illustrated in Figure 7.1. The total IP bandwidth is divided into two parts: the bandwidth reserved or used by the RT traffic over EF PHB; and the remaining bandwidth to transmit the NRT traffic with various priorities by means of different AF and BE PHBs. As mentioned in section 4.4, two cases are considered for the bandwidth allocation between the RT and the NRT traffic: (1) bandwidth partitioning; (2) bandwidth sharing. In case (1) the bandwidth reserved by the RT traffic will not be used by the NRT traffic. It can be realized by using bandwidth reservation function and rate limiting functions, which allow the partitioning of link capacity for different types of traffic. In this case, the available bandwidth for the NRT traffic is calculated by subtracting the bandwidth assigned for the RT traffic from the total link bandwidth. In case (2), it is assumed that the NRT traffic can completely share the bandwidth of the RT traffic, which is the case considered in this thesis. In this case the available bandwidth for the NRT traffic is obtained by subtracting the mean traffic load of the RT traffic from the total link bandwidth. Both cases have been explained in section 6.3.9 in ATM-based UTRAN.

It is assumed that there are a number of service classes (priorities) of NRT traffic, each is served by one certain RAB service and carried on one AF or BE PHB. Then the application QoS performance of each service class strongly depends on its defined priority (corresponding to the weight in WFQ), the resultant available IP bandwidth that can be used for this service class, the data rate of the RAB to transmit this service class, and offered traffic of this service class. The essential idea of the following proposed analytical approach to estimate the application performance of every service class is to apply the M/G/R-PS model per service class, but taking into account of the potential multiplexing gain of bandwidth sharing among different service classes. It can be observed from Figure 7.1 that the total available IP bandwidth for each service class contains not only the bandwidth allocated for that service class by the WFQ scheduling function according to its weight calculated with equation (3.1); but also the bandwidth that has not been utilized by all other PHBs. Because according to the theory of the WFQ scheduling, whenever one PHB does not need to use its assigned bandwidth, its spare bandwidth shall be fairly shared by other PHBs.

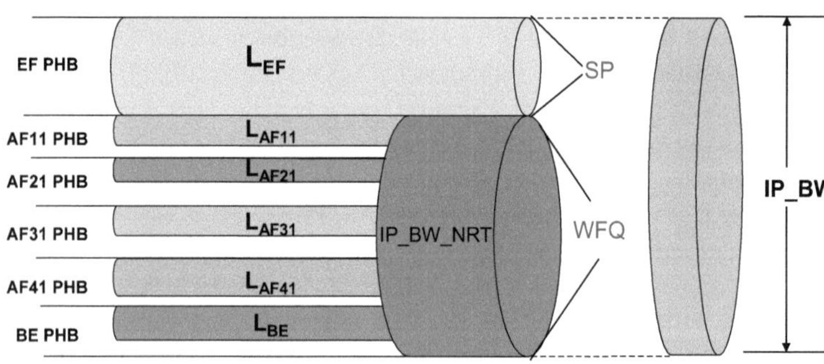

Figure 7.1: *Dimensioning for IP DiffServ-based QoS Structure*

The following presents the proposed analytical approach to estimate the application performance of each service class and then in turn to be used for dimensioning the Iub link bandwidth for satisfying desired application QoS criteria of all service classes. Firstly the following parameters are defined:

- Let *IP_BW* denote the total IP bandwidth;
- Let *IP_BW_NRT* be the total bandwidth allocated for NRT traffic, i.e. AF and BE PHBs;
- Let L_{EF} denote the bandwidth used or reserved by the EF PHB depending on the bandwidth allocation scheme as mentioned above;
- Let r_{peak_k} denote the peak rate of the bearer service (i.e. RAB service) for the service class k that is carried on the kth PHB of a AF or BE PHB;

- Let N_k denote the maximum allowed number of flows for the service class k;
- Let w_k denote the WFQ weight of the service class k;
- Let L_k be the mean offered traffic of the service class k;
- Let x_k denote the average file size of the service class k.

The following explains the detailed steps of calculating the average application delay with the M/G/R-PS model for the elastic traffic of the service class k that is carried on the kth PHB of an AF or BE PHB:

(1) The total available capacity for transmissions of all NRT elastic service classes (denoted as IP_BW_NRT) is obtained by subtracting the bandwidth used for the RT traffic from the total IP bandwidth, i.e. $IP\_BW\_NRT = (IP\_BW - L_{EF})$.

(2) C_k denotes the available IP bandwidth that can be used for the service class k (carried with the kth PHB), which consists of not only the bandwidth allocated by the WFQ scheduling for the kth PHB, but also the bandwidth which has not been utilized by all other PHBs. It is calculated with the following equation:

$$C_k = IP\_BW\_NRT \cdot \frac{w_k}{\sum_k w_k} + \left[IP\_BW\_NRT \cdot (1 - \frac{w_k}{\sum_k w_k}) - \sum_{j \neq k} L_j \right] \tag{7.1}$$

In the equation above, the first term $IP\_BW\_NRT \cdot \frac{w_k}{\sum_k w_k} = (IP\_BW - L_{EF}) \cdot \frac{w_k}{\sum_k w_k}$ represents the allocated bandwidth by the WFQ for the kth PHB accordion to its weight w_k. It is the minimum available bandwidth that can be attained by the kth PHB in case that all other PHBs are fully utilizing their allocated bandwidth determined by the WFQ. The second term $\left[IP\_BW\_NRT \cdot (1 - \frac{w_k}{\sum_k w_k}) - \sum_{j \neq k} L_j \right]$ corresponds to the spare bandwidth that may not be utilized by other AF or BE PHBs. Here $IP\_BW\_NRT \cdot (1 - \frac{w_k}{\sum_k w_k})$ defines the total allocated bandwidth for other AF and BE PHBs and $\sum_{j \neq k} L_j$ is the total offered traffic to be transported on other AF and BE PHBs. The expression of C_k can be also simplified to equation (7.2) with the condition that C_k needs to be above or at least equal to the allocated bandwidth from the WFQ scheduling according to its weight w_k.

$$C_k = \max \left\{ \left(IP\_BW\_NRT \cdot \frac{w_k}{\sum_k w_k} \right), \left(IP\_BW\_NRT - \sum_{j \neq k} L_j \right) \right\} \tag{7.2}$$

It is seen from the above simplified equation that C_k should have a minimum bandwidth that equals to the allocated bandwidth by the WFQ according to its weight w_k, i.e.

$\left(IP\_BW\_NRT \cdot \frac{w_k}{\sum_k w_k} \right)$. In this proposed approach, it shall be noticed that the traffic of all other NRT service classes over other AF/BE PHBs can as well share the bandwidth of C_k as taking into account the potential multiplexing gain of bandwidth sharing among all AF and BE PHBs.

(3) The normalized traffic load of the service class k, denoted as ρ_k, can be derived with $\rho_k = L_k / C_k$, given the mean offered traffic L_k and the calculated available IP bandwidth C_k for the service class k.

(4) In order to apply **M/G/R-PS model** to estimate the application performance, **R** needs to be determined. For the service class k, $R_k = C_k / r_{peak\_k}$ (see section 6.3.3).

(5) For the service class k on the kth PHB, the expected delay of transferring a file of length x_k can be derived from the M/G/R-PS model. The following equation (7.3) applies the basic form of M/G/R-PS formula (based on equation (6.18) in section 6.3.5).

$$E_{M/G/R}\{T(x_k)\} = \frac{x_k}{r_{peak\_k}}\left(1 + \frac{E_2(R_k, R_k\rho_k)}{R_k(1-\rho_k)}\right) = \frac{x_k}{r_{peak\_k}} f_k \tag{7.3}$$

As explained in section 6.3.5, in the above formula E_2 represents Erlang's second formula (Erlang C formula), and f_k is defined as the delay factor which represents the increase of the transfer time or decrease of the throughput caused by the link congestion. In general, the M/G/R-PS model is applied when there is no restriction on the number of concurrently transmitted traffic flows.

(6) If it is assumed that a maximum of **N** **(N≥R)** elastic traffic flows are transported simultaneously over the link, then the **M/G/R/N-PS model** can be applied to calculate the average file transfer delay. The calculation of the mean sojourn time of the M/G/R/N-PS model is presented in Chapter 6.3.7.2. Given the maximum number of flows N_k for the service class k, the state probability equation is given in equation (7.4) which is based on the equation (6.28).

$$p(j) = \begin{cases} \dfrac{(1-\rho_k)\dfrac{R_k!}{j!}(R_k\rho_k)^{j-R_k} E_2(R_k, R_k\rho_k)}{1 - E_2(R_k, R_k\rho_k)\rho_k^{N_k-R_k}\rho_k} & (j < R_k) \\[2ex] \dfrac{E_2(R_k, R_k\rho_k)\rho_k^{j-R_k}(1-\rho_k)}{1 - E_2(R_k, R_k\rho_k)\rho_k^{N_k-R_k}\rho_k} & (R_k \le j \le N_k) \end{cases} \tag{7.4}$$

With the given state probability, the average number of connections can be derived from $E\{W\} = \sum_{j=0}^{N} j \cdot p(j)$. By applying Little's law, the average transfer delay is obtained by $E_{M/G/R/N-PS}\{T(x_k)\} = \dfrac{E\{W\}}{\lambda(1-p(N_k))}$ according to equation (6.29). Here λ denotes the average flow arrival rate of the service class k in Erlang and $p(N_k)$ represents the probability of having above N connections in the network.

It is noted that the above approach is for calculating the average application delay per AF or BE PHB. It can be also applied to a single AF class with different drop levels. Because each drop level belongs to one specific PHB.

The above proposed extension on the M/G/R-PS model for the dimensioning of the IP-based Iub with DiffServ QoS structure was published by the author in [LBG+08].

7.3.2 Validation of Analytical Dimensioning Models

This section validates the proposed analytical dimensioning approach for transport the elastic traffic on the AF or BE PHB, to guarantee certain application delay QoS of a service class by means of simulations.

The simulation model is based on the simplified IP-based UTRAN model described in section 5.5.2. The simulation scenario consists of seven traffic types in UMTS. Table 7.1 describes the used service classes and their mapping to the DiffServ PHBs. In the investigated scenario, each Node B transmits 10 elastic traffic flows over AF11, 10 flows on AF21, 10 flows on AF31, 6 flows on AF41, 5 flows over BE PHB, in addition there are 5 video and 11 voice user connections both are transmitted over EF PHB. The simulation scenario is composed of one Node B and one RNC (single Iub scenario), and between them there are two IP DiffServ routers, which implements the DiffServ functions as explained in section 3.4.2.1. The QoS structure is shown in Figure 3.13. Table 7.2 gives the configuration of the WRED and WFQ parameters for different AF and BE PHBs.

Service class	Mapped to DiffServ PHB
Conversational (voice)	EF
Streaming (video)	EF
Interactive –NRT RAB 64kbps	AF11
Interactive – NRT 128kbps	AF21
Interactive – NRT RAB 256kbps	AF31
Interactive – NRT RAB 384kbps	AF41
Background – HSPA (2 Mbps)	BE

Table 7.1: *Traffic classes and their mapped PHBs*

Their traffic models are described below:

(1) Interactive and Background Traffic Model – web application

- Inactive period (reading/thinking time period): geometric distribution (mean = 5s)

- File/Page size: constant distribution of 25kbyte

(2) Conversational Traffic Model – voice application
- AMR Codec, with traffic model defined in Table 5.2
- Transport over UDP

(3) Streaming Traffic Model – video application
- 800bytes frame size
- 10 frames /second
- 64kbps coding rate
- Transport over UDP

	WRED Parameters				WFQ
PHB	min_{th}	max_{th}	max_p	exp.w.	weight
AF11	4	10	10%	9	20
AF21	4	10	10%	9	30
AF31	5	10	10%	9	40
AF41	6	10	10%	9	50
BE	3	10	10%	9	10

Table 7.2: *WRED and WFQ Parameters*

The M/G/R/N-PS model is used in this case as the analytical model for dimensioning. Figure 7.2 compares the calculated application delay per PHB or RAB service using the M/G/R/N-PS model against the simulation results for the given scenario. It shows that for each PHB (AF or BE) and its serving NRT RAB service, given a certain Iub link utilization, the calculated application delay matches well with the delay values from the simulations. It proves that the proposed analytical model gives an accurate estimation of the application performance for various service classes each with different RAB service and priorities. And moreover, it is observed that when the Iub link is highly utilized, e.g. above 90% link utilization, the achieved delay performance is worse for the lower weight queue or service due to a less share of total bandwidth. For example, AF41 PHB has the highest weight so its delay performance is the best whereas BE PHB experiences the most delay degradation under the congestion.

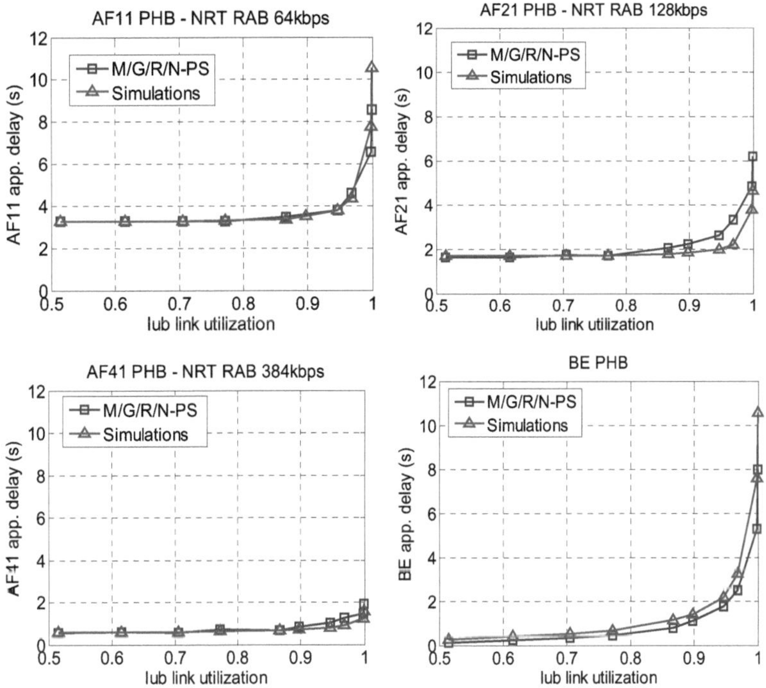

Figure 7.2: *Validation of M/G/R/N-PS Model – transfer delay*

Figure 7.3 presents for each PHB or RAB service the derived IP bandwidth allocated for all NRT traffic (i.e. *IP_BW_NRT*) using the M/G/R/N-PS model under different application delay requirements of that RAB service, and compares them against the simulation results. It shows that for all RAB services, when the delay QoS requirement is higher, i.e. the desired application delay need to be smaller, then a higher IP bandwidth is required. As explained in the above dimensioning steps in section V, if each RAB service has a specific delay requirement, then for each service we can calculate the required total IP bandwidth for all NRT traffic (i.e. *IP_BW_NRT*) under which the delay boundary of that service is satisfied, and we should take the maximum of the calculated bandwidth as the total IP bandwidth for carrying the NRT traffic (i.e. *IP_BW_NRT*). By comparing the simulation results and the derived IP bandwidth from the proposed model in Figure 7.3, it shows that in general the proposed analytical approach based on the M/G/R/N-PS model can give a relatively good approximation of the required IP bandwidth for the dimensioning results, so it is proven to be a proper dimensioning approach. The presented simulation results demonstrate that the proposed analytical methods can appropriately estimate the application performances of different service classes with a DiffServ QoS framework. Hence it can be applied to efficiently dimension IP bandwidth of the Iub link and promise multi-service QoS provisioning in the UTRAN.

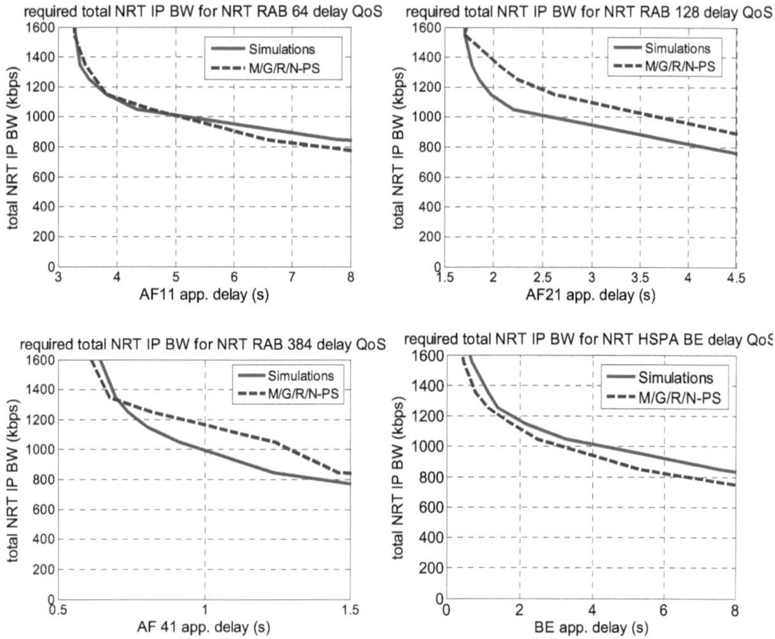

Figure 7.3: *Validation of M/G/R/N-PS Model – IP bandwidth*

7.3.3 Dimensioning Procedure

The above analytical approach is able to estimate the achieved average application performance in terms of mean transfer delay or application throughput for each AF or BE PHB. Given a specific application QoS per service class, the required IP bandwidth for meeting the QoS of that service class can be derived with the presented analytical model. For the dimensioning process, the total required IP bandwidth on the Iub link needs to fulfill the QoS requirements of all service classes. It can be derived with the following steps.

1) For one service class of elastic traffic, the required total IP bandwidth can be derived numerically by performing delay calculations for a range of configured IP bandwidths until the resultant average transfer delay from a certain IP bandwidth reaches the desired application delay QoS of that service class.

2) Repeat the above step 1) for all service classes. Then for each service class *k*, there is one dimensioned IP bandwidth that satisfies the delay QoS of that service class *IP_BW_NRT (k)*.
3) Take the maximum IP bandwidth from all *IP_BW_NRT (k)* to be the dimensioned bandwidth for that link, as it will satisfy the QoS requirements of each service class.
4) If there are additional IP bandwidths reserved for *Real Time* (RT) (derived from MD-Erlang) or signaling traffic (e.g. DCCH, CCCH), then these extra bandwidths also need to be added to compute the total required IP bandwidth for the Iub link.

7.3.4 Dimensioning Results

The following gives an example of applying the above dimensioning procedure to calculate the required IP bandwidth for a single Iub link. Figure 7.4(a) shows the estimated IP bandwidths over different target delay factors of BE PHB (in this example BE PHB is the worst case) and compared with the simulated bandwidths. In Figure 7.4(b) the relative error is given. In this example, there is pure elastic traffic, which is carried by AF and BE PHBs. The following configurations are applied in the simulations:

- there are five PHBs: AF 11 (RAB 64 kbps), AF 21 (RAB 128 kbps), AF 31 (RAB 256 kbps), AF 41 (RAB 384 kbps)
- all NRT service classes use the same web traffic model: page size is 50 kbyte of a constant distribution
- each Node B carries 6 elastic traffic flows per service class

It can be seen from Figure 7.4 that the bandwidth requirement is reduced with a higher target delay factor (i.e. lower application QoS requirement). And through comparing the analytical results with the simulation results, it can be seen that the suggested analytical approach can give proper estimations on the required IP bandwidths (with relative error around 10%).

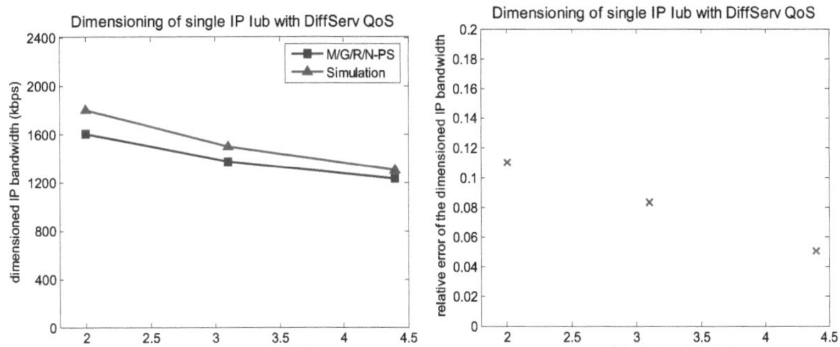

(a) IP bandwidth (b) relative error

Figure 7.4: *Dimensioning of a single IP Iub link*

After demonstrating the analytical approach for dimensioning a single IP Iub link, the analytical dimensioning approach can be used to investigate dimensioning rules. Figure 7.5 presents required normalized capacity (definition is given in section 4.3) over traffic load (in kbps) derived from the analytical approaches. Figure 7.5(a) shows the required capacity under different QoS requirements. It is seen that with a higher QoS requirement, the bandwidth demand is higher. It can be also seen that there is an increasing multiplexing gain with larger traffic load, and thus the required normalized capacity approaches one. Figure 7.5(b) demonstrates the influence of the circuit-switched traffic (voice in this example) on the required capacity. It shows that more percentage of voice traffic requires less bandwidth, similar conclusions are obtained in the mixed traffic scenario with a single ATM-based Iub (presented in section 6.3.10).

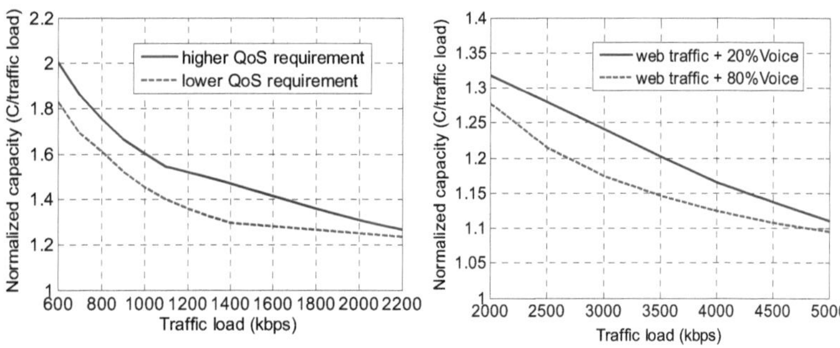

(a) different QoS requirements (b) different voice portions

Figure 7.5: *Capacity vs. traffic load*

7.4 Comparing the ATM and IP-Based Iub

When comparing the ATM and the IP transport, their individual transport network overheads, applied QoS or multiplexing schemes are different. This results in significant differences on overall network and end user performances and in turn the network dimensioning results. This section compares the achieved performances and bandwidth requirements of the ATM and IP based UTRAN.

At first, *Bandwidth Efficiency* of both networks is compared in Figure 7.6. Effective bandwidth utilization in the UTRAN is directly related to the transport costs. High bandwidth efficiency means low transport network costs. In the context of this thesis, it is defined as the ratio of the data payload size (without transport network layer protocol headers) to the total bits of transmitting this payload on the physical link. In Figure 7.6, the bandwidth efficiency is calculated by estimating the total protocol layer overheads in both ATM and IP transport. In the following comparisons, the IP transport does not apply any multiplexing scheme. That means, each radio frame (i.e. FP PDU) is encapsulated into one UDP/IP packet. In Appendix A.22, two optional multiplexing schemes *Composite IP* (CIP) and *Lightweight IP Encapsulation* (LIPE) can be used to improve the bandwidth efficiency of IP based UTRAN. They allow multiplexing multiple radio frames (FP PDU) into one UDP/IP packet, and thus reduce the relative UDP/IP transport overheads.

It can be seen in Figure 7.6 that for AMR voice and low RAB rate bearers such as RAB 64 kbps, the ATM-based UTRAN achieves higher efficiency than the IP-based UTRAN. This is because that the radio frame sizes of low RAB rate bearers are relatively small, and thus the added UDP/IP/Ethernet overhead of the IP transport network is larger compared to the ATM transport. On the other hand, the bandwidth efficiency of the IP-based UTRAN increases for higher RAB rate and even exceeds the efficiency of the ATM-based UTRAN when the RAB rates are 128kbps or above. This is because that with higher RAB rate, the radio frame sizes are larger and thus the overhead of UDP/IP/Ethernet is relatively smaller than the one in the ATM-based UTRAN. Appendix A.22 shows that when multiplexing scheme is applied in the IP-based UTRAN, with either CIP or LIPE, the bandwidth efficiency is significantly better compared with not using multiplexing scheme in the IP-based Iub, and even better than the ATM-based one, by combining several frames into one IP packet.

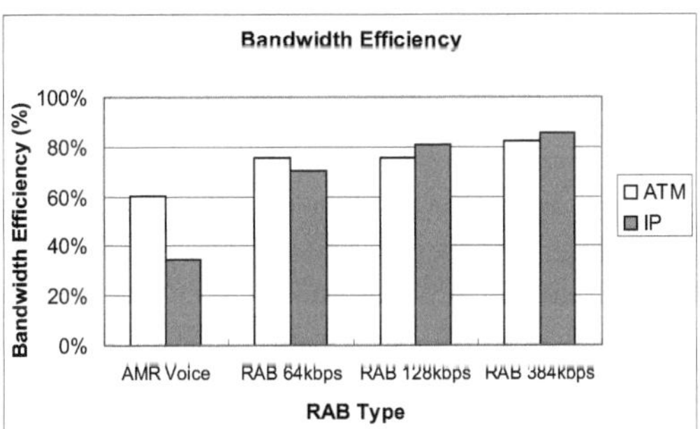

Figure 7.6: *Bandwidth Efficiency: ATM vs. IP*

The following results compare the dimensioning of a single Iub link in the ATM and IP-based UTRAN. In the given examples, the considered user-relevant QoS for elastic traffic is the end-to-end application delay indicated by a target delay factor, and for circuit-switched traffic the connection reject ratio as a result of CAC. Figure 7.7 compares the required Iub link bandwidths of the ATM and the IP (without multiplexing scheme) over different voice traffic loads, for meeting 1% and 5% CAC connection reject ratio (blocking). The dimensioning results are derived from Erlang-B formula while considering the individual transport overheads of both transport networks.

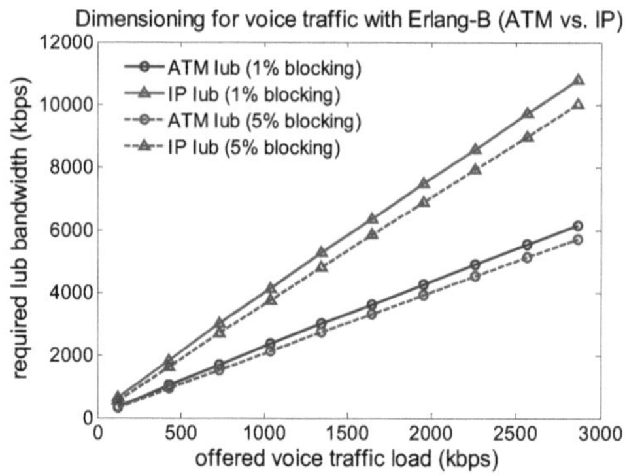

Figure 7.7: *ATM vs. IP (without multiplexing) link bandwidth over traffic load voice scenario: QoS target: 1% and 5% connection reject ratio*

It can be seen that the IP transport requires much higher bandwidth than the ATM, as the IP has a much lower bandwidth efficiency for 12.2 kbps AMR voice bearer as shown in Figure 7.6 when not using any multiplexing scheme. In order to reduce the transport costs of IP, it is suggested to apply multiplexing schemes to improve the bandwidth efficiency. First investigations of various multiplexing schemes are presented by the author in [LC+07] and the importance of optimization is addressed.

The following several examples compare the Iub dimensioning for web traffic to guarantee a target end-to-end delay factor of 1.25 for transferring a 50 kbyte web page. Figure 7.8 compares the dimensioning for RAB 128 kbps, and Figure 7.9 for RAB 64 kbps and RAB 384 kbps in the ATM and IP-based Iub.

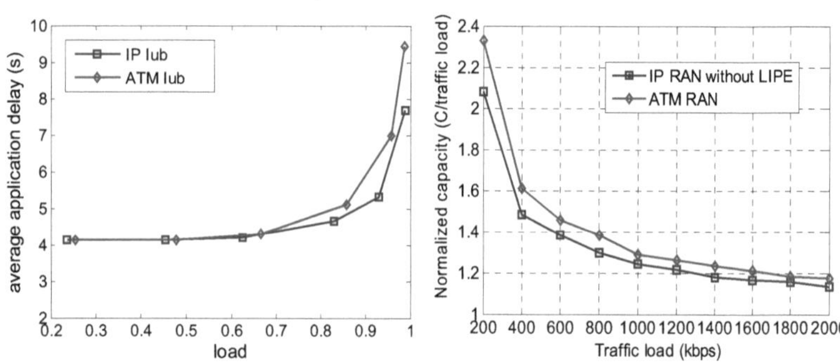

(a) delay over load (link utilization) (b) capacity over traffic load

Figure 7.8: *ATM vs. IP: web scenario, RAB 128 kbps, no CAC, constant page size of 50 kbyte, target delay factor f =1.25*

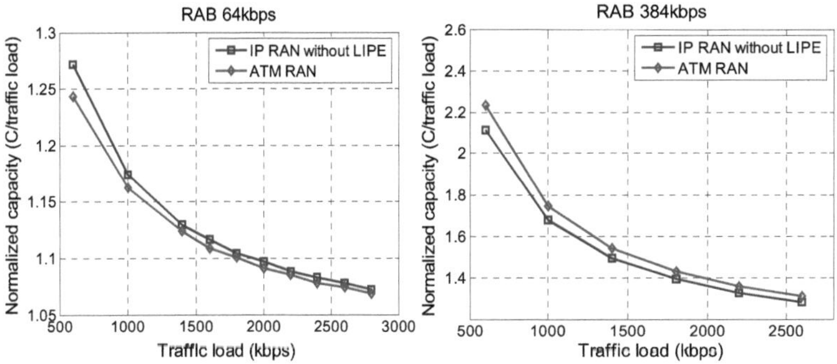

Figure 7.9: *ATM vs. IP (capacity over traffic load): web scenario, RAB 64 kbps (left) and RAB 384 kbps (right), no CAC, page size =50 kbyte, target delay factor f =1.25*

The left graph in Figure 7.8 shows the average transfer delay with RAB 128 kbps over different loads (link utilizations). The results in the graph are taken from simulations with the ATM and IP based UTRAN simulation model. It is seen in Figure 7.8 that when the load is low, the obtained average application delays are same in both networks. As the minimum application delay is dependent on the RAB rate. But when the load gets higher, it can be observed that the delay performance of the ATM-based Iub gets worse than the IP-based one. This is because that with RAB 128 kbps the ATM based transport network generates more transport network overheads (lower bandwidth efficiency) for the same amount of application load, and in turn for the same link rate the resultant **R** in the **M/G/R-PS** model is smaller. This result also implies that the ATM transport requires more link bandwidth than the IP transport for achieving the same target application delay. This conclusion is proven with the right graph in Figure 7.8, where the dimensioning results are derived from the analytical models. In this graph, the dimensioned link bandwidth is represented by the normalized capacity (defined in section 4.3) over different traffic loads. The result shows that with RAB 128 kbps, the required bandwidth for the ATM-based Iub is higher than IP-based Iub since it has relatively lower bandwidth efficiency. Similar conclusion is obtained for RAB 384 kbps in Figure 7.9 (right diagram), due to higher bandwidth efficiency with the IP transport than the ATM (see Figure 7.6). While for RAB 64 kbps as given in the right diagram of Figure 7.9, the dimensioning result is slightly better in the ATM-based Iub as a result of higher bandwidth efficiency with the ATM transport (see Figure 7.6). In Appendix A.22, more comparisons of the ATM and IP based transport are given. Additionally, based on [LC+07] some examples of applying different multiplexing schemes are given in Appendix A.22. The results show that a balance should be given between the bandwidth efficiency and the transport network delay by optimizing the multiplexing scheme parameters.

7.5 Summary

In this chapter, analytical dimensioning approach was proposed for dimensioning the IP bandwidth for the IP-based Iub interface applying the DiffServ QoS, for elastic NRT traffic and RT streaming individually. The proposed dimensioning approach for elastic traffic is based on the extension work of the M/G/R-PS model. The analytical approach was validated by the simulations. The simulation results demonstrate that the proposed analytical methods can appropriately estimate the application performances of the different service classes, hence we can efficiently plan bandwidth capacity of the network links and promise multi-service QoS provisioning in the IP-based UTRAN.

At the end of this chapter, the performance and the dimensioning of the IP-based Iub is compared with the ATM-based Iub. The comparisons show that the IP transport has more gain for the higher RAB rate bearers due to a relatively lower transport overhead and thus a higher bandwidth efficiency. However, generally it is suggested to apply multiplexing schemes to improve the transport efficiency, especially for the lower RAB rate bearers. But a balance should be given between the bandwidth efficiency and the transport network delay by optimizing the multiplexing scheme parameters.

8 Dimensioning for Multi-Iub RAN Scenario

After the investigation of dimensioning for the single link scenario in the previous chapters, this chapter focuses on the dimensioning for the multi-Iub *Radio Access Network* (RAN) considering a star network topology. In this chapter, individual dimensioning approaches are presented for dimensioning the circuit-switched traffic and the elastic traffic, each of them tailored individually for ATM and IP based RAN, which allows dimensioning of the last mile link for each Node B and the backbone link individually.

One related issue on dimensioning of the Iub backbone link is *overbooking*. It is a popular technique used by network operators to allocate less bandwidth on the backbone link than the total bandwidth request by all connected last mile links. For a cost-efficient dimensioning of the backbone link, an optimum overbooking is desired to provide balance between the costs and the QoS objectives. In this chapter, the impact of overbooking is presented and various important aspects that have influence on overbooking are investigated. For a given radio access network, through dimensioning the backbone link, an appropriate overbooking can be derived. Thus, the analytical approaches which are proposed in this chapter for dimensioning the multi-Iub RAN are applied to calculate optimum overbooking for the Iub backbone link. By comparing with simulation results, the proposed analytical approaches are proven to give accurate estimations on the overbooking and the related dimensioning results.

8.1 Overview and Objectives

The structure of a star topology multi-Iub RAN is presented in Figure 1.2 (b). It consists of a number of last mile links each connected to one Node B and a backbone link that is connected to the RNC with an aggregate router or switch. For dimensioning such a radio access network, the bandwidths, which need to be assigned to each last mile link and the backbone link, have to be determined. The goal of dimensioning is to guarantee the QoS requirements with the maximum utilization of transport resources on both last mile link and backbone link. Thus, the overall network costs can be minimized.

This chapter mainly presents the dimensioning of a multi-Iub transport network to guarantee *user QoS requirements* as the objective of dimensioning. As specified in section 4.1.2.1, the relevant user QoS for the elastic traffic is end-to-end application throughput or delay. For the circuit-switched traffic the considered user QoS is the connection reject ratio.

In the previous chapters for a single Iub link scenario, processor sharing models have been presented to estimate the application throughput or the delay of the elastic traffic; and the Erlang-B or multi-dimensional Erlang-B formula has been used to calculate the connection reject ratio of the circuit-switched traffic. These two models have been demonstrated to be appropriate analytical models for dimensioning a single Iub transport link in both ATM and IP based UTRAN. In this chapter, they are proposed to be applied for dimensioning the multi-Iub RAN as well, but some extensions are required to consider both the last mile link and the backbone link for calculating an end-to-end QoS. For the dimensioning process, two issues are discussed and investigated in this chapter. The first issue is to develop methodologies to properly dimension the last mile links and the backbone link individually in order to fulfill the user QoS requirements on an end-to-end basis. The second one is to perform overbooking on the backbone link, given a number of last mile links each carries a certain amount of traffic.

8.1.1 Dimensioning of Individual Links

The objective of this task is firstly to develop correct methods to estimate the end user QoS, taking into account of the impact of both last mile link and the backbone link on the end-to-end quality of service. To analytically estimate the end user QoS, the analytical models which were proposed to estimate the end user QoS in the single Iub link scenario can be applied here in a way that the impact of the last mile link and the backbone link are considered. Then in the second step based on the relations of the user QoS with the individual links, appropriate dimensioning approaches need to be suggested to determine the bandwidth for the last mile link and the backbone link separately, which shall result in a minimum overall network cost while fulfilling the desired end-to-end user QoS requirements.

8.1.2 Overbooking of the Iub Backbone Link

Overbooking means reserving more resources than are actually available. The nature of network traffic is usually "bursty". It allows active connections to utilize the bandwidth during the idle period of other connections. Hence a statistical multiplexing gain can be achieved in the aggregated traffic, especially for bursty traffic. Due to the potential multiplexing gain, overbooking becomes possible to increase the utilization of the available resources. It has become an efficient way to improve bandwidth utilization and network efficiency, thereby increasing the revenue generated by a network infrastructure. However, for a cost-efficient network dimensioning, overbooking should be performed appropriately: too much overbooking causes degradation of QoS for the existing customers whereas too little overbooking leads to underutilization of available bandwidth. Thus, finding an optimum overbooking is an essential task for network dimensioning.

Overbooking has been studied for ATM and IP networks. Pazos and Gerla studied overbooking in the use of ATM ABR services on Internet backbones [CM98a]. [CM98b, TS96] demonstrated overbooking is a reliable alternative to increase link utilization in ATM networks. The IETF draft [CI01] addresses the notion of overbooking and its application to IP/MPLS traffic engineering. Besides, there are studies focused on finding approaches of determining the optimum overbooking factor for the Internet backbone networks [RM[+]04, HDB[+]04].

In the star topology multi-Iub radio access network, overbooking is used to allocate less bandwidth on the backbone link than the total bandwidth request of all the connected last mile links. The Iub backbone links can be overbooked by taking advantage of statistical multiplexing gain among traffic from different Node Bs. The achievable statistical multiplexing is dependent on the network scale (number of Node Bs/last mile links), trunk types, traffic types and user traffic characteristics. Thus, these aspects will be considered in setting an appropriate overbooking for a cost-efficient dimensioning. In this thesis, the overbooking factor (*OB_factor*) is used to specify the degree of overbooking. Assuming there are n Node Bs in the radio access network, each Node B has a last mile link rate C_{ac_i} and the Iub backbone link has a capacity of C_{bb}, then the overbooking factor is calculated as follows.

$$OB\_factor = \frac{\sum_{i=1}^{n} C_{ac\_i}}{C_{bb}} \qquad (OB\_factor \geq 1) \tag{8.1}$$

The total requested bandwidth by N Node Bs is expressed as $\sum_{i=1}^{n} C_{ac\_i}$. The value of *OB_factor* should be greater than or equal to 1. If *OB_factor* equals to 1, that means no overbooking is applied as the backbone link offers the same bandwidth to each last mile link as its requested bandwidth C. If *OB_factor* is larger than 1, the backbone link is overbooked as the total requested bandwidth by N Node Bs is above the given backbone link capacity. A higher value of the *OB_factor* means a more aggressive or higher overbooking, i.e. less capacity is offered on the backbone link.

In this chapter, overbooking will be investigated for both ATM and IP RAN. Their impacts and various important aspects that have influence on the overbooking will be discussed. More importantly, analytical approaches will be presented, which can estimate the optimum overbooking factor for network dimensioning. The estimation is based on the developed analytical models for dimensioning the multi-Iub RAN. And the estimated overbooking factor is validated with simulations.

8.2 Dimensioning for Circuit-Switched Traffic

As introduced in Chapter 1, the circuit-switched traffic is subject to a *connection admission control* (CAC) function. In the star topology multi-Iub RAN, it is considered that the CAC function is performed through the UTRAN network. That means, a new connection needs to reserve transport bandwidth throughout the radio access network domain and it can be only accepted to transport if both the last mile link and the backbone link provide sufficient bandwidth to accommodate the requested bandwidth of the new connection. However, overbooking is allowed on the backbone link. That means, the CAC guaranteed bandwidth on the backbone link can be scaled adequately by taking into account the effect of overbooking. Nevertheless, the degree of overbooking which can be done on the backbone link depends on the QoS requirements and the achieved possible multiplexing gain as a result of aggregating the traffic from multiple last mile links over a single backbone link. In the framework of this thesis, the network related QoS requirements such as packet delay and packet loss are considered for determining the overbooking for the backbone link, since they are also important QoS for the circuit-switched type of traffic in the UTRAN. As introduced in Chapter 1, the Iub interface needs to satisfy the strict delay requirements to ensure optimal radio resource utilization. Moreover, the packets which experience excessively long delays over the Iub have to be discarded as explained in section 3.3.3.4. Though packet losses usually do not trigger retransmissions for the real time applications, they usually result in service degradation for the end users due to the lost data. Therefore for dimensioning process, it is important to perform appropriate overbooking on the backbone link in order to guarantee a certain packet loss ratio or packet delay over the Iub.

The following subsections present the analytical approach for dimensioning the last mile link and the backbone link for the circuit-switched traffic in a star-structured multi-Iub RAN.

8.2.1 Network Dimensioning Approach

In this thesis, it is proposed to apply the multi-dimensional Erlang-B (MD-Erlang) model to dimension the last mile link to meet the desired CAC connection reject ratio QoS; and apply packet-level queuing models to dimension the backbone link in order to fulfill the transport network QoS requirements, i.e. packet delay and packet loss ratio. In Chapter 6.4.3, the MMPP/D/1 model has been presented and demonstrated for dimensioning a single Iub link for the circuit-switched traffic to satisfy a desired packet delay or packet loss QoS. In the multi-Iub scenario, it is suggested by the author to apply the MMPP/D/1 model as well to dimension the backbone link to meet the objective of a desired packet delay or packet loss ratio in the network. It is assumed that there are neglectable packet losses on the last mile links. As it is supposed that the given CAC guaranteed bit rate per connection on the last mile link is optimum, so that each connection gets a sufficient bandwidth to transport and thus protect from packet losses on the last mile links.

The corresponding analytical approach for dimensioning such star-structured RAN (Figure 8.1(a)) is illustrated in Figure 8.1(b), with the application of the MD-Erlang model for dimensioning the individual last mile links and the MMPP/D/1 queuing model to dimension the backbone link. Assuming there are n Node Bs in the radio access network and each Node B transmits the same circuit-switched traffic, the following parameters are defined:

- Let C_{ac_i} be the last mile link capacity for Node B i;
- Let C_{bb} be the capacity of the backbone link;
- Let n be the number of Node Bs in the radio access network;
- Let r_m be the CAC guaranteed bit rate requested by each connection at last mile;
- Let r_{peak} be the peak data rate of each connection
- Let p_b denote the desired connection reject ratio as the QoS target for dimensioning the last mile link;
- Let a_i be the amount of offered user traffic on the last mile link i in Erlang;
- Let y_i be the carried traffic on each last mile link i as a result of CAC in Erlang;
- Let qos_{bb} denote the desired packet loss ratio as the required network QoS for dimensioning the backbone link. As mentioned above, the packet losses are caused by the extremely long transport delays over the link which exceed a maximum allowed delay threshold. Here the packet loss ratio is equal to the ratio of the amount of packets whose delays exceeding a predefined maximum value to the total amount of packets.

As shown in Figure 8.1, the capacity of each last mile link C_{ac_i} is calculated by using the MD-Erlang model. Given a certain last mile link capacity, based on the offered traffic a_i and the requested CAC guaranteed bit rate r_m, the connection reject ratio can be directly calculated from the MD Erlang Model. Then for bandwidth dimensioning, according to the dimensioning procedure described in section 4.3 the link capacity will be adjusted until its resulting connection reject ratio reaches the desired value p_b. At the end, the link capacity which meets the desired target is given to C_{ac_i}.

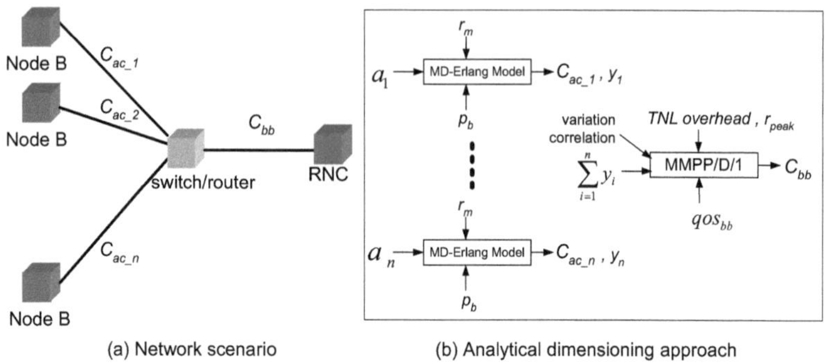

Figure 8.1: *Dimensioning for circuit-switched traffic in a multi-Iub RAN: with MD-Erlang model for the last mile and MMPP/D/1 model for the backbone link*

For the backbone link, the dimensioning is aimed to fulfill the transport network QoS requirements and the MMPP/D/1queuing model is applied to derive the required backbone link bandwidth. The dimensioning for the backbone link takes into account the aggregated traffic on the backbone link, which is the sum of the carried traffic (the traffic admitted by the CAC) from all last mile links. The carried traffic on each last mile link y_i is a result of the CAC. It is calculated as $y_i = a_i \cdot (1 - p_b)$ according to equation (6.7) where the dimensioned last mile link meets the target connection reject ratio p_b. In addition, the transport network layer overhead (*TNL overhead*) and the user peak data rate r_{peak} need to be considered for estimating the total bit rate transmitted on the backbone link. For characterizing the aggregated traffic to be modeled by the MMPP arrival process, the variation and the correlation of the aggregated traffic need to be measured and given as the input parameters of the queuing model as well. With the proper estimation of the aggregated traffic over the backbone link, for a given a link capacity the resultant packet loss ratio can be calculated from the MMPP/D/1 queuing model (see section 6.4.3.2). Then using the dimensioning process described in section 4.3, the optimum backbone link capacity can be determined which satisfies the desired packet loss QoS on the backbone link.

The presented network dimensioning approach is twofold: (1) on each last mile link the desired CAC connection reject ratio can be guaranteed; (2) the backbone link can be properly overbooked by considering possible multiplexing gain from aggregating the traffic from all last mile links; while guaranteeing the strict packet delay and packet loss QoS of the Iub interface to avoid unsatisfied degradation of end user performances.

8.2.2 Applicability of the Dimensioning Approaches

This section presents the results of applying the above proposed analytical approach as shown in Figure 8.1(b) for dimensioning the circuit-switched traffic in a multi-Iub RAN. The bandwidth of the last mile link and the backbone link are calculated separately with the MD-Erlang model and MMPP/D/1 model. The main criterion for the applicability of the proposed dimensioning approaches is that, the required connection rejection ratio needs to be guaranteed on the last mile as well as the multiplexing gain needs to be considered on the backbone link and thus overbooking of the backbone link bandwidth can be achieved. For validation of the proposed analytical approach, simulations were carried out to compare with the theoretic expectations. In the presented example scenario, each last mile link generates the same amount of circuit-switched voice traffic a_i with the same AMR rate r_{peak}, and each connection requests the same CAC guaranteed bit rate r_m at the last mile. The applied voice traffic model is defined in Table 5.2 with AMR voice codec rate of 12.2 kbps. The following parameters and traffic are configured:

- Voice traffic per last mile link: a_i=19.24 Erlang, r_{peak}=12.2 kbps (the AMR codec rate = 12.2kbps), r_m =12.4 kbps (the measured average ATM link throughput for the applied voice traffic model is 12.4 kbps, refer to Figure 5.5)
- Desired CAC connection reject ratio for the last mile link p_b = 1%
- Defined packet loss ratio on the backbone link qos_{bb} = 1 %

For satisfying the desired connection rejection ratio p_b per last mile link, the capacity of each last mile link C_{ac_i} is calculated by using the MD-Erlang model. The calculated last mile link capacity for 1% connection reject ratio is C_{ac_i}= 360 kbps. It is the same as the simulation results for the last mile link. With the 360 kbps last mile link capacity, the resultant last mile link utilization is 60.8%. Figure 8.2 presents the dimensioning results for the backbone link obtained from simulations and calculations. Figure 8.2(a) shows the required backbone link bandwidths over different number of Node Bs or last mile links, which are calculated from the MMPP/D/1 queuing model for the 1% packet loss ratio (qos_{bb} = 1%). And it is compared with the results obtained from the simulations. For a given number of Node Bs each configured with the calculated last mile link capacity of 360 kbps, the simulations were performed for getting an optimum backbone link bandwidth, which need to transmit the total amount of carried traffic from all last mile links with maximum 1% packet loss ratio over the link. Figure 8.2(b) gives the relative error of the analytical results of using the MMPP/D/1 model. As indicated in Figure 8.2, the analytical results derived from the MMPP/D/1 queuing model matches essentially well with the simulation results. Though the calculated results have some underestimation on the required bandwidth due to the fact that the MMPP arrival process cannot give a perfect estimation of the correlation and burstiness of the aggregated traffic, the obtained relative errors of the MMPP/D/1 model are all below 10% as shown in

Figure 8.2(b). Therefore it can be concluded that the MMPP/D/1 model gives relatively accurate dimensioning results and thus it can be used to calculate the backbone link capacity for satisfying a defined packet loss ratio.

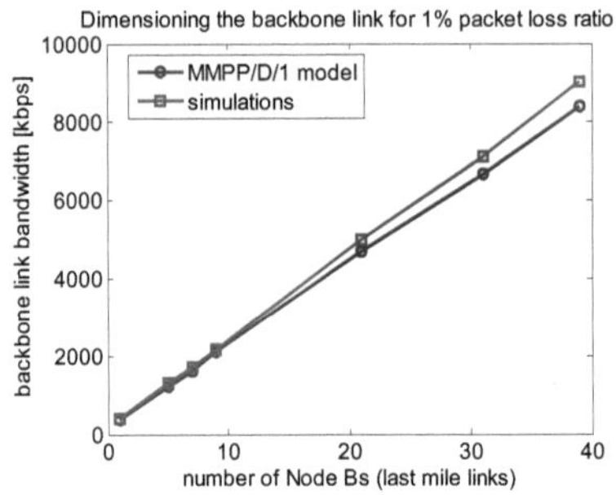

(a) dimensioning results

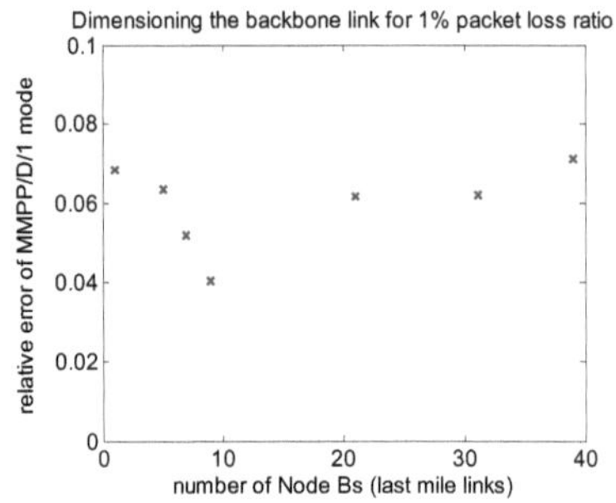

(b) relative error

Figure 8.2: *Dimensioning the backbone link for 1% packet loss ratio*

Moreover the proposed dimensioning approach achieves bandwidth saving over the backbone link, since it takes the multiplexing characteristics of the aggregated traffic into account. This makes the overbooking possible on the backbone link. Figure 8.3 presents the obtained overbooking factor calculated with equation (8.1), given the dimensioned last mile and backbone link bandwidths.

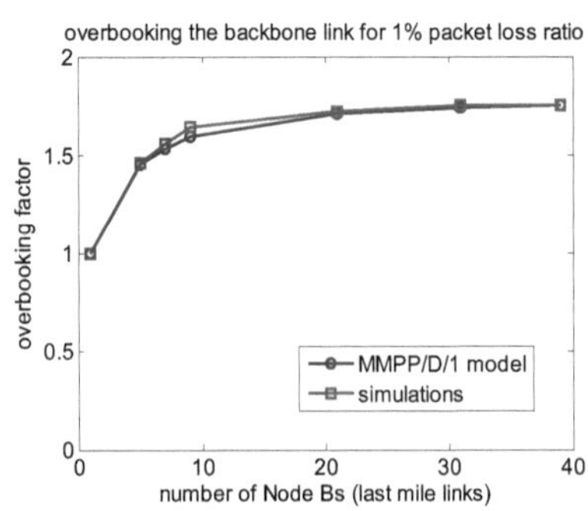

Figure 8.3: *Overbooking the backbone link*

It shows that the obtained overbooking factors are larger than 1 and it increases with the increased number of Node Bs. It can also be seen that the overbooking factor is gradually close to a maximum value when the number of Node Bs continues to increase. This is because that when both the last mile link rate and the offered traffic on each last mile link are fixed, though the aggregated traffic on the backbone link keeps increasing with the increased number of Node Bs, the potential multiplexing gain steadily reaches a maximum value, which is equal to 1/the last mile utilization. In this example, the achieved last mile link utilization is 60.8%, thus the maximum overbooking factor is close to 1/0.68 = 1.64.

8.2.3 Summary

As a summary, the presented results demonstrate that the proposed analytical approach can be applied for dimensioning the circuit-switched traffic in a multi-Iub radio access network. The capacity of the last mile link and the backbone link are calculated separately: the last mile link is dimensioned with the MD-Erlang Model to satisfy a desired CAC connection reject ratio at the last mile; the backbone link is dimensioned with the application of the MMPP/D/1 model for guaranteeing a defined packet loss ratio at the backbone link. In the presented approach, the dimensioning for the backbone link takes the possible multiplexing gain of the aggregated traffic into consideration. Thus the calculated bandwidths from the proposed analytical approaches show overbooking capability on the backbone link.

The suggested dimensioning approaches can be applied for both ATM and IP based transport network, as the effect of the implemented traffic control and resource control functions for the transport of the circuit-switched traffic in both transport networks are quite similar. The main difference lies in the transport overheads due to different transport protocols and underlying transport layers. Therefore, when applying the proposed dimensioning approach, the proper transport overheads corresponding to each transport network need to be taken.

It needs to be noted that other analytical approaches can be applied for dimensioning the multi-Iub radio access network, depending on the QoS requirements of the network operators. For example, if the dimensioning is only focused on the transport network QoS at both last mile link and the backbone link, then each link can be dimensioned with the MMPP/D/1 queuing model individually. However, different approaches will result in different dimensioning results which in turn lead to different network costs. The selection of a most suitable approach is dependent on the QoS requirements of the network operators.

8.3 Dimensioning for Elastic Traffic in ATM-Based UTRAN

As known, processor sharing models are proposed to dimension for the elastic traffic for a guaranteed user application QoS and have been demonstrated for the single Iub link scenario in both ATM and IP RAN. This section presents extended analytical approaches based on processor sharing models for dimensioning an ATM based multi-Iub radio access network transmitting elastic traffic. The next section will present the dimensioning approach for an IP based multi-Iub UTRAN.

As mentioned in section 8.1.1, for setting up an appropriate network dimensioning approach, firstly correct methods need to be developed to estimate the user QoS on an end-to-end basis. The following section 8.3.1 presents the proposed analytical methods for estimating the end-to-end application performance, taking into account the individual impact of last mile link and the backbone link. Based on the proposed methods, several dimensioning approaches are suggested in section 8.3.2 to determine the bandwidths for the last mile link and the backbone link separately.

8.3.1 Estimation of End-to-End Application Performance

The system model of a multi-Iub network is shown in Figure 8.4. The elastic traffic is transmitted through the network between the two ends: the end user (i.e. UE) and the corresponding node in the external networks (e.g. Internet). The required end-to-end QoS requirement is the average throughput D_t, which each flow receives for data transactions, or that an amount of x_t bytes can be transferred within an expected time period T_t. The expected end-to-end delay is mainly influenced by the UMTS radio access network, subject to the assigned *Radio Access Bearer* (RAB) rate and the limited transport capacities over the Iub interface. In the star or tree-structured access network, every link over the transport path can be the bottleneck and cause considerable impact on the end-to-end delay. Thus, in the presented star-structured access network, both the last mile link and the backbone link shall be considered for estimating the end-to-end application delay performance. Moreover, the additional delays within the UMTS core network and the external networks are modeled as an extra delay value added to the TCP *Round Trip Time* (*RTT*).

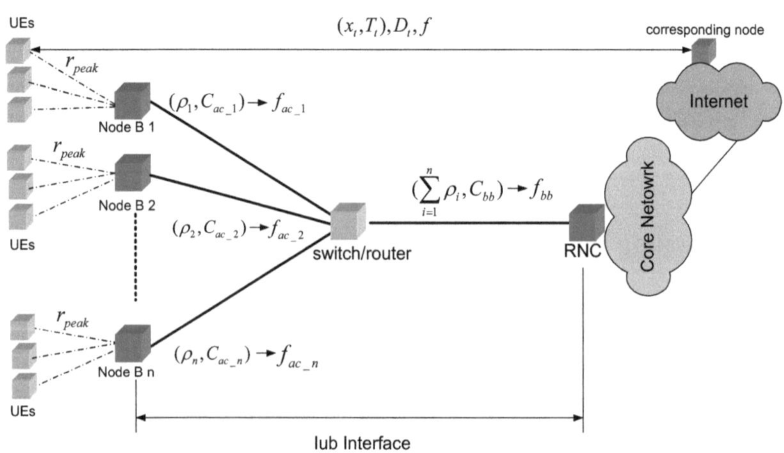

Figure 8.4: *System model for elastic traffic in the multi-Iub RAN*

It is assumed that there are n Node Bs in the radio access network and each Node B serves a number of UEs, as shown in Figure 8.4, and that all flows are given the same RAB rate r_{peak}. Each last mile link has a capacity C_{ac_i} and transmits a certain amount of elastic traffic ρ_i. The backbone link is assigned with link capacity C_{bb}. The total amount of elastic traffic traversing the backbone link is the sum of the traffic from all last mile links. Then the end-to-end transaction time T_t for a file size of x_t is calculated as a function of the requested file size x_t, the assigned RAB rate r_{peak}, the carried amount of traffic over the last mile link and the backbone link as well as their allocated link capacities, and additionally the TCP *RTT* if taking into account the impact of the TCP slow start. The following equation (8.2) gives a general function of calculating the end-to-end delay T_{t_i} for the ith Iub interface:

$$T_{t\_i} = F(x_t, r_{peak}, \rho_i, C_{ac\_i}, \sum_{i=1}^{n} \rho_i, C_{bb}, RTT) \tag{8.2}$$

The required end-to-end delay QoS requirement can be represented as well by the end-to-end delay factor f. As introduced in Chapter 6.3.5, the delay factor represents the increase of the average transfer time (or decrease of the average throughput) imposed by the congestion (see equation (6.20)). For each last mile link and the backbone link, an individual link delay factor can be calculated. The individual link delay factor indicates the additional delay caused by the congestion of the individual link. f_{ac_i} denotes the link delay factor of the last mile link of the ith Iub interface and f_{bb} denotes the link delay factor of the backbone link.

To estimate the end-to-end delay T_t, three basic network cases are considered:

(1) The last mile link is configured with a limited capacity that results in f_{ac_i}>1, while the backbone link is configured with adequate bandwidth that will not cause any traffic congestion, i.e. f_{bb}=1;

(2) The last mile link has a sufficient bandwidth that makes f_{ac_i}=1, while the backbone link is set to a limited bandwidth that cause congestion on the backbone link which leads to f_{bb}>1;

(3) Both last mile link and the backbone link are configured with limited link bandwidths, where congestion happens on both links, i.e. f_{ac_i}>1 and f_{bb}>1.

For the above case (1) and (2), only one of the links has congestion which will influence the end-to-end delay T_t, whereas the other one has a delay factor equaling to 1 and therefore does not contribute additional delays. Therefore, in these two cases, the end-to-end delay is influenced only by the congested link. Thus for estimating the end-to-end delay, the Iub transport network can be seen as a single congested link and the M/G/R-PS model can be applied on the congested link (either last mile or backbone link) to calculate the expected sojourn time. In this way, the delay estimation method in these two cases is the same as the methods used for the single link case (see Chapter 6.3). However, for case (3) the end-to-end delay T_t will be influenced by both last mile and backbone links as both links contribute additional delays to the end-to-end delay performance. Therefore, for estimating T_t the impact of the two concatenated links need to be considered at the same time, and thus the methods for a single Iub case is not suitable anymore. In the following two methods are proposed to calculate the end-to-end delay factor f for case (3).

- Method 1: use the maximum delay factor as the end-to-end delay factor, assuming that the end-to-end delay is only determined by the main congested link (i.e. the bottleneck link) but ignoring the impact of the other one.

$$f = \max.(f_{ac\_i}, f_{bb}) \tag{8.3}$$

In equation (8.3), the delay factor of the last mile link f_{ac_i} is calculated with the basic M/G/R-PS model (with equation (6.20)) given the last mile link capacity C_{ac_i} and the carried traffic amount on the last mile link ρ_i. The delay factor of the backbone link f_{bb} is also calculated with the basic M/G/R-PS model given the backbone link capacity C_{bb} and the aggregated traffic $\sum_{i=1}^{n} \rho_i$. From equation (8.3), the main congested link is determined by having a larger delay factor and this delay factor will be chosen for the end-to-end delay factor f. This method assumes that the average end-to-end delay T_t is only determined by the main congested link, therefore in this method the Iub transport network can be modeled as a single congested link. Thus T_t can be calculated by applying the suggested extended M/G/R-PS models (proposed for single ATM Iub link in Chapter 6.3.7) on the main congested link.

- Method 2: let the end-to-end delay consider the total additional delays caused by the congestion of both links. The end-to-end delay T_{t_i} for the ith Iub interface is calculated with equation (8.4).

$$T_{t\_i} = T_{\min} + (T_{ac\_i} - T_{\min}) + (T_{bb} - T_{\min}) = (T_{ac\_i} + T_{bb}) - T_{\min} \tag{8.4}$$

Here T_{ac_i} denotes the calculated delay assuming that only the last mile link is congested but the backbone link has sufficient bandwidth, which is the same as case (1). T_{bb} represents the calculated delay assuming that only the backbone link is congested while the last mile link is not, same as case (2). The calculations for T_{ac_i} and T_{bb} use the same analytical approach suggested for the case (1) and (2). $T_{\min}$ is the delay when both links are ideal (without congestions in the transport network). It will be equal to the delay of transmitting the file x_t with the peak data rate r_{peak}. Thus, the additional delay caused by the last mile link is calculated as $(T_{ac\_i} - T_{\min})$ and the additional delay caused by the backbone link is $(T_{bb} - T_{\min})$. With equation (8.4), the individual impact of each link on the end-to-end delay is taken into account.

In the following, for each of the three basic network cases, the proposed analytical methods are validated with simulations. In the presented examples, the simulation scenarios are configured as follows:

- all users are assigned to the same RAB rate $r_{peak} = 128$ kbps
- web traffic: page size = constant (50kbyte)
- there is no restrictions on the number of elastic traffic flows

Figure 8.5 presents the delay estimation for the case (1) where the last mile link is congested and the backbone link is ideal (i.e. with sufficient bandwidth). In the example, the backbone link is set to 10 Mbps and the last mile link is 1 E1 (1600kbps used for the user plane). There are 2 Node Bs configured in the radio access network. In Figure 8.5, the application delay is calculated with the M/G/R-PS model that models the last mile link (with equation (6.26) and (6.27) presented in section 6.3.7.1) and compared with the measured average page transfer delay from the simulations. In the figure, the last mile link utilization is the x-axis. The left graph shows the average application delay and the right graph shows the calculated relative error of the analytical models. It can be seen from Figure 8.5 that for different load levels the average end-to-end application delays calculated from the M/G/R-PS model are quite similar to the simulation results where the relative error is below 5%.

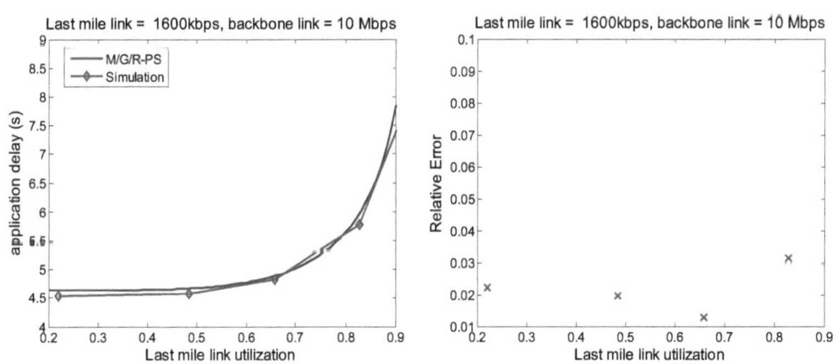

Figure 8.5: *Delay estimations for case (1): congested last mile link and ideal backbone link*

Similar validation results can be obtained for case (2). Figure 8.6 shows the estimated delays from both analytical model and the simulations. In this example, there are 3 Node Bs each connected to a 3.4 Mbps last mile link (which is sufficient for the offered traffic amount on the last mile link that will not cause congestion) and the backbone link is configured with 4 Mbps. In Figure 8.6, the backbone link utilization is the x-axis. The left graph shows the average application delay and the right graph shows the relative error. The presented results validate the applicability of the proposed method for estimating the end-to-end application delay for the case (2).

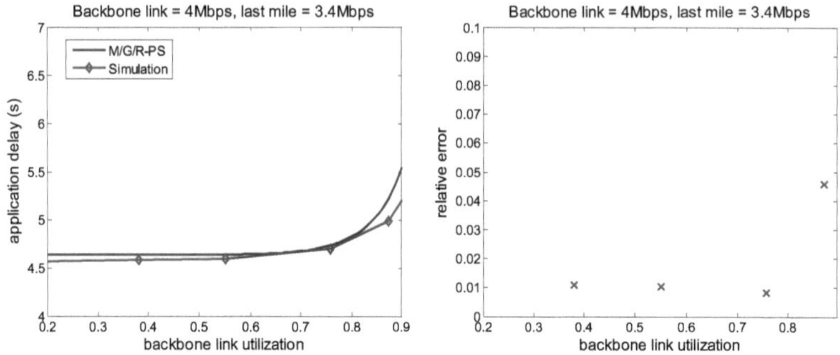

Figure 8.6: *Delay estimations for case (2): ideal last mile link and congested backbone link*

The above two examples demonstrate that when only one of the link is congested in the Iub transport network the corresponding Iub interface can be modeled as a single congested link. Thus the average end-to-end application delay can be estimated by applying the M/G/R-PS model for that congested link.

Case (3) is a more common scenario, as often in the access network both last mile link and backbone link are configured with limited link bandwidths so that the congestion can happen on both links. Figure 8.7 validates the two proposed methods to calculate the end-to-end delay for the case (3). In the given example, there are 4 Node Bs each connected to an E1 link (1600kbps used for the user plane). The backbone link is configured with 4.8 Mbps.

As shown in Figure 8.7, both proposed methods can estimate the end-to-end application delay with below 10% relative error (shown in the right graph). But by comparing the relative error of each method, it is found that the second method given by equation (8.4) offers a better estimation for the end-to-end delay. The first method, which takes the maximum delay factor as the end-to-end delay factor, leaves out the additional delay caused by the other less congested link. Therefore, it will result in an underestimation of the end-to-end delay as shown in the left graph in Figure 8.7. Hence, the second method given by equation (8.4) is recommended to estimate the end-to-end application performance as it gives a better result. This method has been validated in different simulation scenarios and demonstrated that it has the best estimation accuracy. Thus in the following the second method is applied to perform network dimensioning in a multi-Iub RAN. Moreover, it shall be noticed that the second method can also be applied for network case (1) and (2) where the end-to-end delay factor f just equals to the delay factor of the congested link.

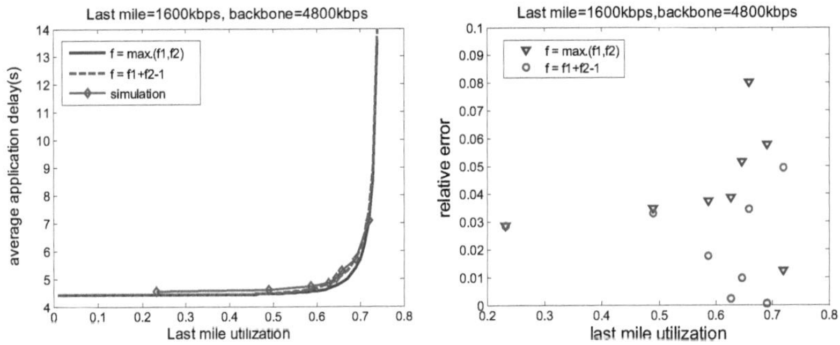

Figure 8.7: *Delay estimations for case (3): congested backbone and last mile links*

8.3.2 Dimensioning Approaches

Based on the above proposed application performance estimation methods, this section is going to propose several dimensioning approaches for a star-structured multi-Iub RAN to determine the suitable bandwidth for the last mile link and the backbone link separately. The objective of dimensioning is to guarantee a certain average transfer time T_t for a file of size x_t or average throughput D_t on an end-to-end basis with the minimum network costs. Equivalently, dimensioning can be driven by satisfying an end-to-end delay factor f which results in the desired average transfer time T_t ($T_t = T_{\min} \cdot f$) or the average throughput D_t ($D_t = r_{peak} / f$) for all file transactions. In this context, the task of dimensioning of a star-structured multi-Iub RAN is to distribute the total allowed congestion (represented by the end-to-end delay factor f) on the individual links over the transport path, i.e. to assign proper target delay factors for the last mile link and the backbone link individually to achieve the desired end-to-end delay factor f. The assignment of a proper target delay factor for each link needs to consider the impact of each link on the end-to-end application QoS and its resultant transport costs.

Let a target end-to-end delay factor $f_{t\arg et}$ represent the required application QoS level for a specific service class on all Iub interfaces. Depending on the value of $f_{t\arg et}$, three basic approaches are proposed in this thesis to assign the separate target delay factor for the last mile link and the backbone link individually:

(1) If the desired end-to-end delay factor $f_{t\arg et}$ is close to 1, which is a practical case for network dimensioning, each link can be dimensioned according to $f_{t\arg et}$. That means, the assigned delay factor for the last mile link and the backbone are set to the same delay factor equal to $f_{t\arg et}$, i.e. $f_{ac\_i} = f_{bb} = f_{t\arg et}$. As the desired delay factor is close to 1, the expected congestion on each link is not high. Therefore the impact of each link

on the end-to-end delay is not especially significant. This allows dimensioning every link in a way that it is the bottleneck while neglecting the impact of the other link. Noticing that throughput reductions induced by the bottleneck cannot be compensated by higher bandwidth values at other links. So optimum network dimensioning is achieved by assigning both link capacities according to the end-to-end delay factor f_{target}.

(2) If the target end-to-end delay factor f_{target} is significantly larger than 1 (e.g. f_{target}>1.3), one possible approach is to dimension each last mile link according to the end-to-end delay factor f_{target} ($f_{ac\_i} = f_{target}$) while setting the target delay factor for the backbone link to a value which is close to 1 (e.g. 1.1). In this way, each last mile link acts as the bottleneck that determines the achievable throughput of an individual flow while the backbone link is allocated to a relative sufficient bandwidth that has considerably lower impact on the end-to-end delay. Thus, the end-to-end delay factor f mainly corresponds to the one of the last mile link, which is the maximum over all delay factors along the way. With this approach, the transport costs on the last mile will be minimized.

(3) Another approach for the case of f_{target} significantly larger than 1 is to assign the same delay factor for the last mile links and the backbone link, i.e. $f_{ac\_i}= f_{bb}$. As the required delay factor is much larger than 1, the impact of each link on the end-to-end delay cannot be neglected as in approach (1). Thus the dependences of the end-to-end delay with both links have to be considered. Their relation has been expressed by equation (8.4). This equation can be converted to equation (8.5) with the use of the delay factor of the ith last mile link f_{ac_i} and the backbone link f_{bb}.

$$T_{t\_i} = (T_{ac\_i} + T_{bb}) - T_{\min} = (T_{\min} \cdot f_{ac\_i} + T_{\min} \cdot f_{bb}) - T_{\min} = T_{\min} \cdot (f_{ac\_i} + f_{bb} - 1) \quad (8.5)$$

With the given end-to-end delay factor f_{target} and the condition of $f_{ac\_i}= f_{bb}$, the assigned target delay factor for the ith last mile link and the backbone link can be calculated with equation (8.6).

$$\begin{aligned} &T_{t\_i} = T_{\min} \cdot f_{target} = T_{\min} \cdot (f_{ac\_i} + f_{bb} - 1) \\ &thus,\ f_{target} = f_{ac\_i} + f_{bb} - 1 \\ &if\ (f_{ac\_i} = f_{bb}),\ then\ f_{ac\_i} = f_{bb} = \frac{(f_{target} + 1)}{2} \end{aligned} \quad (8.6)$$

Approach (1) is only suitable for the case with small end-to-end delay factors, i.e. the application QoS requirement is relative high. Approach (2) and (3) can be used for the case with large target delay factors. If the network operators try to minimize the transport costs spent on the last mile links, approach (2) is suggested. As in a real network usually the number of last mile links is much larger than the number of backbone links, so if every last mile link is assigned a minimum bandwidth the overall network costs can be low. The main drawback of the approach (2) is that there will be low overbooking gain over the backbone link, since the last mile links will be highly utilized and the backbone link has a lower target delay factor than the last mile links.

If there are no specific requirements on the transport costs for the last mile and the backbone links, approach (3) can be used as a general dimensioning solution which assigns the same delay factor to both last mile and backbone links. Approach (3) is also valid for the case of a small end-to-end delay factor. When f_{target} is close to 1, the calculated delay factor for the last mile link and the backbone link from the approach (3) will be similar to the ones obtained from the approach (1).

Given a properly assigned delay factor for each link using the above suggested approaches, the minimum bandwidths of the individual links can be calculated with the same processor sharing model as used in the single link scenarios. Thus the developed analytical dimensioning approach for a star-structured multi-Iub RAN is illustrated in Figure 8.8. In this approach, the last mile link and the backbone link are dimensioned individually with a processor sharing model to meet the individual assigned target delay factors f_{ac_i} (for i^{th} last mile link) and f_{bb} (for the backbone link). For each link, given the amount of the offered traffic for that link and the assigned target delay factor, the required minimum link capacity can be calculated with the proposed M/G/R-PS models applied in the single Iub link scenarios in Chapter 6.3. For different UMTS network and traffic configurations, various extensions and adjustments to the basic M/G/R-PS model have been proposed in section 6.3.7 for the ATM-based Iub interface. These proposed extensions can be applied as well in the multi-Iub network scenarios for estimating the individual link capacities, given certain network and user configurations. For instance, when the UTRAN transport network does not apply CAC for the elastic traffic and each user is assigned to the same RAB rate which is fixed throughout the data transmission, then the required minimum capacity of the individual links can be obtained by resolving equation (6.26) and (6.27) iteratively to get the desired bandwidth. In case that multiple RABs are used in parallel, the required link bandwidth can be estimated based on the delay estimation model proposed in section 6.3.7.3. In case of a mixed traffic scenario (elastic traffic mixed with circuit-switched traffic), the solutions suggested in section 6.3.7.5 can be applied to calculate required bandwidth for each individual link in the multi-Iub RAN network.

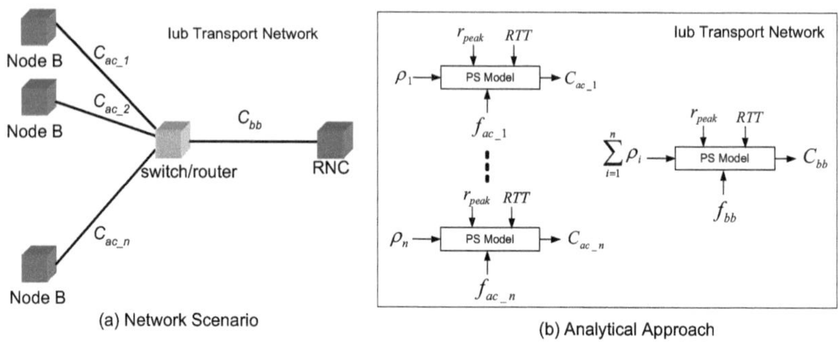

Figure 8.8: *Dimensioning for elastic traffic in a multi-Iub ATM RAN: with processor sharing model for both last mile and the backbone link*

As additional remarks, the above introduced three approaches to assign target delay factors for the individual links can be applied as well in a general multi-Iub RAN, where per Iub there are multiple concatenated links along the transport path to connect the RNC and the Node B. When the desired end-to-end delay factor f_{target} is close to 1, approach (1) can be used so that each link along the path is assigned to the same delay factor equal to f_{target}. For the case of the desired end-to-end delay factor f_{target} significantly larger than 1, if the network operators try to minimize the transport costs on the last mile links then approach (2) can be applied to assign f_{target} to the last mile link whereas set the target delay factors for all other links along the way to a value a little higher than 1. Approach (3) can be used for both cases. Given K concatenated links along the transport path, approach (3) will assign the same delay factor to every link along the path, which is equal to $\frac{(f_{target}+(K-1))}{K}$.

8.3.3 Dimensioning Results

This section presents the dimensioning results of applying dimensioning approach (3) proposed in the last section and validates them with simulations (the dimensioning results for approach (1) and (2) are given in Appendix A.18). Approach (3) is considered as a general approach to be applied for dimensioning regardless of the degree of the desired delay factor. In the following presented examples, the simulation scenarios are configured as follows:

- all users in the network are assigned to the same RAB rate r_{peak} = 128 kbps
- web traffic model: page size is 50 kbyte of a constant distribution
- there is no restrictions on the number of elastic traffic flows
- there are 4 Node Bs (last mile links) in the RAN, each generating the same amount of the offered traffic

In the presented examples, the target end-to-end delay factor f_{target} changes from 1.1 to 1.5. With approach (3), the assigned target delay factors for the last mile and the backbone link are equal to $(f_{target}+1)/2$ according to the formula (8.6). The offered traffic amount on the last mile is around 1700 kbps. The validation of the applicability of approach (3) is by comparing with simulations. For each link, the bandwidth will be set and adapted until the assigned target delay factor for that link is reached in simulations. In addition to the simulations, the theoretical bandwidths for each individual link for the assigned target delay factor are calculated with the proposed extended M/G/R-PS model and they are compared with the simulated bandwidths. Figure 8.9 presents the required last mile link capacities obtained from both simulations and theoretical calculations using the proposed extended M/G/R-PS model. Figure 8.10 presents the dimensioning results for the backbone link.

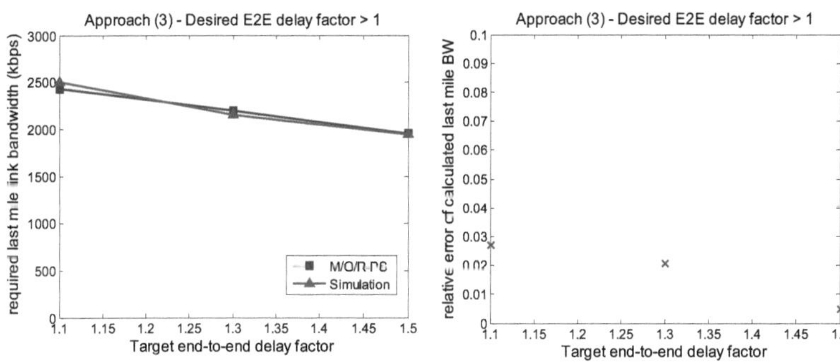

(a) last mile link bandwidth (b) relative error

Figure 8.9: *Validating approach (3): dimensioned last mile link bandwidths over different target delay factors*

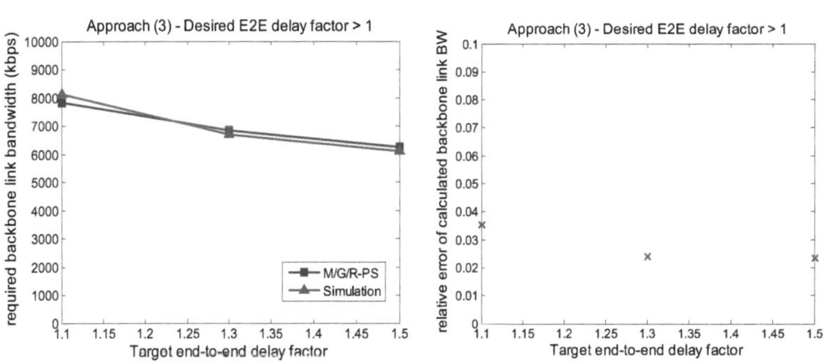

(a) backbone link bandwidth (b) relative error

Figure 8.10: *Validating approach (3): dimensioned backbone link bandwidths over different target delay factors*

In both figures (a) shows the required last mile link capacities over different target end-to-end delay factors as the QoS criteria, while (b) shows the relative error of the calculated bandwidths. It can be seen in both figures that for the last mile link and the backbone link, the bandwidths given from simulations comply with the theoretical bandwidths derived from the proposed extended M/G/R-PS model. This demonstrates that the proposed M/G/R-PS approach can provide sufficiently exact predictions for the required bandwidth for the elastic traffic flows.

The graph in Figure 8.11 (a) shows the average end-to-end delays T_t for different target end-to-end delay factors as the QoS criteria. The curve represents the mean end-to-end transfer delays measured from simulations for the corresponding target end-to-end delay factors, and the expected end-to-end delays (target delays). In addition, graph (b) evaluates the deviation of the simulated delays and the expected end-to-end delay. The results prove that approach (3) is capable of providing QoS guarantee for different ranges of delay factors and thus can be applied generally for dimensioning networks for elastic traffic. Furthermore, from Table 8.1 which gives the dimensioned bandwidths out of simulations and the resultant overbooking factors, it can be seen that approach (3) additionally achieves an efficient overbooking on the backbone link.

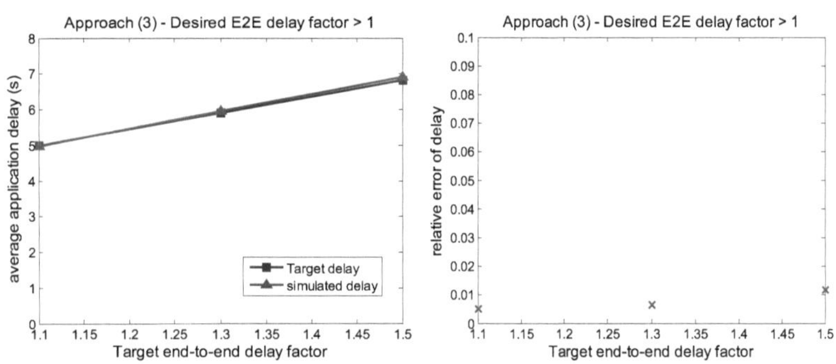

(a) End-to-end delays (b) relative error

Figure 8.11: *Validating Approach (3):E2E delays over different target delay factors*

Target end-to-end delay factor f_{target}	Last mile link bandwidth (kbps)	Backbone link bandwidth (kbps)	Overbooking factor for backbone link
1.1	2500	8125	1.23
1.3	2150	6700	1.28
1.5	1950	6100	1.28

Table 8.1: *Dimensioning results for different target delay factors with approach (3)*

8.3.4 Summary

In section 8.3, the dimensioning for elastic traffic in an ATM based multi-Iub radio access network was presented. The analytical methods, which are based on the M/G/R-PS model, are developed to estimate the application performance on an end-to-end basis, taking into account of the individual impact of the last mile link and the backbone link. Based on that, three dimensioning approaches are proposed to determine the necessary bandwidths for the last mile link and the backbone link for a desired end-to-end delay factor as the QoS requirement. The applicability of the proposed three approaches for network dimensioning is demonstrated through simulations. Overall, the simulation results comply with the theoretical expectations sufficiently well. That means, the suggested three dimensioning approaches can work quite well for assigning appropriate bandwidths for the individual links in the network for satisfying a given delay factor. Moreover, the simulation results again verify that the applied M/G/R-PS model in the dimensioning approaches can predict the achievable application QoS accurately and thus can be used for dimensioning links and networks for elastic traffic for a guaranteed user relevant application QoS.

The selection of a proper dimensioning approach for mobile network operators depends on the required application QoS level and the overall network costs. The suggested approach (1) is only suitable for the case with small end-to-end delay factors. Approach (2) can be used for the case with large target delay factors, but it has low overbooking gain over the backbone link. Approach (3) is proven to be a general dimensioning solution, which can work sufficiently well for different ranges of target delay factors and is also capable of attaining overbooking on the backbone link. Therefore considering its wide applicability, good accuracy, and ability of overbooking for optimizing network cost-efficiency, approach (3) is recommended in this thesis to be applied in general for dimensioning networks.

8.4 Dimensioning for Elastic Traffic in IP-Based UTRAN

This section presents analytical approaches for dimensioning an IP based multi-Iub UTRAN transport network for elastic traffic. In the considered IP Iub transport network, a DiffServ-based QoS architecture is deployed for QoS support and traffic management. The fundamental idea of the proposed dimensioning approaches are the same as the ones developed for the ATM based UTRAN presented in section 8.3.2, but the applied processor sharing models are based on the ones extended for a DiffServ-based IP RAN which are presented in section 7.3 for the single Iub link dimensioning. The following section 8.4.1 describes the detailed analytical approaches and the complete procedures for dimensioning the IP based multi-Iub UTRAN. The corresponding validation results are presented in section 8.4.2.

8.4.1 Dimensioning Approaches

Figure 8.12 illustrates a star-structured multi-Iub IP RAN, where a DiffServ QoS scheme is applied to classify different service classes and guarantee QoS requirements of the individual service classes. To deploy the DiffServ QoS architecture throughout the IP RAN, every Node B and the RNC as well as the connecting IP routers need to provide the DiffServ QoS function for both uplink and downlink transmissions. As can be seen from Figure 8.12, the DiffServ QoS function is performed on each last mile link and the backbone link individually.

Assuming that each Node B serves a group of elastic traffic flows of the UEs, on the last mile link the applied DiffServ QoS scheme firstly maps the multiple elastic traffic flows to a number of service classes, i.e. *Per Hop Behaviors* (PHBs), and then forwards them to the corresponding buffers before sending them down to the link. The applied scheduler decides at what time which buffer can be served, according to the given priority of each PHB. As a result, the last mile link carries several service classes, each being allocated to a share of the complete IP bandwidth that is assigned to the last mile link according to its defined priority.

When the traffic flows of the different Node Bs are aggregated on the backbone link, the total number of the elastic traffic flows carried by the backbone link will be the sum of the number of flows of each Node B, and the total amount of elastic traffic traversing on the backbone link will be the sum of the amount of traffic from all last mile links. It is assumed that throughout the IP RAN, the provided DiffServ PHBs and the related DiffServ QoS functions such as mapping, marking, dropping, and scheduling are configured the same at every DiffServ node (i.e. at the Node B, the RNC, and the IP routers). Thus, the backbone link allows the same types of service classes (PHBs) and experiences the same DiffServ functions as each last mile link. In the context of this thesis, the applied scheduler at each last mile link and the backbone link is a combination of *Weighted Fair Queue* (WFQ) and *Strict Priority* (SP) scheduling. The corresponding QoS structure is presented in section 3.4.2.3. At the end, the backbone link transmits the same set of service classes (PHBs) as each last mile link and each of the service classes will be allocated a share of the entire IP bandwidth that is assigned to the backbone link according to its priority level and WFQ weight.

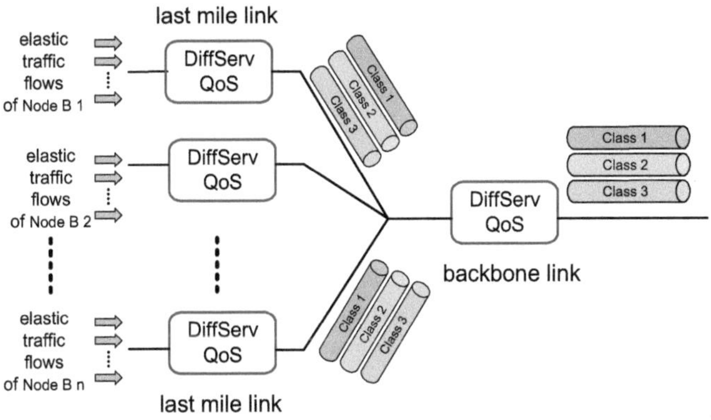

Figure 8.12: *A star topology multi-Iub IP RAN using DiffServ QoS for IP transport*

The analytical models for dimensioning a single IP-based Iub link for elastic traffic have been presented in Chapter 7.3, which considers the impact of the DiffServ QoS function applied at the IP transport layer. As seen in section 7.3.1, for elastic traffic the analytical dimensioning models are extensions of the **M/G/R-PS model** in case that there is no restriction on the number of concurrently transmitted elastic traffic flows, and the **M/G/R/N-PS model** in case that a maximum of **N** (N≥R) elastic traffic flows are transported simultaneously. Section 7.3.3 explains the detailed procedures for dimensioning a single IP-based Iub link with the proposed analytical models from section 7.3.1. In this proposed dimensioning approach, it is assumed that *Assured Forwarding* (AF) and *Best Effort* (BE) PHBs carry elastic traffic and each PHB (AF or BE) serves one *Non Real Time* (NRT) RAB service. The task of dimensioning is to derive a minimum required IP bandwidth that guarantees the QoS requirement of each DiffServ service class.

To apply the previously developed analytical models used for a single link to dimension the star-structured multi-Iub IP RAN, first of all the QoS requirements for the individual links through the Iub transport network need to be defined. It is proposed to use the same ideas of dimensioning the ATM based multi-Iub radio access network which are presented in section 8.3.2. In section 8.3.2, three basic approaches are proposed to allocate the separate target delay factors for the last mile link and the backbone link individually, according to the requested application QoS level and the required network costs. All of these three proposed approaches can also be applied for dimensioning a star-structured IP RAN. In this thesis approach (3) is presented for dimensioning the multi-Iub IP RAN, due to its wide applicability, fine accuracy, and capable of overbooking as validated in section 8.3 for the ATM based multi-Iub RAN.

For a certain class of elastic traffic flows that is carried by one of the AF or BE PHBs, the target end-to-end QoS requirement for this service class is an expected time period *for* transmitting a certain amount of bytes, or simply a value for the target end-to-end delay factor f_{target}. It is assumed that the QoS requirement of this service class is applied throughout the network. According to the approach (3) in section 8.3.2, the assigned target delay factor for each last mile link f_{ac} and for the backbone link f_{bb} will be calculated with equation (8.6).

After allocating a proper target delay factor for the last mile link and the backbone link for each service class, the dimensioning of the IP RAN transport network (which is carrying several service classes) is performed by calculating the required minimum IP bandwidth for each link individually, which needs to guarantee the target delay factor of every service class on that link. The calculation for the required minimum IP bandwidth of each link in this case can be according to the proposed the dimensioning approach for a single link presented in section 7.3.1 and section 7.3.3. The resultant dimensioning approach is illustrated in Figure 8.13.

In Figure 8.13, it is assumed that there are n Node Bs in the radio access network and each Node B servers a group of elastic traffic flows of the UEs. The IP transport network defines $\boldsymbol{K}$ service classes (PHBs) for the DiffServ QoS, which is applied at every last mile link as well as the backbone link. For the service class k (carried by the kth PHB), the following parameters are defined:

- Let r_{peak_k} denote the peak data rate of the RAB service for the service class k;
- Let w_k denote the WFQ weight of the service class k on the kth PHB queue;
- Let x_k denote the average file size of the service class k;
- Let f_{target_k} denote the target end-to-end delay factor for the service class k;
- Let f_{ac_k} be the target delay factor of the *last mile link* for the service class k;
- Let f_{bb_k} be the target delay factor of the *backbone link* for the service class k.

The settings of the above parameters for the service class k are used for all Iub interfaces in the multi-Iub RAN. In addition, $\rho_k(i)$ denotes the mean offered traffic of the service class k carried on the ith Iub interface corresponding to the Node B i, and $N_k(i)$ defines the maximum allowed number of flows for the service class k on the ith Iub interface.

As can be seen from Figure 8.13, the dimensioning for each last mile link and the backbone link is carried out separately with the proposed processor sharing model that is extended for DiffServ for a single link presented in section 7.3.1. For the dimensioning process, the above defined parameters for each service class need to be given as the input parameters. According to the dimensioning procedure described in section 7.3.3, for the individual link, for each service class a required minimum IP bandwidth, which can satisfy the assigned target delay factor of that service class on that link, will be derived numerically with the proposed analytical model presented in section 7.3.1. Then after calculating the required IP bandwidth per service class, the maximum one will be taken to be the dimensioned bandwidth for that link, as it will satisfy the assigned target delay factor of every service class on that link. If there are additional IP bandwidths reserved for *Real Time* (RT) traffic or signaling traffic, then these extra reserved bandwidths need to be added as well for the dimensioning of the total link capacity. At the end, the calculated dimensioned capacity for the ith last mile link is denoted as C_{ac_i} and the one for the backbone link is represented by C_{bb}.

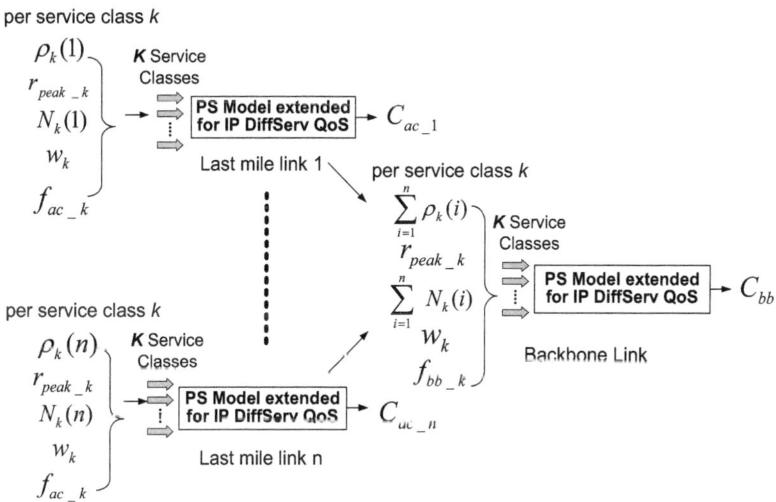

Figure 8.13: *Dimensioning for elastic traffic in a multi-Iub IP RAN with* ***K*** *service classes using DiffServ QoS Structure*

8.4.2 Dimensioning Results

This section presents the dimensioning results, which are derived from the proposed analytical dimensioning approach presented in the last section 8.4.1. And they are validated by simulations. The applied simulation model is the simplified IP-based UTRAN model (see section 5.5.2). In the following presented dimensioning examples, the simulation scenarios consist of five NRT service classes each carried by one AF or BE PHB. Table 8.2 characterizes these five NRT service classes and their mapping to the DiffServ PHBs. The configurations of *Weighted Random Early Detection* (WRED) function and WFQ parameters for each AF and BE PHB are given in Table 8.3. In addition, the following configurations are applied in the simulations.

- there are 4 Node Bs in the RAN, structured as a star topology network architecture.
- each Node B transports the five NRT service classes concurrently using the defined DiffServ QoS functions. Each Node B has the same traffic demand, and the total amount of traffic transmitted on each last mile link is around 1.25 Mbps in average.
- all NRT service classes use the same web traffic model: page size is 50 kbyte of a constant distribution.
- each Node B carries 8 elastic traffic flows per service class.

The QoS objectives of dimensioning are defined in Table 8.4. For each service class the required application end-to-end QoS is specified in terms of a desired end-to-end delay factor. In the investigated examples, three scenarios are simulated each with a different setting of QoS requirements for the five service classes.

Service Class	DiffServ PHB
Interactive – NRT RAB 64 kbps	AF11
Interactive – NRT RAB 128 kbps	AF21
Interactive – NRT RAB 256 kbps	AF31
Interactive – NRT RAB 384 kbps	AF41
Background – HSPA (2 Mbps)	BE

Table 8.2: *Service classes and their mapped DiffServ PHBs*

	WRED Parameters				WFQ
PHB	min_{th}	max_{th}	max_p	exp.w.	weight
AF11	4	10	10 %	9	20
AF21	5	10	10 %	9	30
AF31	6	10	10 %	9	40
AF41	7	10	10 %	9	50
BE	3	10	10 %	9	10

Table 8.3: *WRED and WFQ Parameters*

Figure 8.14 presents the resultant dimensioning results for the last mile link. Every point in the curves represents one scenario in Table 8.4 with a different setting of QoS requirements for dimensioning. For each scenario, in order to dimension the last mile link and the backbone link individually, the target delay factor of each link needs to be determined. Based on the proposed approach (3) for QoS allocations presented in section 8.3.2, given the desired end-to-end delay factor per service class defined in the Table 8.4, the individual target delay factor of the last mile link and the backbone link for each service class can be derived from equation (8.6).

Service Class	Target end-to-end delay factor as QoS requirement per service class		
	Scenario 1	Scenario 2	Scenario 3
AF11- RAB 64 kbps	1.4	1.4	1.4
AF21- RAB 128 kbps	1.3	1.3	1.3
AF31- RAB 256 kbps	1.2	1.2	1.2
AF41- RAB 384 kbps	1.1	1.2	1.3
BE - HSPA	3	5	7

Table 8.4: *QoS requirements per service class for IP RAN dimensioning*

In Figure 8.14, the graph (a) presents the dimensioned last mile link capacities for different QoS requirements derived from the analytical calculations and in addition the ones obtained from simulations. The analytical solutions are achieved according to the proposed analytical dimensioning approach presented in the preceding section (see Figure 8.13). With the calculated per link target delay factors of the last mile link, the dimensioning of each last mile link will be derived with the proposed processor sharing model that is extended for DiffServ for a single link presented in section 7.3.1. The M/G/R/N-PS model is used in this case as the analytical model for dimensioning. The final derived last mile link bandwidth is the bandwidth satisfying the target delay factor of every service class on that link. To validate the analytical results, simulations are performed. In the simulations, for each scenario the bandwidth of the last mile link will be adapted independently from the backbone link (i.e. configuring a sufficient bandwidth for the backbone link so that f_{bb}=1) until the required target delay factor of each service class on the last mile link can be met. Figure 8.14(b) evaluates the deviations of the calculated last mile link bandwidths with the simulated values. It can be seen clearly that for the last mile link the analytical dimensioning results based on the M/G/R/N-PS model fit quite well with the simulation results. The observed gap in the graph (b) is fairly small, with less than 10% relative error.

Similar to the dimensioning of the last mile link, the M/G/R/N-PS model is also applied for dimensioning the backbone link, only that the maximum number of the elastic traffic flows on the backbone link is the sum of the maximum number of flows of each Node B, and the total amount of elastic traffic per service class traversing on the backbone link is the sum of the amount of traffic of that service class of each Node B (see Figure 8.13). For each scenario, a required minimum backbone link bandwidth is calculated, which satisfies the assigned target delay factor of each service class assigned to the backbone link. The analytically derived backbone link bandwidths are presented in Figure 8.15(a). To validate them, the dimensioning of the backbone link is also performed by simulations. In the simulations, for each scenario, the last mile links are configured with the dimensioned bandwidth obtained from the previous simulations (see Figure 8.14(a)), the bandwidth of the backbone link will be adjusted until the desired end-to-end delay factor of each service class is met. It can be seen in Figure 8.15(a) that the analytically derived backbone link bandwidths based on the M/G/R/N-PS model match well with the simulation results. Figure 8.15(b) shows the corresponding relative errors of the analytical results.

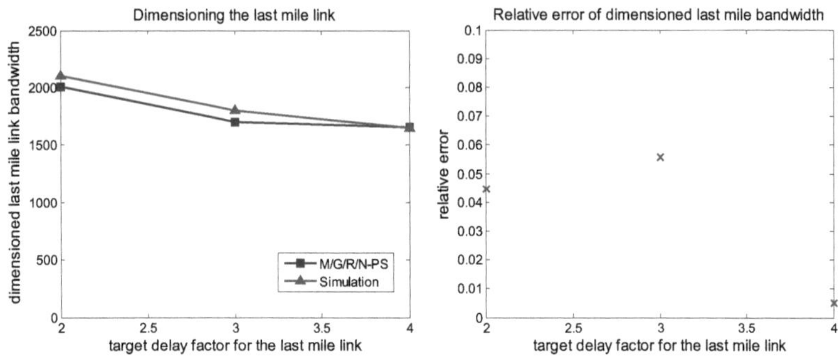

(a) last mile link bandwidth (b) relative error

Figure 8.14: *Dimensioned last mile link bandwidths over different target delay factors in the star-structured IP-based UTRAN with DiffServ QoS scheme*

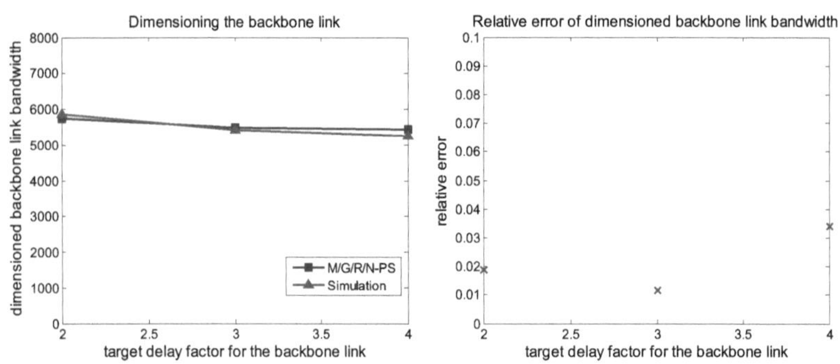

(a) backbone link bandwidth (b) relative error

Figure 8.15: *Dimensioned backbone link bandwidths over different target delay factors in the star-structured IP-based UTRAN with DiffServ QoS scheme*

In addition, the achieved end-to-end delay of each service class for the above three QoS scenarios are given in Table 8.5. For each scenario, the target end-to-end delays are shown, which are calculated according to the desired end-to-end delay factors as defined in Table 8.4. Besides, the corresponding end-to-end delays measured from the simulations (which are configured with the dimensioned last mile link and backbone link bandwidths in the IP transport network) are given. In Table 8.5, through comparing the target delays to the simulated values, it is found that the target end-to-end delay QoS of every service class is guaranteed in the simulations. The second observation is that the obtained end-to-end delays from the simulations for the BE HSPA service class are close to the target delays, but for the other four service classes the simulated delays are far lower than their target delays (i.e. the end-to-end delays of these four service classes are overachieved with the dimensioned bandwidths). That means, in the configured scenarios the BE HSPA service class is the worst case for dimensioning, which requires the most bandwidth to satisfy its defined QoS requirement. According to the proposed dimensioning approach described in section 7.3.3, the dimensioned bandwidth needs to satisfy the desired target delay factor of every service class. Therefore, in the given example, the dimensioning of the individual links are according to the bandwidth requirements of the BE HSPA service class. As a result, the delay performances of the BE HSPA service reach the desired HSPA end-to-end delay targets while the ones of other four service classes are overachieved.

Service Class	End-to-end application delay per service class (seconds)					
	Scenario 1		Scenario 2		Scenario 3	
	target	simulated	target	simulated	target	simulated
AF11- RAB 64 kbps	9.0	6.43	9.0	6.44	9.0	6.45
AF21- RAB 128 kbps	4.25	3.28	4.25	3.30	4.25	3.34
AF31- RAB 256 kbps	2.04	1.72	2.04	1.76	2.04	1.79
AF41- RAB 384 kbps	1.30	1.20	1.42	1.24	1.54	1.30
BE - HSPA	1.02	1.02	1.71	1.68	2.39	2.37

Table 8.5: *Target end-to-end delay QoS vs. simulated delay*
target: *target end-to-end delay according to the target end-to-end delay factor*
simulated: *measured end-to-end application delay from simulations*

Figure 8.16 presents the resultant degree of overbooking on the backbone link from the above dimensioning results in both simulations and analytical calculations, where the graph (a) shows the derived overbooking factors and the graph (b) shows the deviations of the overbooking factors derived from the simulations and the ones from the proposed analytical approach. With the dimensioned last mile link bandwidth and the backbone link bandwidth, the overbooking factor of the backbone link is calculated with equation (8.1). It is observed from Figure 8.16 that the obtained overbooking factors are larger than 1. That means the proposed dimensioning approach is capable of attaining overbooking on the backbone link. Figure 8.16 also shows that the overbooking factors derived from the analytical approach comply with the simulations.

From the presented results, the following conclusions can be drawn. From the presented dimensioning results for the last mile link and the backbone link as shown in Figure 8.14 and Figure 8.15, it is seen that the analytical dimensioning results match well with the simulation results. Thus it can be demonstrated that the proposed processor sharing model that is extended for DiffServ for a single link (presented in section 7.3.1) can be applied well for dimensioning the individual links for elastic traffic in a star-structured multi-Iub IP RAN. As it can provide sufficiently accurate estimations for the required bandwidth for each link. From Table 8.5, it is observed that for each QoS scenario using the obtained dimensioning results from simulations, the resultant QoS performance of each service class meets the individual defined delay requirement. It proves that the proposed dimensioning approach is able to guarantee the QoS of every service class in the DiffServ scenario for different QoS conditions. Furthermore, from Figure 8.16 it can be clearly seen that the proposed dimensioning approach is capable of attaining an optimistic overbooking on the backbone link.

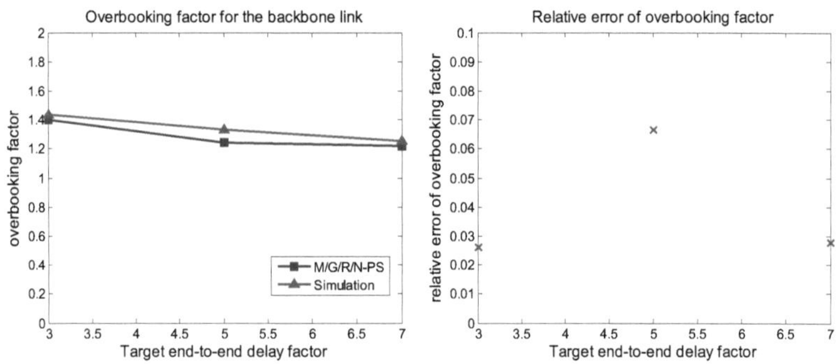

(a) overbooking factor (b) relative error

Figure 8.16: *Overbooking factor over different target delay factors in the star-structured IP-based UTRAN with DiffServ QoS scheme*

8.4.3 Summary

Section 8.4 discussed the dimensioning for elastic traffic in an IP based multi-Iub radio access network, where the IP DiffServ QoS scheme is deployed individually at each last mile link as well as at the backbone link. The dimensioning for the star-structured IP RAN transport network applies the same essential ideas of dimensioning the ATM based multi-Iub RAN which are presented in section 8.3.2. That is, for a desired end-to-end delay factor, an appropriate target delay factor needs to be assigned to each last mile link and the backbone link individually, which will be used to determine the necessary bandwidths of the individual links with the proposed processor sharing models. In this thesis, approach (3), which is a general dimensioning solution proposed in section 8.3.2, is demonstrated to dimension the multi-Iub IP RAN. And the applied processor sharing models for the IP RAN are based on the ones extended for DiffServ for a single link presented in section 7.3.1. The complete dimensioning approach is shown in Figure 8.13. The applicability of this dimensioning approach is demonstrated through simulations. The presented results show that the proposed analytical dimensioning approach can be applied well for dimensioning the individual links for elastic traffic in the multi-Iub IP RAN, and moreover it can calculate a factor for efficient overbooking on the backbone link.

8.5 Overbooking of the Iub Backbone Link

After presenting the dimensioning approaches for the multi-Iub RAN networks for both ATM and IP transport, this section will investigate the overbooking, i.e. to derive an efficient and optimum overbooking for the Iub backbone link, given a fixed load and last mile link capacity per Node B and a number of Node Bs in the UTRAN (see detailed introduction in section 8.1.2).

In this section, the proposed analytical dimensioning approaches, which are presented in the preceding sections for dimensioning the multi-Iub RAN for different types of traffic and transport networks, are applied to analytically estimate an optimum overbooking factor for dimensioning the backbone link to meet the user-relevant QoS targets. The calculated overbooking factors will be validated through simulations. After validations, the impacts of various aspects such as the network size (the number of Node Bs), last mile utilization, traffic scenario, etc., on the overbooking are investigated by using the analytical approach as well as through simulations. At the end, the rules of setting appropriate overbooking factor are summarized. Section 8.5.1 presents the overbooking for the ATM-based UTRAN and section 8.5.2 for the IP-based UTRAN.

8.5.1 Overbooking of ATM-Based Iub Backbone Link

In Appendix A.19, the impact of overbooking on the overall network and user performance is shown. Moreover, intensive investigations were carried out by extensive simulation to study the overbooking by considering both user-relevant and transport network-relevant QoS requirements in the ATM-based UTRAN, looking at different traffic scenarios, last mile loads, last mile link capacities and the number of Node Bs. The presented simulation results in Appendix A.19 show that they all have strong influences on the achieved overbooking: (1) the overbooking is decreased when the last mile utilization increases; (2); the overbooking usually increases with a larger last mile link rate; (3) when there is more circuit-switched traffic in the network, a higher overbooking can be obtained; (4) the overbooking increases with the increase of the number of Node Bs. Detailed results can be found in Appendix A.19.

This section focuses on studying the applicability of the analytical approach for deriving the overbooking to satisfy user-related QoS in an ATM-based multi-Iub RAN, and further apply the analytical approach to analyze the impact of different network and traffic settings on the overbooking. Section 8.2 and section 8.3 have presented analytical approaches to dimension the last mile link and the backbone link individually in the multi-Iub ATM RAN. In this section, they are applied to estimate proper overbooking factor for the Iub backbone link. The analytical results are validated by simulations.

Figure 8.17 presents an example of transmitting circuit-switched traffic. The simulated scenario and parameters are:

- There are 4 Node Bs in the RAN, structured as a star network architecture (see Figure 8.1 (a)), each last mile link has a bandwidth of 1.3 Mbps
- Source traffic: voice traffic with AMR codec (traffic model is described in Table 5.2), transmitted with the peak data rate r_{peak} =12.2 kbps and the configured CAC guaranteed bit rate at last mile r_m =12.4 kbps
- On the backbone link, the desired QoS is 1% packet loss ratio ($qos_{bb} = 1\ \%$)

Figure 8.17(a) presents the achieved overbooking for different amounts of the offered traffic per last mile link (represented by last mile utilization in the figure). In the graph, in addition to the calculated values obtained from the analytical approach, the simulated overbooking factors are also shown for comparisons. For each offered traffic level, the analytical approach as presented in section 8.2.1 sums up the offered traffic from each Node B and applies the MMPP/D/1 queuing model to calculate the required backbone link bandwidth for reaching 1% packet loss ratio QoS target. With the simulation approach, the dimensioned bandwidth is obtained by performing a number of simulations with different bandwidth settings and choosing the one which meets the 1% packet loss ratio in the simulations. Figure 8.17(b) shows the relative error of the calculated overbooking factor compared to the simulated ones.

Both graphs of Figure 8.17 demonstrate that the proposed MMPP/D/1 model is certainly capable of deriving an appropriate overbooking factor for the backbone link for the 1% packet loss ratio QoS. Same as the investigations presented in Appendix A.19, given a fixed last mile link when the last mile utilization is increasing (i.e. more traffic coming from each Node B), the possible multiplexing gain on the backbone link is decreasing and thus the resultant overbooking factor declines.

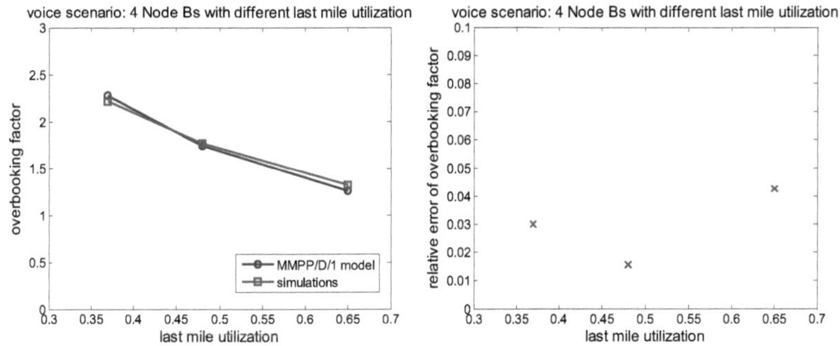

(a) overbooking factor (b) relative error

Figure 8.17: *Validating the calculated Overbooking factor for a voice traffic scenario*

In the following, an example of elastic traffic is validated shown in Figure 8.18. The required QoS for dimensioning is the user-related application QoS, i.e. in this context the end-to-end delay factor.

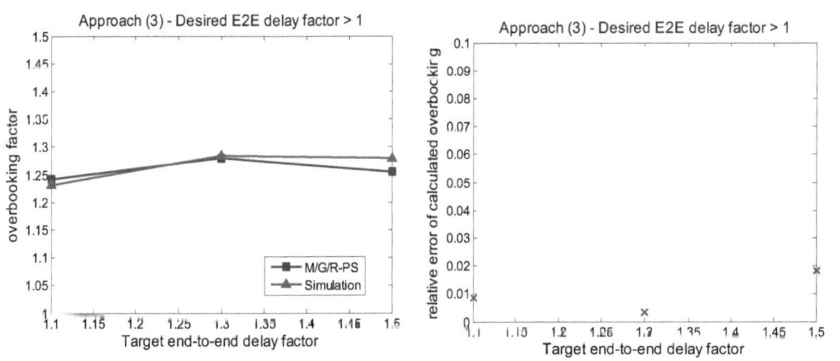

(a) overbooking factor (b) relative error

Figure 8.18: *Validating the calculated Overbooking factor for a web traffic scenario*

The presented results are taken from the scenario given in section 8.3.3 for validating approach (3). There, the target delay factors for the individual links are assigned according to the suggested approach (3). For the assigned delay factors, the bandwidths for the last mile link and the backbone link are derived separately with the proposed extended M/G/R-PS model. Then based on the dimensioned bandwidths, the overbooking factor can be calculated with equation (8.1). Figure 8.18 compares the overbooking factor derived from the analytical approach and the one from simulations. Based on the results shown in Figure 8.9, Figure 8.10, and Figure 8.18, it can be concluded that the analytical dimensioning approaches presented in section 8.3.2 for elastic traffic can give quite accurate predictions on the bandwidth demands for the individual links and in turn derive a correct overbooking factor for the backbone link, for different desired end-to-end delay factors.

After validating the proposed analytical approach for deriving the optimum overbooking factors for the Iub backbone link under different network scenarios, the analytical approach can also be applied to analytically evaluate the impact of different network and traffic configurations on the overbooking. In the following results, the overbooking is calculated for elastic traffic to satisfy a desired application QoS (end-to-end delay factor). Figure 8.19 shows the derived optimum overbooking factor over a different number of Node Bs under different last mile loads, and Figure 8.20 presents the overbooking factor over different delay factors for different offered last mile loads.

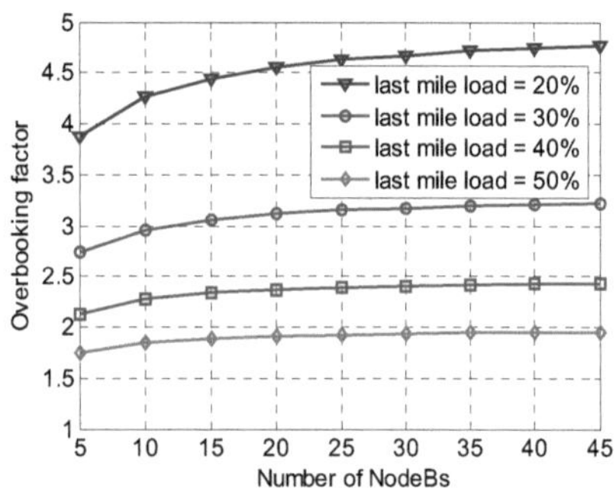

Figure 8.19: *Overbooking factors over different number of Node Bs (RAB=128, f=1.2)*

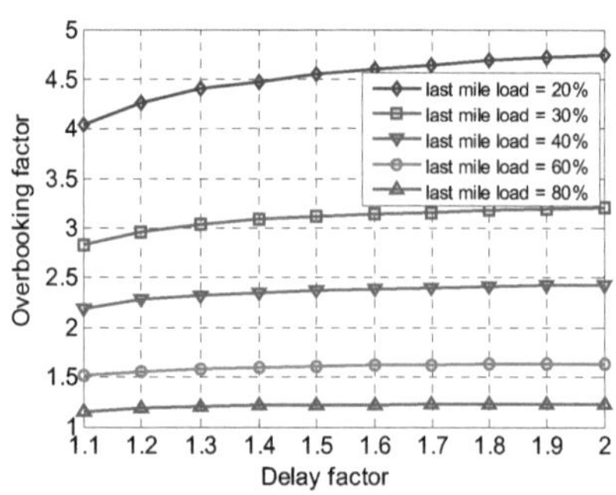

Figure 8.20: *Overbooking factors over different QoS requirements (RAB=128)*

It can be observed in Figure 8.19 and Figure 8.20 that the optimum overbooking factor is influenced by the network size (number of Node Bs), given traffic demands (last mile utilizations) and QoS requirements. The overbooking factor is usually reduced when the last mile utilization increases. When there are more Node Bs in the UTRAN, the overbooking capability is increased due to a higher multiplexing gain. The desired delay factors also have influence on the overbooking results. The higher the delay factor, the higher overbooking can be achieved, but it steadily approaches a maximum value.

8.5.2 Overbooking of IP-Based Iub Backbone Link

This section investigates the overbooking in the IP-based multi-Iub RAN where the DiffServ QoS scheme is used. Same as in the last section, different network configurations are investigated through simulations to find out what are the main factors that will have significant impact on overbooking in the IP-based UTRAN and how their influences are. In addition, the analytical approach, which is proposed in section 8.4.1 for dimensioning the backbone link for elastic traffic, is applied to estimate the appropriate overbooking for the Iub backbone link which needs to satisfy the defined the QoS of every service class. Figure 8.21 presents overbooking factor related to a different number of Node Bs. In this example, the configured simulation settings are:

- There are several Node Bs in the RAN, structured as a star topology network architecture. Each last mile link is configured with 1.6 Mbps for the user plane.
- Each Node B carries five NRT service classes as defined in Table 8.2. Each Node B generates the same traffic demand, having a 30% last mile utilization.
- All NRT service classes apply the same web traffic model: page size is 50 kbyte of a constant distribution.
- The maximum number of elastic flows per service class is given in Table 8.6.
- The desired end to end delay factor per service class is also defined in Table 8.6.
- The configurations for WRED and WFQ functions are given in Table 8.3.

Service Class	Maximum number of elastic traffic flows	Target end-to-end delay factor (QoS criteria)
AF11- RAB 64 kbps	2	1.4
AF21- RAB 128 kbps	3	1.3
AF31- RAB 256 kbps	3	1.2
AF41- RAB 384 kbps	3	1.1
BE - HSPA	3	3

Table 8.6: *Settings for different service classes (overbooking)*

For satisfying the required QoS of each service class defined in Table 8.6, the recommended optimum overbooking factors for setting the backbone link are shown in the graph (a) of Figure 8.21 for different number of Node Bs. Both the analytical and simulated results are presented. First of all, from graph (a) it can be observed that for a fixed 30% last mile utilization the optimum overbooking factor gets larger with higher number of Node Bs. This observation complies with the rule summarized from the last section for investigating the influence of a different number of Node Bs on the optimum overbooking in the ATM-based UTRAN. The improved overbooking capability is due to a higher multiplexing gain caused by a higher degree of aggregation with the increased number of Node Bs. Secondly, to validate the applied analytical approach, graph (a) also compares the calculated overbooking factor with the simulated values and graph (b) gives the corresponding relative error of the calculated overbooking factor compared to the simulated results. Both graphs of Figure 8.21 demonstrate that the proposed analytical approach is able to provide fairly precise estimations on the optimum overbooking factor for the backbone link in the IP based multi-Iub UTRAN.

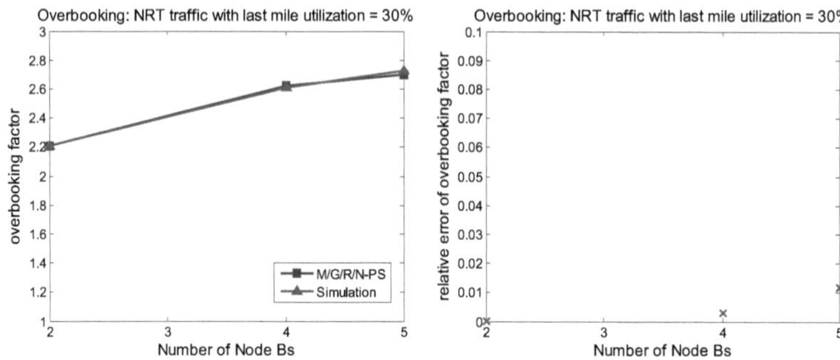

(a) overbooking factor (b) relative error

Figure 8.21: *Validating overbooking factor vs. the number of Node Bs (web traffic)*

Figure 8.22 presents the impact of different last mile utilizations on the obtained optimum overbooking factor. In the given example, the simulation setup is same as the one above, only that the number of Node Bs in the RAN is fixed to 4 and the offered traffic per Node B (i.e. last mile utilization) is varying by means of changing the maximum allowed flows per service class.

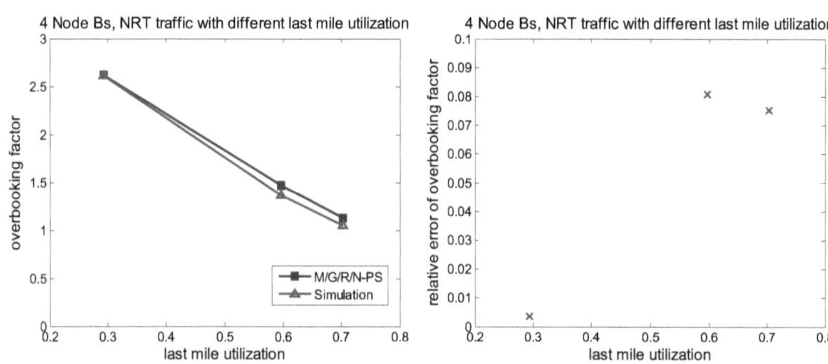

(a) overbooking factor (b) relative error

Figure 8.22: *Overbooking factor vs. last mile utilizations (4 Node Bs) - web traffic*

Figure 8.22(a) shows that the overbooking factor decreases when the utilization of each last mile link gets higher. This rule is also consistent with the one observed from Appendix A.19 for investigating the impact of different last mile utilizations in the ATM-based UTRAN. Again Figure 8.22 validates the accuracy of the analytical approach for predicting optimum overbooking factors.

Figure 8.23 presents a mixed traffic scenario. In this example, the reserved bandwidth by the RT traffic has been derived separately and the RT traffic has higher priority. So the task below is mainly to dimension a proper bandwidth for transporting all NRT traffic, whose QoS criteria is the average application QoS. And then the total required bandwidth at the backbone link will be the sum of the bandwidth reserved for the RT traffic and the one calculated for the NRT traffic. The desired end-to-end delay factors for different NRT service classes are taking the same values as given in Table 8.6. The network and parameter settings in the investigated simulations are:

- There are a number of Node Bs in the RAN, structured as a star network architecture. Each last mile link is configured with 1.6 Mbps for the user plane.
- Each Node B carries seven service classes as same as the ones defined in Table 7.1 in section 7.3.2. The applied QoS structure is illustrated in Figure 3.13. Each Node B generates the same traffic demand, containing 74% Non Real Time (NRT) traffic and 26% Real Time (RT) traffic and in total having around 680 kbps traffic transmitted on the last mile link (equal to 43% last mile utilization).
- All NRT service classes apply the same web traffic model: page size is 50 kbyte of a constant distribution.
- The RT traffic includes the voice and the video services, both sharing one *Expedited Forwarding* (EF) PHB. The traffic models for the RT service classes are same as the ones used in section 7.3.2.
- In this example, each Node B carries 2 voice connections, 2 video connections, and 3 elastic traffic flows per NRT service class.
- The configurations for WRED and WFQ functions are given in Table 8.3.

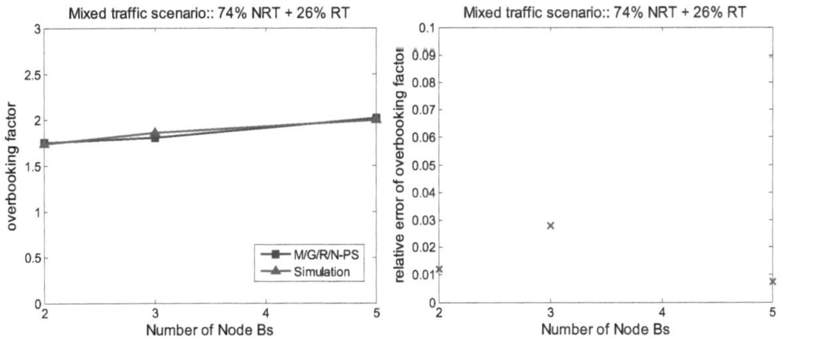

(a) overbooking factor (b) relative error

Figure 8.23: *Overbooking factor vs. the number of Node Bs (mixed traffic scenario)*

In Figure 8.23, the graph (a) shows the derived overbooking factors for different number of Node Bs from both simulation and analytical calculations. The applied analytical dimensioning approach as described in section 8.4.1 is based on the analytical method proposed in section 7.3.1 for dimensioning a single IP-based Iub link. As can be seen in section 7.3.1, the proposed method already considers the situation of mixing with RT traffic that has a higher priority. Thus, the proposed analytical approach can also be applied quite well here for the mixed traffic scenario. Graph (b) evaluates the deviations of the calculated overbooking factor to the simulated ones. The graph (a) of Figure 8.23 shows the similar conclusion as derived from Figure 8.21: the optimum overbooking factor becomes larger when there are more Node Bs in the RAN. In this example, the accuracy of the proposed analytical approach is also demonstrated for the mixed traffic scenario.

The above presented results of different scenarios demonstrate that the analytically derived overbooking factors match well with the ones derived from simulations. The relative errors are all below 10%. It demonstrates the analytical approach proposed in section 8.4.1 can give quite accurate dimensioning results and thus can derive a proper setting for the overbooking for various network scenarios. Moreover, from the above results it can be concluded that the overbooking is strongly related to the number of Node Bs, last mile utilizations, and the traffic types. The relation of these factors with the setting of optimum overbooking factor are similar to conclusions obtained in the ATM-based multi-Iub UTRAN.

After validating the proposed analytical approach for deriving the optimum overbooking factors for the Iub backbone link under different network scenarios, the analytical approach can be applied to analytically evaluate the relation of overbooking factor and different network or traffic configurations (such as network size and traffic demands). Figure 8.24, Figure 8.25 and Figure 8.26 show examples of pure elastic traffic scenario. Figure 8.27 shows an example of a mixed traffic scenario. The values of optimum overbooking factor shown in these graphs are derived from the analytical approach. In Figure 8.24, the relation of the overbooking and the number of Node Bs is investigated for different last mile loads. In Figure 8.25, the overbooking over different last mile utilizations is shown for two different QoS requirements. Figure 8.26 shows the impact of different last mile link capacities on the overbooking, given the same last mile utilization of 50%. Figure 8.27 presents the influence of the percentage of mixed voice traffic on the overbooking.

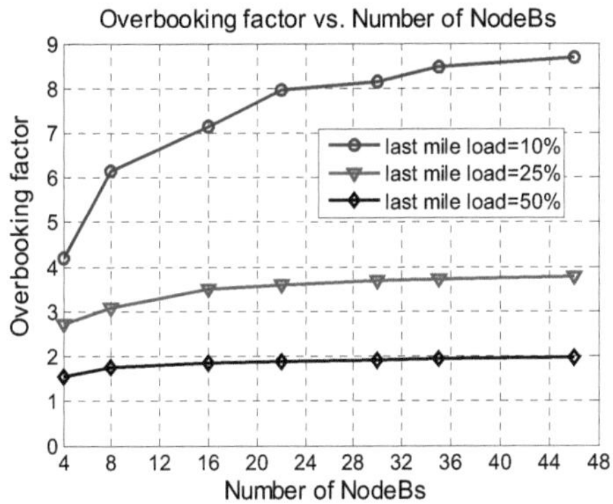

Figure 8.24: *Overbooking factor over number of Node Bs (with various last mile loads)*

It can be seen from Figure 8.24 and Figure 8.25 that they have a similar conclusion as obtained in the ATM-based UTRAN (presented in section 8.5.1). In general the setting of optimum overbooking factor is strongly dependent on the network sizes (number of Node Bs), the given traffic demands (last mile utilizations), and the QoS requirements. When there are more Node Bs in the UTRAN, the overbooking capability is increased due to achieving a higher multiplexing gain. With the increased last mile utilization, the overbooking factor is usually decreased as a result of less room for multiplexing with larger traffic loads. Certainly, the desired QoS of different traffic types will have direct influence on the overbooking results. The higher the QoS requirements, the lower the overbooking that can be achieved. Figure 8.26 demonstrates that the assigned last mile link capacity has also impact on the overbooking. With the same utilization of the link, the overbooking factor is higher with a larger last mile link bandwidth. This is because that a larger link capacity results in more multiplexing gain.

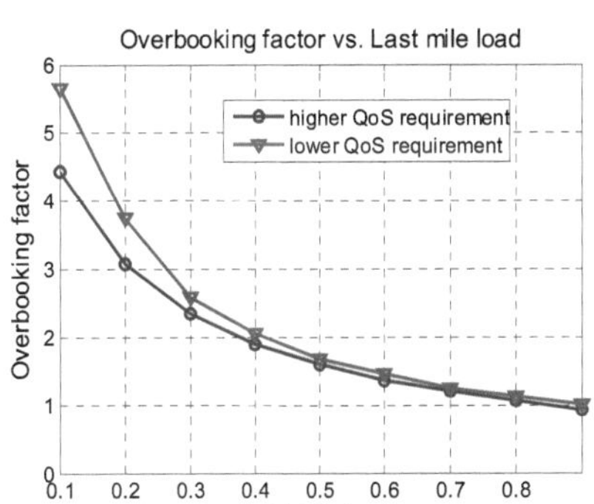

Figure 8.25: *Overbooking factor over last mile loads (with different QoS requirements)*

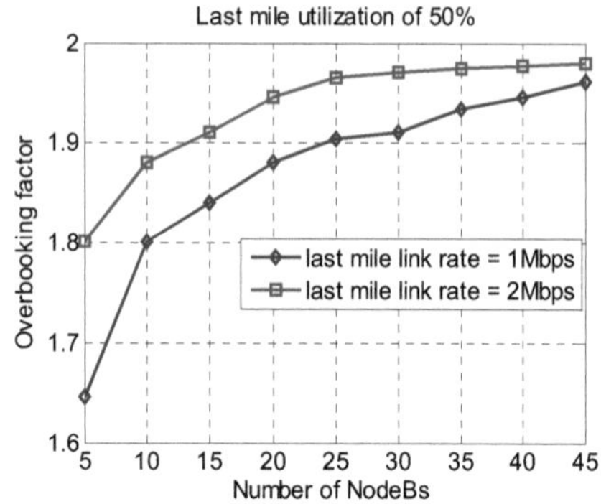

Figure 8.26: *Overbooking factor over number of Node Bs (various last mile capacities)*

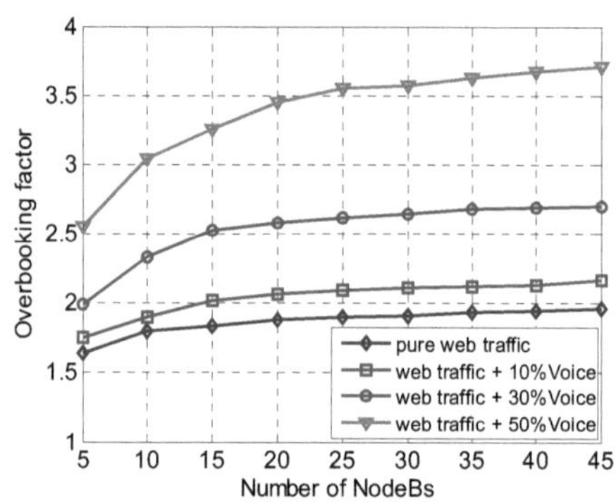

Figure 8.27: *Overbooking factor over number of Node Bs (with different voice portions)*

Figure 8.27 shows that given the same amount of mean offered traffic, with a higher percentage of voice traffic, the achieved optimum overbooking is higher by taking advantage of a lower bandwidth demand of the voice traffic. The more the amount of voice traffic, the more bandwidth saving that can be achieved.

8.5.3 Comparing ATM and IP-Based Multi-Iub RAN

In this section, an example is given to compare the dimensioning of the ATM and IP-based multi-Iub RAN. Since for both ATM and IP based UTRAN, the analytical dimensioning approaches have been already validated by simulations. The following shown dimensioning results will be derived from the individual analytical dimensioning approach presented for the ATM and the IP transport separately. In this example, in both ATM and IP based UTRAN the following parameters are configured:

- there are 4 Node Bs (last mile links) in the RAN, each generating the same amount of the application traffic
- all user flows are assigned to the same RAB rate r_{peak} = 128 kbps (only one service class is considered in this example)
- web traffic model: page size is 50 kbyte of a constant distribution
- there is no restrictions on the number of elastic traffic flows
- target end-to-end delay factor f = 1.3, the target delay factor for each link is assigned according to the suggested approach (3) as presented in section 8.3.2.

Figure 8.28 shows the derived dimensioning for the last mile link (left diagram) and the backbone link (right diagram) over different traffic loads per Node B. The dimensioning results are represented by the normalized capacity (defined in section 4.3). It is seen that in Figure 8.28 that for both last mile link and the backbone link, with RAB 128 kbps the IP transport requires less link capacity than the ATM transport for meeting the same QoS target, because the RAB 128 kbps can achieve a higher bandwidth efficiency in the IP transport. It is the same conclusion that is obtained for the single Iub link scenario presented in section 7.4. It can be further observed in Figure 8.28 that the backbone link can achieve higher cost efficiency due to a higher multiplexing gain is achieved from aggregating the traffic from different number of Node Bs. Figure 8.29 shows the overbooking factor calculated from the dimensioned last mile link and backbone link bandwidths. It can be observed that the obtained overbooking gain is larger than 1. Moreover the ATM-based UTRAN has higher overbooking factor than the IP-based UTRAN, because the ATM transport has lower last mile utilization as shown in Figure 8.28 and thus achieves a higher multiplexing gain over the Iub backbone link. If given the same capacity for the last mile link in both ATM and IP based multi-Iub RAN, it can be expected that for the service class of RAB 128kbps the overbooking factor for the IP-based UTRAN will be higher due to a higher bandwidth efficiency of RAB 128kbps in the IP transport.

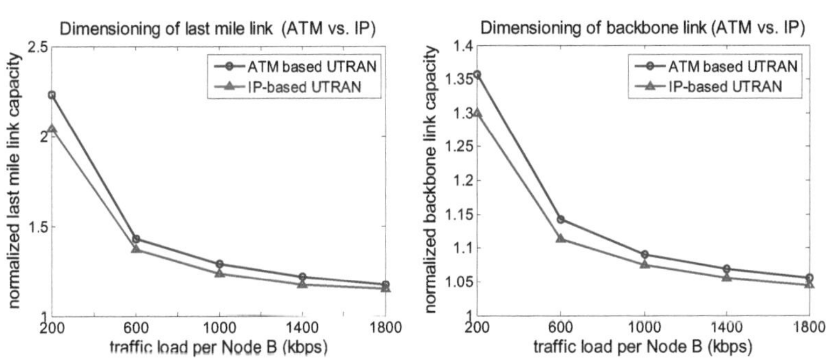

Figure 8.28: *ATM vs. IP- dimensioning of last mile link and backbone link*

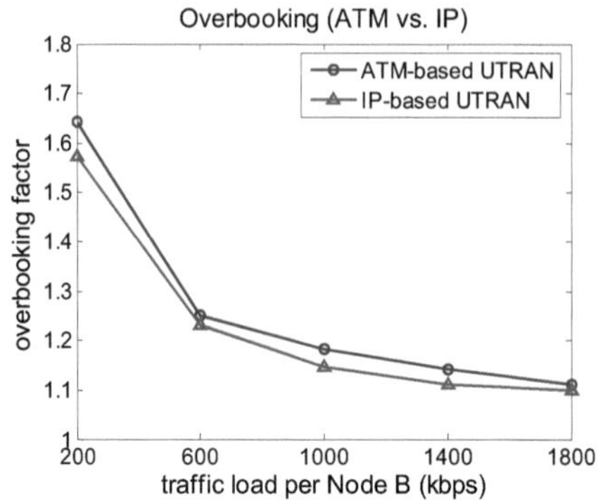

Figure 8.29: *ATM vs. IP- overbooking factor over traffic load*

From this example and the conclusions drawn from the single Iub link dimensioning comparison in section 7.4, it can be expected that for lower RAB rates (e.g. RAB 32 kbps, RAB 64 kbps) with a lower bandwidth efficiency in the IP transport, the IP-based UTRAN will require more bandwidth than the ATM-based UTRAN on both last mile and backbone links, and therefore will have more overbooking than the ATM transport. On the other hand, for higher RAB rates (e.g. RAB 384 kbps or HSPA) the conclusion will be similar to the results of RAB 128 kbps as shown in the given example, only that the achieved link utilization and overbooking gain is dependent on the bandwidth efficiency of each RAB type in different transport networks.

8.5.4 Summary

In section 8.5 overbooking is investigated in detail in both ATM and IP based UTRAN transport networks. It is related to the task of dimensioning the backbone link in the star-structured multi-Iub RAN, given the configurations of a certain number of Node Bs located in the RAN and each Node B carrying a certain amount of traffic on an assigned last mile link. Through analyzing the impacts of the number of Node Bs, given traffic profiles, last mile utilizations, and last mile link capacities on the attained optimum overbooking factor at the backbone link by simulations as well as analytically, the following general rules can be summarized for both ATM and IP transport networks:

- Normally when there are more Node Bs aggregating to the backbone link, the overbooking capability is increased due to a higher multiplexing gain;
- The overbooking usually shall be reduced when the last mile utilization increases. But if the last mile link already reaches the maximum utilization, the optimum overbooking factor will approach 1 independent of the number of Node Bs in the RAN;

- The overbooking can also be dependent on the configured last mile link capacities. Typically, the higher the last mile link bandwidth, the more overbooking is possible.
- The carried traffic characteristics and their QoS priorities will have considerable impact on the achievable overbooking level. And certainly, the desired QoS requirements of different traffic types will have direct influence on the overbooking results. The higher of the QoS requirements, the lower overbooking can be achieved.
- In the mixed traffic scenarios, the percentage of RT traffic will have influence on the achieved overbooking gain. The higher percentage of RT traffic, the higher the overbooking that can be achieved. This is due to that RT traffic is given higher priority and as well it has lower bandwidth demands than the NRT traffic. Thus when there is more portion of RT traffic, there will be more bandwidth saving and thus higher overbooking gain.

In addition to finding out the above important factors and their individual influences on the overbooking, analytical approaches, which are proposed in the preceding sections for dimensioning the multi-Iub RAN for different types of traffic and transport networks, are applied in this section to analytically derive an appropriate overbooking factor for dimensioning the backbone link for a defined QoS. The validations are performed for different network and traffic scenarios for both ATM and IP based UTRAN. For all investigated network scenarios, the presented validation results demonstrate that all proposed analytical approaches are able to derive the optimum overbooking factor for the backbone link and the calculated overbooking factors are quite accurate compared to the ones obtained from the simulations.

At the end, the overbooking of the ATM and IP based UTRAN is compared. The comparison results show that the achieved overbooking gain is dependent on the bandwidth efficiency of each RAB type in the individual ATM and IP transport network.

9 Conclusion and Outlook

An important way for improving the cost-efficiency of UMTS networks is efficient design and use of existing network infrastructure. The radio access network is considered as one of the most important economic aspects for the network planning and bandwidth dimensioning due to its limited and expensive transport resources. With the rapid expansion of the radio access network as well as the fast growing number of mobile users and traffic volume, there is a significantly increasing demand for transport capacities in the radio access network. To achieve a cost-efficient design of the UMTS network, the radio access networks have to be dimensioned appropriately for the many types of services which are to be offered.

This thesis is dedicated to investigate the dimensioning for the Iub interface of different UMTS radio access networks, by simulations as well as analytically. In the framework of this thesis, different UMTS simulation models were developed for the OPNET simulator, according to the 3GPP specifications. Within this work, the simulation models of UMTS Rel99 with the ATM-based UTRAN and UMTS with the IP-based UTRAN were developed. The simulations of HSPA (HSDPA and HSUPA) were using HSPA simulation model established in the context of another project of the Communication Networks working group at the University of Bremen. In this work, these simulation models are used to evaluate the performances of the UMTS networks, and further to provide dimensioning of the transport bandwidths for the Iub interface. Moreover, they are used to derive the dimensioning rules and analytical approaches and further to validate the proposed analytical approaches.

This thesis proposes a number of novel and elaborate analytical models, algorithms and methodologies, which let UMTS network operators dimension their radio access network bandwidths in order to reduce expenditures, while still being able to offer a desired quality of service. For the dimensioning process, different traffic scenarios, evolutions of UMTS radio access networks, transport solutions, QoS mechanisms and network topologies are studied. In this thesis, two fundamental types of traffic are distinguished: elastic and circuit-switched traffic, which are associated with data applications and real-time services, respectively. In this thesis for each traffic type different QoS requirements are investigated. The QoS requirements determine the objective of dimensioning and the selection of suitable analytical dimensioning models. In this thesis two kinds of QoS are considered: *user-relevant QoS* and *network-relevant QoS*. User-relevant QoS refers to the QoS related to the individual user flow; while network-relevant QoS measures are used to evaluate the quality of the transport network such as packet delay and packet loss ratio. Since the characteristics and QoS requirements of different traffic types are essentially different, it is necessary to come up with distinct approaches for bandwidth dimensioning.

Dimensioning for Elastic Traffic for the user-relevant QoS

Elastic traffic is associated with data applications, which use TCP for data transfer. The characteristic property of an elastic traffic flow is its rate adaptability, which is caused by the feedback mechanism of TCP. The transmission rate of a sender is adjusted to the available bandwidth along the route towards the receiver. As a consequence, the overall throughput, which is the user relevant QoS measure for elastic traffic, depends not only on the link capacities, but also on the number of concurrent flows. The proposed dimensioning model for guaranteeing the application throughput for the elastic traffic is the processor sharing model.

The theory of processor sharing is applied to the dimensioning of networks for elastic traffic. In the context of this thesis, processor sharing is interpreted as ideal bandwidth sharing among flows, whose arrival instances form a Poisson process and whose service times are generally distributed. The basic processor sharing model, i.e. the M/G/R-PS model, which has been proposed in the literature, is investigated and deficiencies are identified. The main contributions of this thesis is to apply the processor sharing model for dimensioning the UMTS networks to meet the desired per user application layer throughput, and propose extensive extensions of process sharing model to incorporate important UMTS radio and transport functions and network topologies.

The objective of the dimensioning process is to minimize the total link bandwidth in the network, while still being able to provide a certain average throughput for elastic traffic flows. The presented results show how the processor sharing models, which in principle only relate to a single link, can be applied to a multi-Iub RAN network.

Dimensioning for Circuit-switched Traffic for the user-relevant QoS

Circuit-switched traffic is related to applications and services with real-time critical demands, which require Connection Admission Control (CAC) in order to guarantee the desired QoS. The objective of the dimensioning process is also the minimization of the total bandwidth in the network. Capacities need to be assigned in a way that call blocking probabilities do not exceed a certain threshold and that packet-level QoS of accepted flows is guaranteed throughout their duration. The presented corresponding dimensioning model is based on the Multidimensional Erlang (MD-Erlang) model.

Dimensioning for the network-relevant QoS

For dimensioning to meet the desired network-relevant QoS, the proposed analytical models are based on queuing models, which are focused on the packet level taking the characteristics of the aggregated Iub traffic into account. The proposed queuing models define an appropriate arrival process to capture the important characteristics of the aggregated traffic on the link, and an accurate server process considering the packet length distribution and the given link capacity. In the mix traffic scenario, scheduling function is considered in the queuing models taking into consideration of the potential multiplexing gain.

In this thesis, the MMPP and BMAP arrival process model are proposed and investigated for the ATM-based RAN. For modeling a single Iub link, MMPP/D/1 or BMAP/D/1 queuing models are suggested for the single traffic scenario. When combining elastic traffic with the circuit-switched traffic where a strict priority scheduling is applied, the proposed analytical queuing model is MMPP/D/1 or BMAP/D/1 with non-preemptive priority queuing.

Dimensioning for a star-structured multi-Iub RAN network

Furthermore, the above analytical dimensioning models, which are proposed for the single link scenario, are extended to be applied to a star-structured multi-Iub RAN network. One important issue for dimensioning a multi-Iub RAN network is overbooking the backbone link, which is to allocate less bandwidth on the backbone link than the total bandwidth request by all the connected last mile links. Usually, the backbone link can be overbooked by taking advantage of statistical multiplexing gain among traffic from different Node Bs. For a cost-efficient dimensioning of the backbone link, an optimum overbooking is desired to provide balance between the costs and the QoS objectives. In this thesis, overbooking is investigated for a wide variety of scenarios through extensive simulations as well as by using the proposed analytical approach, with the objective to find an optimum overbooking. From the investigated results, important factors that have great impact on the overbooking are found. And furthermore the analytical models are applied to derive the optimum overbooking factor (which represents the degree of overbooking) for various network scenarios and validated with simulations.

For all proposed analytical models in this thesis, the validations are performed through extensive simulations using the developed UMTS simulation models. From the presented validation results throughout the thesis, it can be demonstrated that for all relevant network scenarios, the proposed analytical approaches work sufficiently well. That means, the proposed analytical models are able to capture relevant characteristics and provide accurate dimensioning results, and thus can be applied for network dimensioning. At the end, to summarize the proposed analytical dimensioning approaches in this work a dimensioning tool is implemented in this thesis containing all proposed analytical models. The developed dimensioning tool can be used by network operators to perform dimensioning of various traffic scenarios, UMTS radio access networks, transport solutions, QoS mechanisms and network topologies.

Overall, all investigations and analytical dimensioning models presented in this thesis help network service providers to optimize their network infrastructure to reduce the transport costs while still can provide the desired quality of service to the users.

Outlook

The analytical models and approaches proposed in this thesis can serve as a basis for further researches in the area of access network dimensioning. One of the most present radio access network technology, the *Long Term Evolution* (LTE) as introduced in section 2.3.5, has addressed a potentially much higher demand on the transport bandwidth in the access network than UMTS Rel99, HSDPA and HSUPA networks. Thus, to derive new dimensioning approaches for the LTE network to consider future traffic and network scenarios are now an important task. It would be an interesting topic to investigate if the current applied traffic models and dimensioning models are still applicable for the future mobile networks such as a LTE network.

Appendix

A.1 ATM Cell Format

The format of the ATM Cell is given in Figure A.1 for User-Network Interface (UNI).

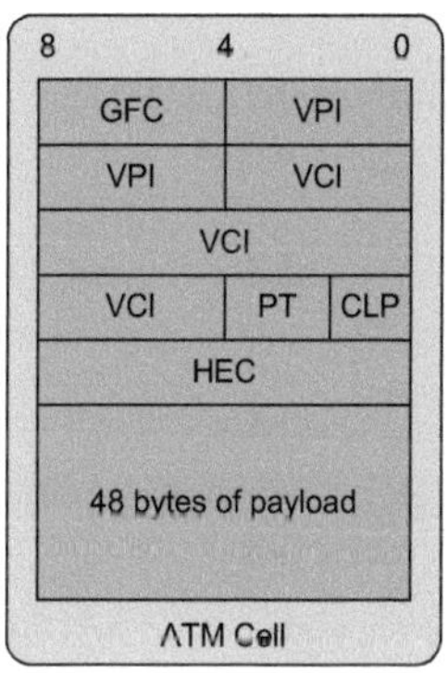

GFC: Generic Flow Control (4 bits)

VPI : Virtual Path Identifier (8 bits UNI) or (12 bits NNI)

VCI: Virtual channel identifier (16 bits)

PT: Payload Type (3 bits)

CLP: Cell Loss Priority (1 bit)

HEC: Header Error Correction

Figure A.1: *Format of the ATM Cell*

A.2 AAL2 Packet Formats

The format of the CPS packet is shown in Figure A.2 [Mcl97].

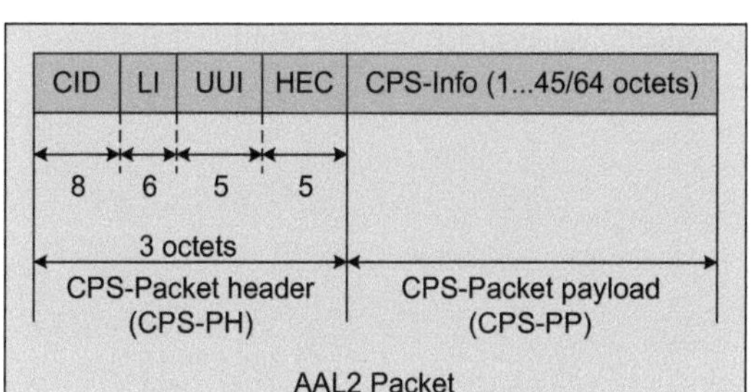

Figure A.2: *Format of the AAL2 CPS Packet*

Key fields of the CPS packet are the Channel Identifier (CID), the Length Indicator (LI), and the User-to-User Indication (UUI) fields. These are defined below:

CID Field:
Uniquely identifies the individual user channels within the AAL2, and allows up to 248 individual users within each AAL2 structure. Coding of the CID field is shown below:

Value	Use
0	Not Used
1	Reserved for Layer Management Peer-to-Peer Procedures
2-7	Reserved
8-255	Identification of AAL2 User (248 total channels)

LI Field:
Since AAL2 allows variable packet length, LI field is used to indicate the actual length of the CPS packet payload associated with each individual user. The value of the LI can vary from 1 to 45 octets (or 64 octets depending on the implementations).

UUI Field:
Provides a link between the CPS and an appropriate SSCS that satisfies the higher layer application. Different SSCS protocols may be defined to support specific AAL2 user services, or groups of services. The SSCS may also be null.

Value	Use
0-27	Identification of SSCS entries
28, 29	Reserved for future standardization
30, 31	Reserved for Layer Management (OAM)

The format of an AAL CPS-PDU block is shown in Figure A.3 [Mcl97]. The Offset Field identifies the location of the start of the next CPS packet within the CPS-PDU. For robustness, the Start Field is protected from errors by the party bit (P) and data integrity is protected by one bit sequence number (SN) used for identifying cell loss.

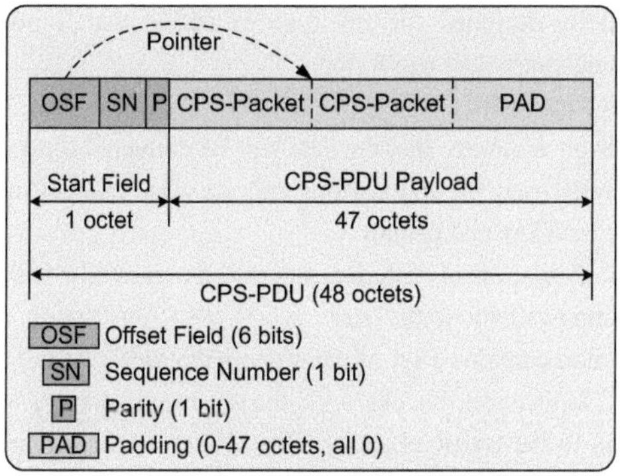

Figure A.3: *Format o the AAL2 CPS-PDU*

A.3 ATM Service Categories and QoS Parameters

The ATM forum defines five different service categories. They are listed as follows:

- Constant Bit Rate (CBR)
  - provides a deterministic bit rate of flow. The amount of bandwidth is characterized by the *Peak Cell Rate* (PCR)
  - supports traffic sensitive to delay and loss, such as video conferencing, telephony (voice services)
  - emulates circuit switching
- Real-Time Variable Bit Rate (rt-VBR)
  - transports traffic at variable bit rates
  - Carries traffic like compressed voice, video, and audio
  - rt-VBR connections are characterized by a *Peak Cell Rate* (PCR), *Sustained Cell Rate* (SCR), and *Maximum Burst Size* (MBS).
- Non-Real-Time Variable Bit Rate (nrt-VBR)
  - transport variable bit rate traffic, for which there are no strict real-time accurate timing requirements
  - carries traffic like data and buffered voice and video
- Available Bit Rate (ABR)
  - similar to nrt-VBR, it is also used for connections that transport variable bit rate traffic for which there is no reliance on time synchronization between the traffic source and destination, but for which no required guarantees of bandwidth or latency exist
  - provides a best-effort transport service, in which flow control mechanisms are used to adjust the amount of bandwidth available to the traffic originator

  - is primarily designed for any type of traffic that is not time sensitive and expects no guarantees of service.
- Unspecified Bit Rate (UBR)
  - provides no assurance that the data will be delivered (best effort only)
  - is generally used for applications that are very tolerant of delay and cell loss, like file transfers and E-mail
  - has no flow-control mechanisms to dynamically adjust the amount of bandwidth available to the user

Each ATM connection contains a set of parameters that describe the traffic characteristics of the source. The ATM network checks with the source to determine the specific traffic requirements according to the traffic characteristics. When each connection is set up, traffic parameters are determined and a traffic contract is made. Then the ATM network shall provide QoS that falls within the agreed traffic contract. Their parameters are:

- *Peak Cell Rate* (PCR): The maximum allowable rate at which cells can be transported along a connection in the ATM network. The PCR is the determining factor in how often cells are sent in relation to time in an effort to minimize jitter. PCR is generally coupled with the CDVT (Cell Delay Variation Tolerance), which indicates how much jitter is allowable.
- *Sustained Cell Rate* (SCR): A calculation of the average allowable, long-term cell transfer rate on a specific connection.
- *Maximum Burst Size* (MBS): The maximum allowed burst size of cells that can be transmitted contiguously on a particular connection.
- *Minimum Cell Rate* (MCR): The minimum allowed rate at which cells can be transported along an ATM connection.

ATM defines a number of QoS parameters to measure the QoS of a connection and quantify end-to-end network performance at the ATM layer. These parameters are negotiated when a connection is set up. The network should guarantee the negotiated QoS by meeting certain values of these parameters.

- *Cell Transfer Delay* (CTD). The delay experienced by a cell between the time it is transmitted by the source and is received by the destination.
- *Cell Delay Variation* (CDV). The variation between the maximum and minimum CTD experienced during the connection.
- *Cell Loss Ratio* (CLR). The percentage of cells that are lost in the network due to error or congestion and are not received by the destination.

When an ATM connection is set up, it needs to specify an appropriate service category, agree a traffic contract with the ATM network and negotiate the QoS parameters that the network shall guarantee.

A.4 Ethernet

Ethernet is a popular transport technology in the IP network. It is a frame-based computer networking technology which could operate at many speeds for *Local* (LANs). A well-known technology CSMA/CD (*Carrier Sense Multiple Access with Collision Detection*) is used in Ethernet to govern the way the computer shared the channels. In this thesis, Ethernet is used for the transport in an IP-based UTRAN due to its low cost, flexibility and scalability.

Ethernet physical link speed could vary from 1M bps to 10G bps, when the physical medium has a range from bulky coaxial cable to optical fiber.

Possible Ethernet Physical Link Rate	Name	Standard
1M	1Base5	802.3
10M	10Base5/2/T/F/FB/FL	802.3
100M	100BaseT/TX/T4/T2	802.3
1G	1GBaseT/SX/CX/BX10	802.3
10G	10GBaseSR/LX4	802.3ae
100G	100GbE	802.3ba

Table A.1: *Ethernet Physical Link Rate and Standard*

100 Gigabit Ethernet or 100GbE is an Ethernet standard presently under early development by the IEEE. The fastest existing standard is 10 gigabit Ethernet. In late November 2006, an IEEE study group agreed to target 100 Gbps Ethernet as the next version of the technology.

A.5 Random Early Detection and Weighted RED

Random Early Detection (RED) is a congestion control algorithm to be used in network routers and gateways that may drop (or mark) packets randomly before buffer overflow. Packets are dropped in a probabilistic fashion before the queue reaches an overflow state. By randomly dropping packets prior to periods of high congestion, RED tells the packet source to decrease its transmission rate. RED was proposed by Floyd and Jacobson [FJ93] as an effective mechanism to control congestion in the network routers or gateways. Currently it is recommended as one of mechanisms for so-called active queue management by IETF (see Braden et al. [BBC+98].) and it has been implemented in vendor products. In addition, RED has been incorporated in various drafts for Differentiated Services for the Internet in such a way that RED operates on different flows with different parameters depending on the flows' priority, i.e. RED is used to provide different classes of services.

RED is the probabilistic discard or marking of packets as a function of queue fill before overflow conditions are reached. Random dropping takes place when the algorithm detects signs of permanent congestion. The congestion control variable in RED is the average queue size that is calculated from the instantaneous queue size by using exponential averaging. Persistent congestion results in an increase of the average queue size, whereas transient bursts in buffer have only little influence.

Following describes the RED algorithm in detail. Let q_n denote the current queue length, i.e. the number of packets in the system (including the one in service) at the time of *n*th packet arrival. Let K_n denote the average queue length at the time of *n*th arrival. The *n*th packet will be discarded with probability p_n that is determined RED drop function, which is a linear function that is determined by the following parameters:

- max_p: maximum dropping probability.
- min_{th}: minimum average queue length threshold for packet dropping.
- max_{th}: maximum average queue length threshold.
- The average queue length K_n at the time of *n*th arrival. For each arriving packet we compute the average (smoothened) queue length. The average queue size is based on the previous average K_{n-1} (the old average queue length at the time of *(n-1)*th arrival) and the current size of the queue q_n. The formula is given following:

$$K_n = K_{n-1} \cdot (1 - \frac{1}{2^w}) + q_n \cdot \frac{1}{2^w} \tag{A-1}$$

Here w is the exponential weight factor, a user-configurable value. For high values of w, the previous average becomes more important. A large factor smoothes out the peaks and lows in queue length. The average queue size is unlikely to change very quickly, avoiding drastic swings in size. The RED process will be slow to start dropping packets, but it may continue dropping packets for a time after the actual queue size has fallen below the minimum threshold. The slow-moving average will accommodate temporary bursts in traffic. But if the value of w gets too high, RED will not react to congestion. Packets will be transmitted or dropped as if RED were not in effect. For low values of w, the average queue size closely tracks the current queue size. The resulting average may fluctuate with changes in the traffic levels. In this case, the RED process responds quickly to long queues. Once the queue falls below the minimum threshold, the process will stop dropping packets. If the value of w gets too low, RED will overreact to temporary traffic bursts and drop traffic unnecessarily.

The packet drop probability is based on the minimum average queue length threshold min_{th}, maximum average queue length threshold max_{th}, and maximum dropping probability max_p. Figure A.4 summarizes the packet drop function.

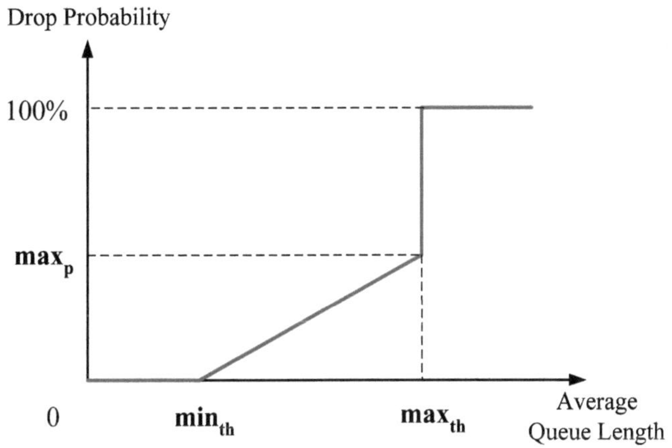

Figure A.4: *RED Packet Drop Function*

When the average queue length calculated according to equation (A-1) is above the minimum average queue length threshold min_{th}, RED starts dropping packets. The probability of packet drop increases linearly as the average queue size increases until the average queue size reaches the maximum average queue length threshold max_{th}. When the average queue size is at the maximum average queue length threshold max_{th}, the probability of packet drop reaches the maximum dropping probability max_p. When the average queue size is above the maximum threshold, all packets are dropped. The minimum threshold value should be set high enough to maximize the link utilization. If the minimum threshold is too low, packets may be dropped unnecessarily, and the transmission link will not be fully used. The packet drop probability P_n of the nth packet arrival is calculated in equation (A-2).

Weighted Random Early Detection (WRED) uses the same parameters as RED, but it has the ability to perform RED on traffic classes individually. WRED generally drops packets selectively based on IP precedence. Edge routers assign IP precedences to packets as they enter the network. WRED uses these precedences to determine how it treats different types of traffic. Packets with a higher IP precedence are less likely to be dropped than packets with a lower precedence. Thus, higher priority traffic is delivered with a higher probability than lower priority traffic. However, WRED can be also configured to ignore IP precedence when making drop decisions so that non-weighted RED behavior is achieved.

$$P_n = \begin{cases} p_n = 0, & if \quad K_n < \min_{th} \\ p_n = 1, & if \quad K_n > \max_{th} \\ p_n = \max_p \cdot \dfrac{K_n - \min_{th}}{\max_{th} - \min_{th}} & if \quad \min_{th} < K_n < \max_{th} \end{cases} \tag{A-2}$$

A.6 AMR

The AMR speech coder [3GP99a] consists of the multi-rate speech coder, a source controlled rate scheme including a *voice activity detector* (VAD) and a comfort noise generation system, and an error concealment mechanism to combat the effects of transmission errors and lost packets.

The speech encoder takes its input as a 13-bit uniform *Pulse Code Modulated* (PCM) signal either from the audio part of the UE or on the network side, from the *Public Switched Telephone Network* (PSTN) via an 8-bit *A-law* or *μ-law* to 13-bit uniform PCM conversion. The encoded speech at the output of the speech encoder is packetized and delivered to the network interface. In the receive direction, the inverse operations take place.

The multi-rate speech coder is a single integrated speech codec with eight source rates ranging from 4.75 kbps to 12.2 kbps, and a low rate background noise encoding mode, as listed in Table 3-1. Here, the speech coder is capable of switching its bit-rate every 20 ms speech frame on command.

Codec mode	Source codec bit-rate
AMR_12.20	12,20 kbps (GSM EFR)
AMR_10.20	10,20 kbps
AMR_7.95	7,95 kbps
AMR_7.40	7,40 kbps (IS-641)
AMR_6.70	6,70 kbps (PDC-EFR)
AMR_5.90	5,90 kbps
AMR_5.15	5,15 kbps
AMR_4.75	4,75 kbps
AMR_SID	1,80 kbps

Table A.2: Sou*rce codec bit-rates for the AMR codec [3GP99a]*

In each 20 ms speech frame, 95, 103, 118, 134, 148, 159, 204 or 244 bits are produced, corresponding to a bit-rate of 4.75, 5.15, 5.90, 6.70, 7.40, 7.95, 10.2 or 12.2 kbps. The following table shows the bit allocation of the AMR codec modes for 12.2 kbps speech.

Mode	Parameter	1st subframe	2nd subframe	3rd subframe	4th subframe	total per frame
	2 LSP sets					38
12.2 kbps	Pitch delay	9	6	9	6	30
(GSM EFR)	Pitch gain	4	4	4	4	16
	Algebraic code	35	35	35	35	140
	Codebook gain	5	5	5	5	20

Table A.3: *Bit allocation of the AMR coding algorithm for 20 ms frame*

The coder operates on speech frames of 20 ms corresponding to 160 samples at the sampling frequency of 8000 sample/s. It performs the mapping from input blocks of 160 speech samples in 13-bit uniform PCM format to encoded blocks of 95, 103, 118, 134, 148, 159, 204, and 244 bits and in reverse from encoded blocks of 95, 103, 118, 134, 148, 159, 204, and 244 bits to output blocks of 160 reconstructed speech samples.

The coding scheme for the multi-rate coding modes is the so-called *Algebraic Code Excited Linear Prediction Coder* (ACELP). At each 160 speech samples, the speech signal is analyzed to extract the parameters of the CELP model (LP filter coefficients, adaptive and fixed codebooks' indices and gains). These parameters are encoded and transmitted. At the decoder, these parameters are decoded and speech is synthesized by filtering the reconstructed excitation signal through the LP synthesis filter.

The AMR codec uses a VAD method to recognize the active periods, and the data are only transferred within this period. For 12.2 kbps coding rate, each 20ms voice frame is 244 bits (12.2kbps * 20ms). During the inactive period, the data volume is rapidly reduced, and only a *Silence Insertion Descriptor* (SID) packet is sent at the beginning of the inactive period, an AMR_SID frame has only 36 bits.

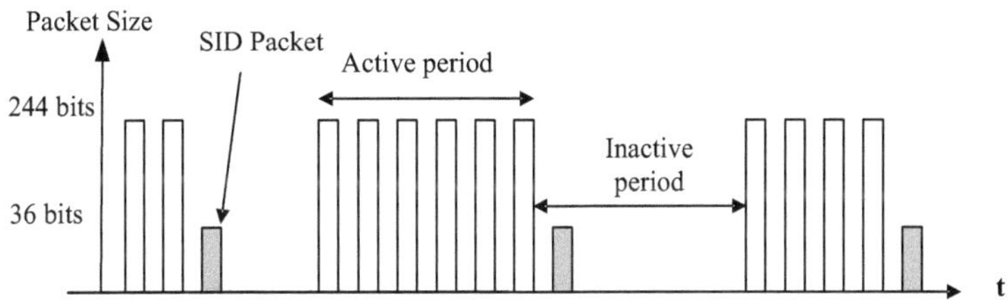

Figure A.5: *AMR Speech Codec Model for 12.2 kbps speech traffic*

A.7 Self-Similarity

There are a number of different, not equivalent, definitions of self-similarity. This definition tightly follows reference [KLL03]. Consider a weakly stationary discrete time stochastic process $X = (X_1, X_2, X_3, ...)$ with constant mean, finite variance and autocorrelation function $r(k)$. Further, let $X^{(m)}$ denote a new time series obtained by averaging X over non overlapping blocks of size m. That is, for $k = 1,2,3...$ the time series $X_k^{(m)} = (1/m)(X_{km-m+1} + \cdots + X_{km})$ and let $r^{(m)}$ denote the autocorrelation function of $X^{(m)}$. The process X is called strictly self-similar with self-similarity (or Hurst) parameter H if $mX^{(m)}$ has the same finite dimensional distributions as $m^H X$ for all $m \geq 1$. The above is the mathematical definition of self-similarity. The physical meaning of self-similarity is that the traffic rates, i.e. number of packets or bytes per time unit, has the burst behavior within a small and large time scale.

Since self-similarity is believed to have a significant on network and queuing performance, the measurement of self-similarity is important. There are several methods to represent the degree of self-similarity such as variance-time plot, R/S analyze and periodogram-based analysis. Here propose R/S analyze to measure the self-similarity.

Given an empirical time series $(X_i : i = 1,2,...,N)$ of length N, the rescaled adjusted range statistics $R(u)/S(u)$ of values u is given by

$$\frac{R}{S}(u) = \frac{1}{S(u)}[\max\{0, W_1, W_2, ..., W_u\} - \min\{0, W_1, W_2, W_u\}] \quad \text{(A-3)}$$

with

$$W_k = (X_1 + X_2 + ... + X_k) - k\overline{X}(u), (k = 1,2,...,u, 0 < u \leq N) \quad \text{(A-4)}$$

Here $\overline{X}(u)$ is the sample mean. $S^2(u)$ is the sample variance.

Hurst found that many naturally occurring empirical records are well represented by relation below.

$$E\left[\left(\frac{R}{S}(u)\right)\right] \rightarrow c \cdot u^H \quad \text{(A-5)}$$

The positive constant c is independent with u. For $u \rightarrow \infty$, the Hurst parameter $H \approx 0.73$. It has been shown that the above relation holds for short-range dependent processes with $H = 0.5$ and for increment processes of self-similarity with $0.5 < H < 1$. This is generally referred as the Hurst effect. Now we make logarithm of this relation in equation (A-6).

$$log\left(E\left[\left(\frac{R}{S}(u)\right)\right]\right) \rightarrow log\left(c \cdot u^{H}\right) \Rightarrow log\left(E\left[\left(\frac{R}{S}(u)\right)\right]\right) \rightarrow H \cdot log(u) + log(c) \qquad \text{(A-6)}$$

The Hurst parameter H can be obtained by the skewness of the line ($\log(u), \log(R/S)$).

A.8 ATM Workstation and Server

The OPNET ATM workstation model represents an ATM node with client-server applications running over TCP and UDP. This model features an application layer that resides over the IP layer, which in turn resides over the ATM layer. The following summarizes the process modules in the ATM workstation/server model.

- The Application module generates different application traffic, like ftp, http, e-mail, voice, etc.
- The *Transport Adaptation Layer* (TPAL) module suite presents a basic, uniform interface between applications and different transport protocols. TPAL also provides a global registry of servers.
- The TCP and IP modules are working as TCP/IP protocol suite.
- The IPAL Module transports IP datagram across the ATM network; establishes and releases AAL connections as required.
- The ATM Layer module encapsulates and forwards data from the AAL layer; decapsulates and forwards data to the AAL layer; forwards cells from the ATM Management module.
- The AAL module implements the encapsulation and decapsulation of SAR_PDUs following the AAL1, AAL2, AAL3/4 and AAL5 protocol, and also creates and invokes signaling and connection processes to handle new connections, incoming signals and data.
- The ATM_call_control module accesses interfaces between the data plane and the control plane.
- The ATM_sig module handles signaling to establish and release an ATM connection
- The ATM Switch module switches cells to output ports. It also implements buffer management, based on QoS and the ABR feedback mechanism.
- The ATM Translation module receives incoming ATM cells from network; translates VPI/VCI values for outgoing cells; forwards cells to appropriate ATM module.

The ATM server model is usually used together with the ATM workstation. The server can support all of the application models.

A.9 Introduction for HSPA (HSUPA/HSDPA)

As mentioned in chapter 2, the evolution of the mobile communication market brings demands for both larger system capacity and higher data rates. To boost the support for the packet switched services, *High Speed Downlink Packet Access* (HSDPA) and *High Speed Uplink Packet Access* (HSUPA) are introduced by the 3GPP Release 5 and 6 respectively as the evolution of UMTS which strongly enhance the transmission of data packet traffic on the downlink and uplink separately. They are jointly referred to as High-Speed Packet Access (HSPA). Following gives a detailed introduction of HSDPA and HSUPA protocol architectures, the impact of HSPA on the UTRAN Iub interface, and specific traffic control functions to be employed in the UTRAN for HSPA.

HSDPA and HSUPA User Plane Protocol Architecture

HSDPA and HSUPA are the extensions of the UMTS Rel99 network, so their basic protocol layers are similar to the Rel99 system. But both introduce new protocol entities in the architecture to support the enhanced features such as fast scheduling, fast HARQ, etc. Figure A.6 and Figure A.7 illustrate the user plane protocol architecture of HSDPA and HSUPA [3GP05b, 3GP06a], highlighting the HSDPA/HSUPA-specific additions and their location in the protocol stacks.

As shown in Figure A.6, HSDPA introduces a new MAC entity, MAC-hs (hs for high speed), at the base station (Node B). The RNC retains the MAC-d (d for dedicated), but the only remaining function is transport channel switching and all other MAC functionalities such as scheduling and priority handling are moved to the MAC-hs located in the Node B which resides close to the radio interface. The main reason is that the scheduler is able to exploit rapid variations in the interference level and the channel conditions, and HARQ with soft combining also benefits from the rapid retransmissions as this reduces the cost of retransmissions. The RLC layer, however, stays mainly unchanged. It is worth mentioning that even though HSDPA has introduced fast physical layer retransmissions using HARQ at the Node B, there is still the RLC layer handling the retransmissions in case the physical layer operation fails transmission in RLC acknowledge mode (AM). Using RLC unacknowledged mode (UM), retransmissions only take place in the physical layer. An example is a VoIP call where the RLC layer transmissions from the RNC would take too long.

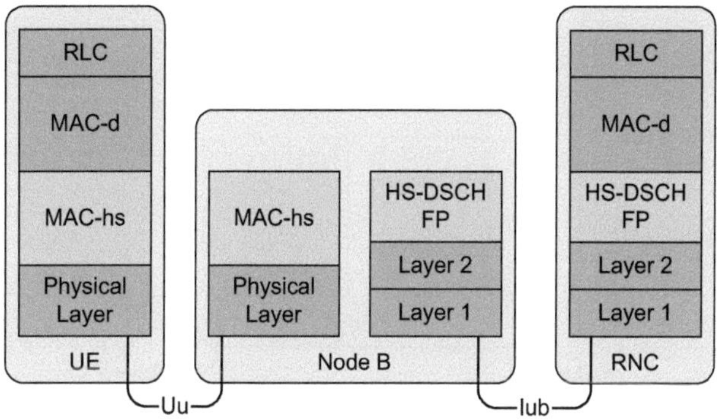

Figure A.6: *HSDPA protocol architecture*

Figure A.7 shows the HSUPA protocol architecture and highlights the main modifications done in the MAC layer compared to the previous release. To allow UMTS system to support the HSUPA enhancements, a new MAC entity, the MAC-e, is introduced in the UE and Node B. In the Node B, the MAC-e is responsible for support of fast hybrid ARQ retransmissions and scheduling, while in the UE, the MAC-e is responsible for selecting the data rate within the limits allocated by the scheduler of the Node B MAC-e. For similar reasons as for HSDPA, uplink scheduling and HARQ functionalities are located in the Node B which stays close to the radio interface. Thus the Node B can track the rapid changes of the air interface and make fast decisions concerning the scheduling and retransmissions. Moreover, there is a new protocol entity for the RNC as well. When the UE is in soft handover with multiple Node Bs, different transport blocks may be successfully received by different Node Bs. Then there is a possibility that the packets from different Node Bs arrive such that the order of packets is not retained. In order to ensure in-sequence delivery of data blocks to the RLC layer, a reordering functionality is required in the RNC in the new MAC entity, the MAC-es. If such reordering is handled at the Node B, then an unnecessary delay would be introduced because the Node B would have to wait for missing packets until they could be determined to have been correctly received by another Node B in the same active set. In a similar manner to HSDPA, the RLC layer in HSUPA is involved with the retransmission of packets if the physical layer fails to correctly deliver them after the maximum number of retransmissions is exceeded.

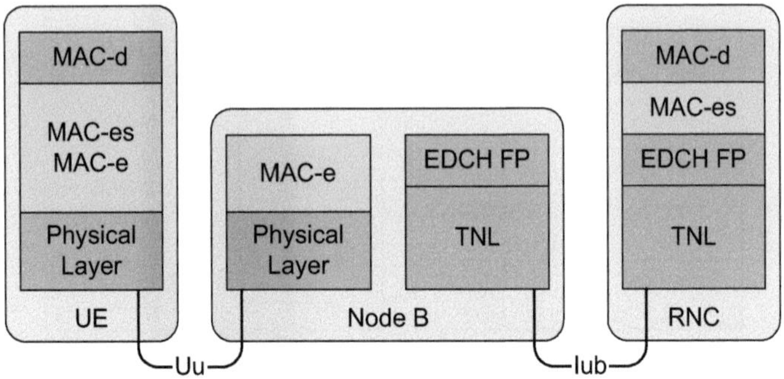

Figure A.7: *HSUPA protocol architecture*

Impact of HSDPA and HSUPA on the UTRAN Iub Interface

With the introduction of HSDPA and HSUPA in the UMTS network, the traffic on the downlink and uplink are expected to increase drastically due to a greatly improved peak data rate over the air interface. HSDPA provides data rates up to 14.4 Mbps over short (2 ms) period in the downlink and HSUPA offers uplink speeds up to 5.76 Mbps. While Rel99 UEs support data rates of at most 384 kbps, and the data rate on different interfaces, including the Iu-PS interface to the packet core network, is equal to the one used for radio interface. With HSPA, the situation has changed. Though a high peak data rate available in the air interface, it does not mean the same data rate being used on the Iub and Iu-PS for that particular user. Because in the HSPA network, the radio resources and code are shared by all users in the cell, thus the average bit rate for a single user in a loaded cell is clearly lower. Thus the average traffic on the Iub interface is usually less than the peak rate on the radio interface. For example, with HSDPA the radio interface peak rate supported by the UE is 7.2 Mbps, but the service data rate over the Iub interface can be limited, e.g. to 1 Mbps. Therefore, the buffering in the Node B is required to allow high peak rate of the air interface (over a short period of time) while limiting the bit rate on the Iub interface in line with the leased transport network bandwidth and the negotiated QoS parameters.

Since HSDPA and HSUPA enable the mobile users to transmit data with high peak data rate over a short time period, this certainly also leads to a bursty nature of the traffic on the Iub interface. Moreover, due to the increased user throughput and system capacity offered by the HSPA to transfer high-speed packet data, the carried traffic amount and traffic variations on the Iub interface are also much higher than that in the UMTS Rel99 system. It is important to have a sufficient, non-congested transport at the Iub interface. To fulfill this target, it is necessary to apply traffic control techniques on the Iub interface to pursue for maximum bandwidth utilization while guaranteeing the QoS to the end user and also reduce the operator cost. The next section introduces HSDPA and HSUPA specific traffic control functions which aim to improve the transport network performance, and prevent overload situations on the Iub interface.

HSPA-Specific Traffic Control Functions

HSDPA and HSUPA introduce two essential traffic control schemes to apply in the Iub interface: flow control (only for HSDPA) and congestion control techniques. They are designed to reduce the burstiness of carried traffic on the Iub interface, and protect the Iub link from congestion collapse and overload.

Flow Control (FC)

In the case of HSDPA, as the traffic is mainly transmitted from the RNC to the Node B over the Iub link, the transmission buffer (MAC-hs buffer) in the Node B requires flow control function to be applied between RNC and Node B for controlling the amount of traffic each user allowed to send to the base station to avoid buffer overflow. The purpose of flow control is to allocate the Iub resources for each individual user according to their available air interface throughput. The user under good radio conditions can get more Iub allocation as the data move fast through the air interface and thus the filling level of the MAC-hs queue remains low. On the other hand, when the buffer starts to get filled due to poor radio conditions, the flow control will try to slow down the transfer speed for that user.

The flow control mechanisms are implemented in MAC-hs and FP layers. The flow control mechanism uses the frame protocol for the purpose of flow control over the Iub interface and adapts the per-user Iub flow rate to the air interface capacity. The MAC-hs user buffers at the Node B have to be monitored continuously to guarantee the data availability and avoid buffer overflow. Flow control is not a part of the 3GPP specifications. So there are a variety of flow control schemes used by different vendors and network operators. Two example flow control schemes are [WTG+06, WL+06]:

1. ON/OFF flow control;
2. Credit based flow control, i.e. Provided Bit Rate (PBR) based Flow Control (also called Enhanced Flow Control).

The ON/OFF flow control mechanism is considered to be the most simple mechanism which can be applied for the purpose of flow control for HSDPA traffic on the Iub. The flow control monitors the filling levels of the MAC-hs buffers for each user flow. It uses two thresholds to control the MAC-d flow rate over the Iub. When the upper limit is exceeded data flow over the Iub link is stopped and when the buffer limit reaches the lower limit, data flow over the Iub is allowed. This protects the MAC-hs buffers from overflow.

The credit based flow control mechanisms can further smooth the HSDPA traffic over the Iub interface, resulting in a reduction of the burstiness of the total Iub traffic. The credit based flow control introduces two new aspects with respect to the ON/OFF flow control algorithm. Firstly, the Iub allocation message shall be sent periodically from Node B to RNC. Secondly, the credit allocation algorithm is based on the *Provided Bit Rate* (PBR) of the per-user priority queues in the Node B. The HS-DSCH PBR measurement is defined 3GPP: for each priority class the MAC-hs entity measures the total number of MAC-d PDU bits whose transmission over the radio interface has been considered successful by MAC-hs in Node B during the last measurement period (cycle time), divided by the duration of the measurement period. The PBR based Flow Control mechanism has an advantage over the simple ON/OFF flow control mechanism that it has a smoothing effect on the data flow over the Iub as it considers the instantaneous rates of the Node B priority queues. Whereas, the ON/OFF flow control mechanism is based on the threshold levels of the Node B priority queues filling level.

Congestion Control (CC)

In the HSPA network when the user traffic over the air interface is greater than the available Iub capacity, the congestion occurs at the Iub interface. Due to Iub congestion, the packet loss probability increases significantly, causing more retransmissions at higher layers such as RLC and TCP. Hence the offered load is further staggered leading the system into congestion collapse and wasting the radio as well as transport resources. To avoid such circumstances, a proper congestion control scheme is required in addition to the flow control mechanism. The basic concept of the Iub congestion control is to reduce the user data rate in case of congestion. The data rate should be reduced adaptively according to the available bandwidth under congestion situations. This can help to improve the effective throughput by gradually reducing the burstiness.

The congestion control algorithm consists of two basic functionalities: congestion detection and congestion control. These basic functionalities are applied individually per MAC-d flow and basically using information contained in the FP frame. The congestion detection function is responsible for detecting any possible congestion on the Iub link. The detection can be based on frame loss, or the frame delay over the Iub link [3GP06d]. Upon detection of congestions, the congestion indication is signaled to the congestion control module. The congestion control function is to adaptively reduce the MAC-d flows Iub rate according to the available Iub bandwidth, so that persistent congestion situations for a MAC-d flow do not mean degraded throughput but just the adaptation of the MAC-d flow rate to the available bandwidth. In fact this action enables the upper layers to recover from congestion

In HSDPA, the congestion detection and the congestion control are performed at the Node B. The result of the CC is the potential reduction of credits in the capacity allocations (CA) of the frame protocol (FP), which are issued by Node B and need to be followed by RNC. For the HSUPA network, the congestion detection is done at the serving RNC, while the congestion control is executed at the Node B. More detailed information on the congestion control algorithms for HSPA can be found in [WTG⁺08].

A.10 Simulation Model of HSPA

The HSPA is the extensions of the UMTS Rel99. In the framework of this thesis, the UTRAN network for the HSPA is based on the ATM. The following gives a brief introduction of the existing HSPA simulation model, which contains the modeling of HSDPA for the downlink and HSUPA for the uplink separately.

I. Modeling of HSDPA

In the HSPA simulation model, the main features, protocol stacks and specific Iub traffic control functions of the HSDPA were modeled detailed according to the introductions given in section 2.3.2 and Appendix A.9. The simulated network model of HSDPA is illustrated in Figure A.8. Compared to the Rel99 simulation model, the way of modeling the UEs, the corresponding nodes in the external networks, and the Iub AAL/ATM based transport network is the same as in the Rel99 simulation model. The main differences are the newly introduced HSDPA protocol entities, the HS-DSCH transport channel, the HSDPA air interface, the fast scheduling at the Node B (also called MAC-hs scheduler), and the Iub congestion control and flow control functions.

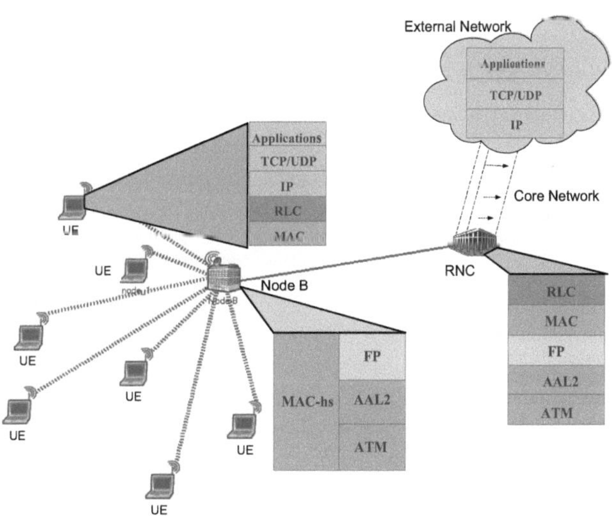

Figure A.8: *HSDPA Simulation Model Architecture [WL+06]*

As highlighted in the figure, the new implemented HSDPA protocol entities are the MAC-d protocol at the RNC and the UE, the MAC-hs protocol at the Node B, and the HS-DSCH FP at both the RNC and the Node B. In addition to these new protocol entities, the RLC acknowledged mode is used for transmitting the HSDPA traffic. It is modeled including the complete RLC functions like flow control, window mechanisms, error detection and retransmission mechanism based on Selective Repeat ARQ.

In the HSDPA, the air interface is different than the one of Rel99 network. As mentioned in section 2.3.2, it offers a much higher peak data rate on the downlink and uses a shorter TTI (2 ms). The modeling the HSDPA air interface is based on deploying the statistical data of the output trace files taken from the dedicated radio simulations provided by the NSN. The trace files provide per-user wireless channel characteristics which are channel dependent and subject to the modeled MAC-hs scheduler in the radio simulations. From the trace files, the statistical data can be obtained for each user on the average user throughput, the probability to transmit n MAC-d PDUs and the probability of retransmissions per TTI. However, the handover UEs are not modeled.

Moreover, different flow control and congestion control algorithms are modeled on the Iub interface. They are used to avoid the link congestion and reduce the losses on the Iub, and therefore can provide guaranteed QoS to the end users while efficiently utilizing the Iub transport resources. The detailed modeling of congestion control and flow control can be found in [*WL*[+]*06*].

II. Modeling of HSUPA

The introduction of HSUPA functions and protocol stacks have been given in section 2.3.3 and Appendix A.9. The HSUPA protocols are implemented on top of the existing HSDPA functions in the HSPA simulation model. The modeling of the HSUPA together with the HSDPA is shown in Figure A.9.

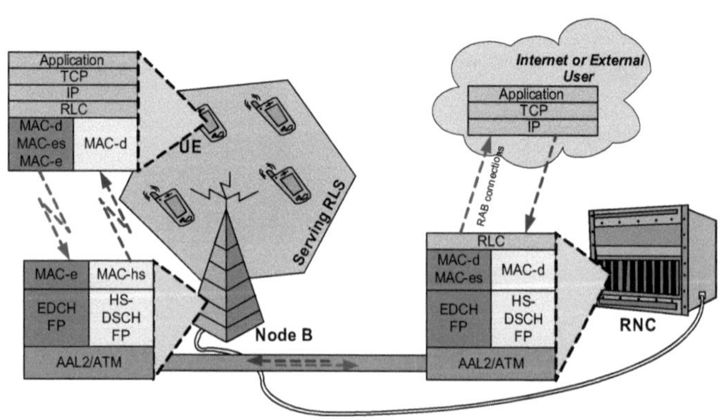

Figure A.9: *HSPA simulator overview [LZK*[+]*08]*

As can be seen, the main enhancements of the model are the new implemented HSUPA protocol entities (MAC-e and MAC-es), the new transport channel E-DCH for each HSUPA UE, and the corresponding E-DCH FP layer over the Iub interface. As well the RLC acknowledged mode is implemented for the uplink including the complete RLC functions for transmitting the HSUPA traffic.

The MAC-e protocol includes HARQ and HSUPA scheduler (also called E-DCH scheduler) functionality. In the HSUPA simulation model, the implemented HARQ is based on the stop and wait protocol. There are four HARQ processes which work in parallel for a single user flow. Whenever the HARQ process is free it allows the one MAC-e PDU to be transmitted with the defined block error rate (BLER) for each retransmission. For the modeling of the HSUPA scheduler, a TTI (Transmission Time Interval) based scheduler is implemented in a way that the interference level is the main criterion for sharing the resources between the HSUPA E-DCH flows. Several HSUPA scheduler implementations and flavors are feasible, but in the current simulation model a simplified scheduler is implemented. The implemented scheduler always gives priority to the secondary UEs, which are transmitting with the SAG, and what is left from the resources is given to the PAG UEs.

Different than the HSDPA modeling, the soft handover users are considered in the HSUPA. The soft handover (SHO) users are about ~20% of the total number of HSUPA UEs in the cell. Those soft handover UEs cause additional packets to be carried over the Iub link, which occupies some of the link bandwidth. Two different cases can be distinguished for the SHO implementation: the first case represents the soft handover which some of *serving radio link set* (S-RLS) UEs are experiencing, in which those UEs will have their packets carried over the *non-serving radio link set* (NS-RLS) Iub link to the RNC (via another Node B). The second case represents the UEs from the neighboring cells that are in soft handover to the own cell. Those UEs are referred to as NS-RLS UEs, and some of their packets have to be carried by the own cell Iub link.

In addition, there are also Iub congestion control functions implemented for the HSUPA. But there is not flow control function in the HSUPA. The HSUPA model has been presented in [LZK⁺08].

A.11 Multiplexing Schemes for the IP-Based UTRAN

With the transport overhead of UDP/IP and Ethernet, bandwidth efficiency in the IP-based UTRAN will be low, especially for small packets due to larger UDP/IP and Ethernet headers. For example for voice traffic, the achievable efficiency in the IP-based UTRAN is only 40.6%. In order to improve the bandwidth efficiency, several multiplexing schemes are proposed in 3GPP TR 25.933 [3GP04a] and [MWIF01], as introduced in section 2.3.4.1. In this thesis, *Composite IP* (CIP) and *Lightweight IP Encapsulation* (LIPE) have been considered. Both schemes allow multiplexing small frames with variable packet sizes into one UDP packet in order to reduce the overhead.

The CIP packet format is shown Figure A.10. The FP PDU is segmented or reassembled to fit the size of the CIP packet payload. Then several CIP packets are encapsulated into one UDP packet payload. The CIP packet header format includes *Context ID* (CID) which is to identify the user connection, payload length which indicates the length of one CIP packet payload, and the sequence number which is used to reassemble the segmented packets.

The LIPE packet format is given in Figure A.11. Each LIPE packet consists of *Multimedia Data Packets* (MDP) and their *Multiplexing Header* (MH). The basic approach of various packets are multiplexed into one UDP/IP packet is similar to CIP scheme. A major difference of CIP and LIPE is the header length. CIP has the fixed CIP header length of 4 bytes, while LIPE allows an extension of the MH header, which can vary between 2 and 3 bytes. Even using the extension of the MH header, LIPE still need 1 byte less for the container overhead. In the multiplexing case, the connection identifier cannot be set in the UDP header using the UDP port because in this case several users share the same UDP header, so we can use the flow ID field in the CIP/LIPE header to identify the user connection.

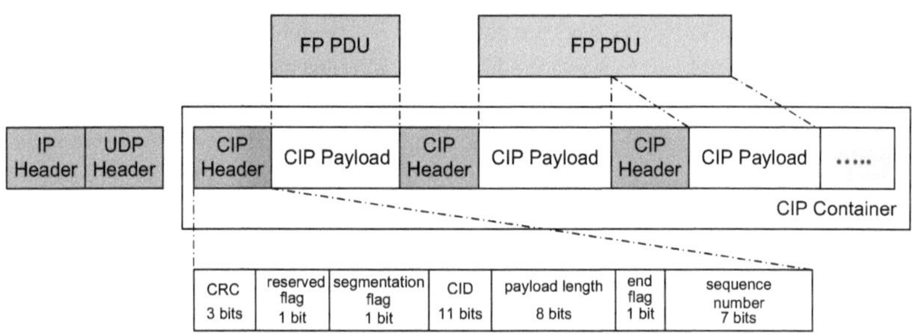

Figure A.10: *CIP packet format*

CIP and LIPE are optional multiplexing schemes for the IP-based UTRAN. Both CIP and LIPE are the multiplexing schemes above layer 3, i.e. between FP and UDP/IP. The multiplexing rule is based on the maximum UDP packet size and the multiplexing timer. The multiplexed packet is only sent out if the maximum size of the packet is reached or the multiplexing timer expires. In order to support multiple services with their QoS requirements, different multiplexers can be implemented to support each QoS class. The 3GPP TR 25.933 [3GP04a] defines three basic types of QoS for multiplexers. The first is voice, which has the most stringent delay requirement, the second is delay sensitive data, which has certain delay requirement less than voice, and the last one is the delay insensitive data, for which link efficiency is more important than a low delay value.

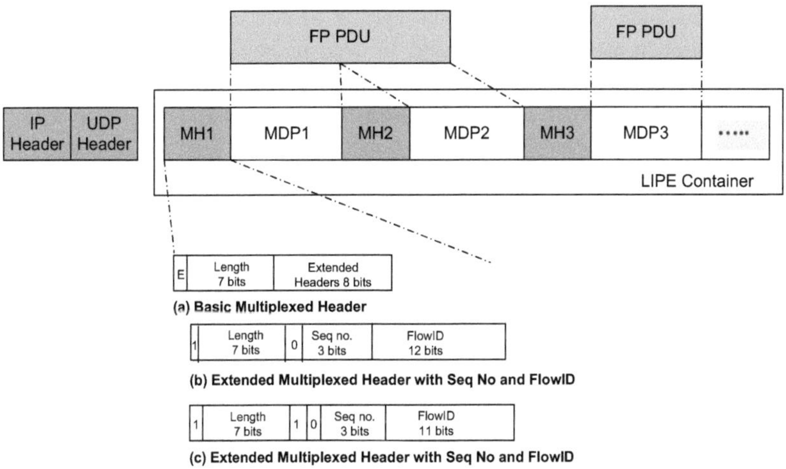

Figure A.11: *LIPE packet format*

For each multiplexer, the number of packets to encapsulate in a LIPE or CIP container is determined by its multiplexing timer and the maximum container packet size. Accordingly, for the above three multiplexers, separate buffers are needed. Timers start when the first FP PDU arrives. If the number of encapsulated packets reaches the maximum number of FP PDUs allowed in one container, then the timer stops and the container packet is encapsulated into the UDP payload and then sent out. But if there are not enough packets in the buffer before the timer expires, the packet will be sent out right away even though the container length does not reach the maximum.

A.12 Exact IP-Based UTRAN Simulation Model

The following gives a detailed description of the modeling of the transport network UDP/IP and Ethernet layers, and the related traffic control and QoS scheme in this exact IP-based UTRAN simulation model, which is briefly introduced in section 5.5.1.

Modeling of UDP/IP

The UDP and IP layers are implemented below the FP layer. The main function of UDP is to manage the transport connection for each user flow and provide a unique identification for the individual user connections. The UDP port numbers in the UDP header field together with the node IP address are utilized to identify different user connections. The UDP layer receives the FP PDUs and encapsulates each FP PDU into one UDP packet by adding 8 bytes UDP header. When a UDP PDU is encapsulated, it is forwarded to the transport IP layer.

In the simulation model, the transport layer IP module implements the functions of encapsulation of IP packets and addressing. In terms of addressing, there are source and destination IP address field defined in the defined 20 bytes IP header. The source or destination IP address is the IP address of the Node B or RNC, which is used for addressing in the UTRAN network. It should be noted that here the IP address is not the IP address of the UE or corresponding node which is outside the UTRAN domain. The routing function is not required for the Node B and RNC as they work as IP host.

Modeling of Ethernet and Physical Layer

Ethernet is the applied layer 2 technology. In the model, the OPNET built-in Ethernet model is used including the *Address Resolution Protocol* (ARP) and MAC modules. The ARP is responsible for translating between the physical MAC address and the logical IP address. The MAC includes the functions of addressing as well as channel access control. On the physical links, a variety of available Ethernet links, such as 10BaseT, 100BaseT, 1GBaseT, etc., can be chosen for building the Iub links.

Modeling of LIPE Multiplexing Schemes

In this IP-based UTRAN model, both Composite IP (CIP) and Lightweight IP Encapsulation (LIPE) multiplexing schemes (introduced in A.11) are modeled. They are located between FP and UDP/IP layers. In order to support multiple services with their QoS requirements, different multiplexers can be implemented to support each QoS class. In the current model, two multiplexers for two different QoS classes are modeled: one is for the delay sensitive (typically for the voice application) and the other is for the delay insensitive services (such as web, ftp applications). These two multiplexers and the packetizers are located above the UDP/IP layer. For each multiplex, the number of FP PDUs to encapsulate into one LIPE or CIP container is determined by its configured multiplexing timer and the maximum container packet size. In order to optimize the performance of each service, the multiplexing timer and the maximum container packet length can be tuned individually for different services. For voice service, usually a small multiplexing timer and smaller packet length is better to ensure a small delay. For the delay insensitive data, a larger packet is more efficient for the bandwidth utilization.

Modeling of QoS Scheme

In this exact IP-based UTRAN model, the QoS provisioning is realized by setting up per QoS class buffer at the Ethernet layer and implementing a scheduler to serve them with different priorities. Currently the model only distinguishes two QoS categories: real time (RT) which is for the delay sensitive services, and non real time (NRT) which is for the non delay sensitive services. Between them a strict priority scheduler is applied: the RT traffic is given higher priority over the NRT traffic.

A.13 Impact of TCP Slow Start

At first, the impact of TCP slow start is investigated with simulations as shown in Figure A.12. It presents the achieved bandwidth utilization efficiency of individual flows as a function of file size, with RAB 64 kbps, 128 kbps and 384 kbps, respectively. The bandwidth utilization efficiency is calculated as the achieved mean rate as a ratio to the offered peak data rate.

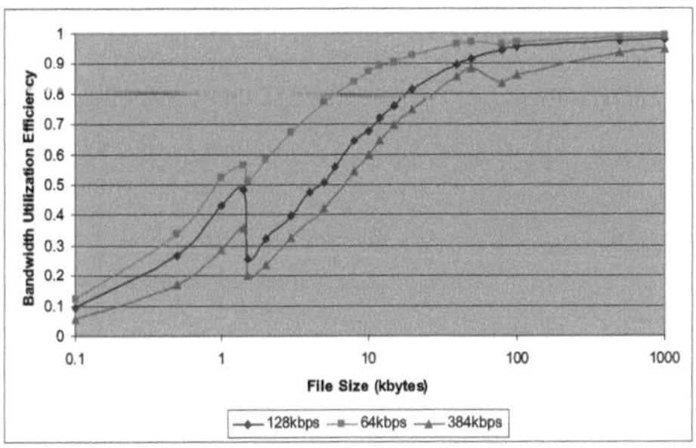

Figure A.12: *Bandwidth utilization efficiency over file sizes*

It can be seen that in general the bandwidth utilization increases with file size. The underutilization of the available bandwidth is due to the TCP slow start at the beginning of the transmission. This underutilization of the available bandwidth is vanishing in case of large file sizes. But for smaller file sizes, especially with web traffic those sizes in the range of 50 kbyte or less, the offered bandwidth cannot be efficiently utilized. It is also noticed that there is a drop for the file size between 1 MSS (here 1460 byte) and 2 MSS, because after the first MSS a secondary IP packet is necessary for transferring the rest of the data, which increases the transfer time by a full RTT cycle. Furthermore, due to the impact of TCP, the relationship between file size and link utilization is non-linear. And it is dependent on the corresponding bearer rate. The lower the radio access bearer rate, the better the bandwidth is utilized. For higher bearer rates, larger file sizes are required to achieve a good utilization of the available bandwidth.

A.14 Processor Sharing among Multiple RABs

Different RAB type traffic dynamically competes for the network resources. Usually higher peak-rate traffic contends more bandwidth before entering the fair capacity sharing phase. Figure A.13 shows a detailed insight of how each RAB competes for the bandwidth resources in a dynamic manner. In this example, there are 3 user connections sharing a 200 kbps link: 2 users with RAB 64 kbps and 1 user with RAB 128 kbps. Each user transfers the same amount of data. In this figure, the green curve shows TCP throughput of RAB 128 kbps, blue and red curves show TCP throughputs of the other two RAB 64 kbps users. It can be seen from Figure A.13 that at the beginning of the transmission, the higher RAB rate gets more transmission resources than the lower RAB rate, and as a result obtains a relatively higher throughput. After some time, the throughput of RAB 128 reduces while the throughput of RAB 64 increases until they both become equal in the stable phase. For RAB 64 kbps, the obtained average throughput on the ATM layer during the stable phase is around 62.5 kbps, and RAB 128 kbps gets around 63 kbps. But considering the overall throughput of the whole connection, RAB 128 kbps obtains a little higher throughput in average.

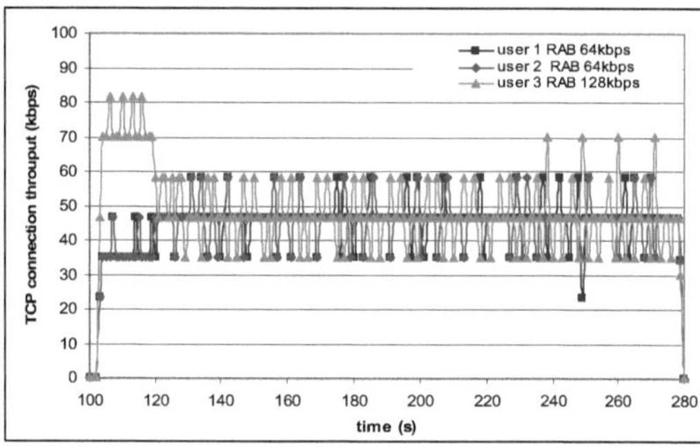

Figure A.13: *TCP connection throughput of each RAB type – multiple RABs scenario*

A.15 Dimensioning for HSPA Traffic

The HSPA traffic is characterized by high peak data rates and high burstiness. To support such HSDPA traffic on the downlink and HSUPA traffic on the uplink, not only the UMTS air interface but also the backhaul of the UMTS access network, will require considerably high capacity for the provisioning of high-speed transmission of packet data. This poses big challenges in the dimensioning of the backhaul network due to the limited and expensive transport resources. In the framework of this thesis, HSPA is still using the ATM as the underlying transport in UTRAN. This chapter presents the dimensioning results for HSPA, which are obtained from simulations using the developed HSPA simulation model (introduced in section 5.6).

Different than Rel99 traffic, the HSPA traffic has low transport delay requirement at the Iub interface. There are three main reasons: (1) HSPA consists of mainly interactive or background services which have much more relaxed requirements on the delay than the real time services; (2) in HSPA network the radio related control functions is shifted from RNC to Node B, such as the fast packet scheduling and fast retransmissions are located at the Node B now, so there is no tight transport delay requirement on the Iub interface as in Rel99; (3) In order to protect from the heavy congestion situation on the Iub interface, a congestion control scheme is applied in HSPA, with which the long buffering delays and the resultant packet losses can be avoided. Therefore, for the HSPA traffic the transport network delay and packet discard are no longer the major QoS issue for the Iub, instead the user-relevant application QoS becomes the main objective of dimensioning.

I. Framework for the Dimensioning of HSPA

This section presents a general framework for dimensioning the Iub interface to transport the HSPA traffic (valid for both HSDPA and HSUPA). In order to properly dimension the backhaul bandwidth for the ATM-based Iub interface, The following aspects need to be determined: (1) how many HSPA cells are supported by one Node B; (2) how many HSPA UEs are active in each cell; (3) how much is the entire air interface capacity of each HSPA cell and the scheduled data rate of HSPA UEs; (4) what are the applications and services to be transferred via HSPA; (5) is there any additional traffic control functions applied on the Iub interface to avoid or diminish the link congestion; (6) at last the desired QoS target for the dimensioning. To include all the above facts into the dimensioning framework, we can summarize them into the following factors to be considered for the Iub dimensioning in HSPA:

(1) The number of HSPA cells supported by the Iub link, depicted by K;
(2) The number of HSPA UEs of cell i, depicted by N_i;
(3) The air interface capacity of cell i, depicted by C_i;
(4) The traffic model used by the HSPA UEs;
(5) The total requested HSPA traffic load in cell i depicted by ρ_i;
(6) The scheduled data rate of HSPA UE j in cell i, depicted by R_{ij} where $j < N_i$; It should be noticed that the data rate of each UE is the decision of the air interface scheduler, which takes into account the assigned cell capacities, the number of active HSPA UEs in each cell, the UE activities based on different traffic models, the UE channel conditions, other cell to own cell interference, soft handover UEs, and the congestion condition on the Iub, etc. The air interface scheduler function is up to the equipment manufacturer implementation. Different scheduler schemes and configurations result in diverse decisions on the per UE uplink data rate.
(7) Any extra congestion control functions used on the Iub interface, which result in a congestion factor φ;
(8) The required QoS for traffic type s: Q_s as the target of the dimensioning

Thus, for supporting a specific service type with Q_s as the desired target of dimensioning, the Iub bandwidth Iub_BW can be estimated according to the following general form given in equation below.

$$Iub\_BW = f(\sum_{i=1}^{K}\rho_i,\ (R_{11}, R_{12}, \ldots\ldots R_{ij}, \ldots\ldots),\ \varphi,\ Q_s) \tag{A-7}$$

If the Iub need to satisfy the QoS of several service types, then the maximum value out of the individual estimated Iub BW of service type s with QoS target Q_s should be taken for dimensioning. In the formula, R_{ij} is the result of the scheduler. In general it is not an easy task to calculate it directly thanks to the complex function of the scheduler itself and also a large amount of information on the UE and network status is required, as described in (6). Hence in this thesis R_{ij} is obtained via simulations.

Usually, the cell capacity, the air interface scheduler, and the Iub congestion control functions are specified and configured by the network operator, therefore this thesis mainly discusses the influence of different traffic models, number of HSPA UEs and various user QoS requirements on the Iub dimensioning, by fixing the scheduler function, cell capacity and the Iub congestion control function, and taking certain statistical assumption of UE channel conditions, other cell to own cell interference, and certain percentage of soft handover UEs. The following sections present the dimensioning results for HSUPA traffic, as an example.

II. Dimensioning for HSUPA Traffic

This section presents the dimensioning results of the Iub interface for carrying HSUPA traffic obtained from simulations. The simulations are performed with the HSUPA simulation model (see Appendix A.10 II). The investigations consist of two parts: (1) the performance under different traffic models, number of HSUPA UEs and their impacts on the Iub bandwidth dimensioning; (2) the estimation of the required Iub bandwidth as a function of user QoS requirements.

Traffic Models and Simulation Scenarios

In the simulations, ftp application is chosen for HSUPA UEs. In order to investigate the impact of different traffic models on the Iub dimensioning in HSUPA network, five ftp traffic models are used to generate a range of uplink traffic loads per HSUPA UE, as given in Table A.4. ftp traffic model II, as a reference model, is defined in 3GPP specification [3GP04f] which generates moderate uplink traffic. ftp traffic model I, III, IV and V are configured with different file sizes while keeping the same interarrival time distribution to produce various traffic load levels. Moreover, for investigating the influence of HSDPA traffic in downlink on the HSUPA performance, the ETSI defined web traffic model is used for HSDPA (Table 5.3). The main configuration parameters for the simulations are given in Table A.5. The CIRtarget vs. user data rates in the simulation are from [Mes06].

Following investigates the scenario of one HSUPA cell and assume a single traffic type used by all HSUPA UEs. The simulated network consists of one Node B and one RNC which are connected with each other via a single Iub link. The simulations scenarios are categorized upon the traffic models and the number of HSUPA UEs in the cell, as well as different QoS levels for the dimensioning targets. All simulation scenarios run for 3600 seconds.

ftp Traffic Model I – 1M 15s	
File size	Constant Distribution, 1 Mbytes
Interarrival time	Exponential Distribution Mean: 15 seconds
ftp Traffic Model II – 3GPP FTP	
File size	Truncated Lognormal Distribution Mean: 2 Mbytes, Max: 5 Mbytes Std. Dev.: 0.722 Mbytes
Interarrival time	Exponential Distribution Mean: 180 seconds
ftp Traffic Model III – 100K 15s	
File size	Constant Distribution, 100 Kbytes
Interarrival time	Exponential Distribution, Mean: 15 seconds
ftp Traffic Model IV – 50K 15s	
File size	Constant Distribution, 50 Kbytes
Interarrival time	Exponential Distribution Mean: 15 seconds
ftp Traffic Model V – 10K 15s	
File size	Constant Distribution, 10 Kbytes
Interarrival time	Exponential Distribution Mean: 15 seconds

Table A.4: *Traffic Models for HSUPA*

Parameters	**values**
TTI	10 ms
SAG	32 kbps
PAG	64 kbps – 1.44 Mbps
NR	6 dB
other to own interference ratio	0.6
SHO	20% of users are in SHO
RLC mode	Acknowledge Mode (AM)

Table A.5: *System Parameter Configurations for HSUPA*

Impact of Different Traffic Models and Number of UEs

This part evaluates the performance under different number of UEs and different traffic models. Figure A.14 shows the average cell throughput measured at MAC-e layer and Figure A.15 depicts the average noise rise. The results show that for the ftp traffic model II-V, the cell throughput as well as the system noise rise increases with increasing the number of

HSUPA users in the cell. It also shows that for these traffic models with 20 HSUPA users the system is still not working in its full capacity since the measured system noise rise is still below the maximum value of the allowed noise rise of 6 dB. But for the ftp traffic model I where each user generates the highest traffic demand among all investigated traffic models, with more than 10 HSUPA UEs in the cell the system is fully utilized where the noise rise reaches the 6 dB upper limit. However, the obtained average cell throughput is not always increasing with the increased number of UEs. It is seen that when the number of users is above 10, the cell throughput starts decreasing. The reason of the reduced cell throughput is because the total requested traffic demand by all users is above the available shared radio resources, therefore there are in average more secondary UEs using 32kbps data rate in the system which results in a lower overall system throughput.

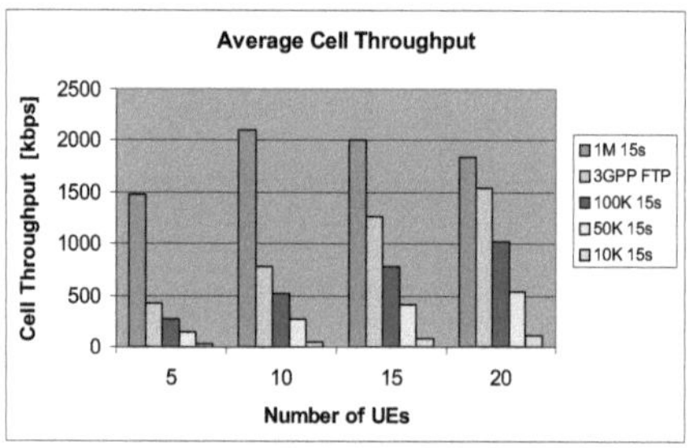

Figure A.14: *Average cell throughput*

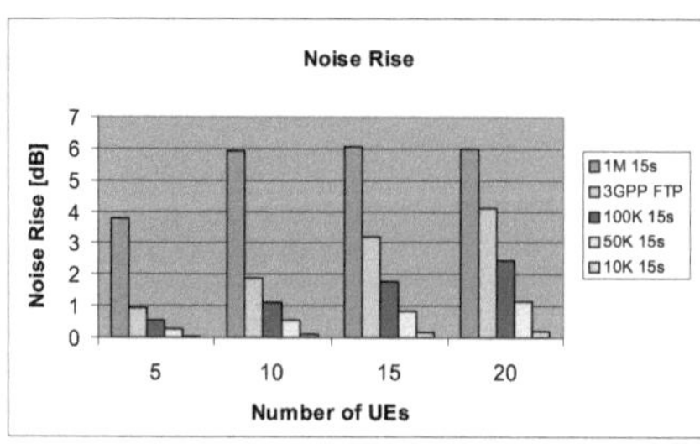

Figure A.15: *Average noise rise*

Figure A.16 gives the results of the overall average ftp application throughput, which is calculated as an average among all HSUPA users and the throughput is measured for uploading a complete file (i.e. dividing the ftp file size by its respective uploading period). It indicates the achieved end user QoS. This figure demonstrates that for all traffic models more UEs in the cell results in a decreased overall application throughput. The reason is straightforward, that is the higher number of users in the cell sharing the radio resources, the less resource each user can get in average. By allocating a lower data rate to each UE, the generated interference by each user is less and thus the total interference level can be controlled below the maximum allowed interference level. It is also seen that the obtained application throughput of using the ftp traffic model I is relatively low compared to the other traffic models, especially for above 10 UEs. This is due to the same reason that the traffic demand and user activity of the ftp traffic model I is rather high, and thus the produced interference by each user is large, so the scheduler need to set lower data rate to control the interference level, therefore in this case SAG grants are sent more frequently to those UEs. Furthermore, the average throughput of each user gets further lower when there are more such users in the cell, and in turn the achieved cell throughput is decreasing as well, as shown in Figure A.14. While observing other traffic models, it shows that as long as the cell capacity is not fully utilized (the average noise rise is below the maximum allowed noise rise of 6 dB), the obtained overall average application throughput get increased if the traffic model generates more uplink traffic.

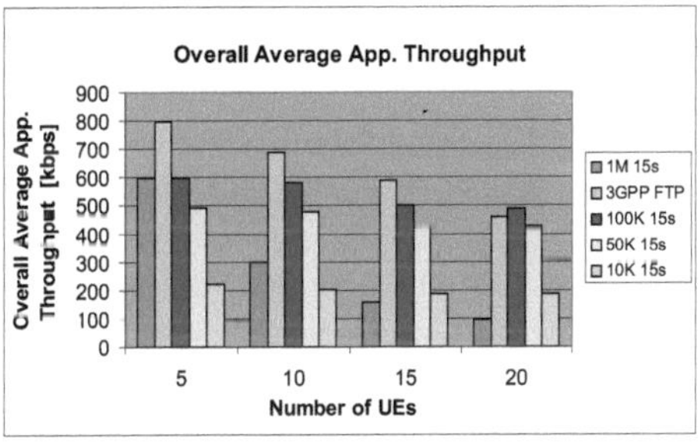

Figure A.16: *Overall average application throughput*

Figure A.17 shows the Coefficient of Variation (COV) of the cell throughput. It implies the variation and burstiness of the traffic. It is seen in Figure A.17 that in general the COV of cell throughput declines when the number of users increases, regardless which traffic model used. And the lower the traffic demand produced by the traffic model, the higher the COV values. It can be seen that the ftp traffic model V, which generates the lowest traffic load level by uploading 10Kbyte every 15seconds per UE, has the highest variation in terms of the highest COV value. The main reason is that with less traffic demands either by less number of UEs or lower traffic loads requested by the UE, there are more resources to allocate for each UE and as a result there is more chance for the user to get PAG for transmissions. The use of PAG grants leads to an abrupt increase of the data rate, which causes a higher variation. Since the maximum number of PAG users is limited in the system, therefore when the number of users increases, the probability of getting the PAG is less for each UE and in turn the COV of cell throughput reduces.

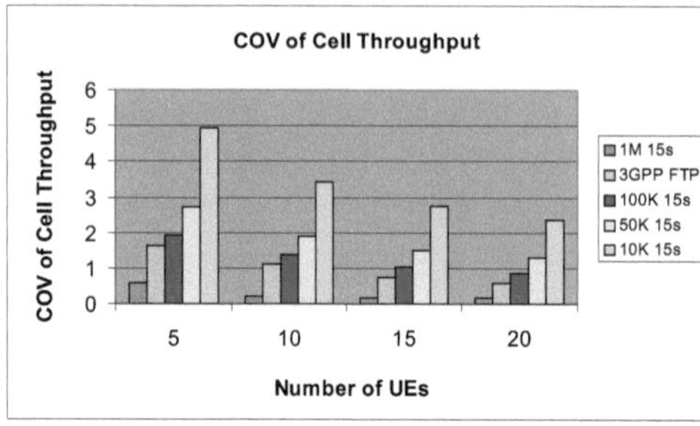

Figure A.17: *COV of cell throughput*

Following discusses the impact of different traffic models, and number of UEs in the cell on the Iub dimensioning results. Figure A.18 shows the Iub dimensioning results for achieving 95% of the overall average application throughput in term of the dimensioning factor, which is calculated as the dimensioned Iub bandwidth divided by the total HSUPA traffic demands of all UEs in the cell, and it indicates the amount of bandwidth required for satisfying QoS requirement as a ratio of the offered traffic load. By comparing the COV of cell throughput given in Figure A.17 with Figure A.18, we can conclude that the dimensioning factor is directly related to the traffic bursty property: a more busty traffic requires relatively higher extra bandwidth on the Iub link for satisfying the same QoS requirements than the traffic with less burstiness. Since with less number of UEs or having lower demand traffic model results in a more bursty traffic characteristic as presented in Figure A.17, hence more additional bandwidth in terms of higher proportion of the offered traffic demand is needed. Figure A.19 illustrates the obtained Iub link utilization calculated as the average Iub link throughput as a percentage of the Iub link rate. It shows that with the increased number of UEs or larger per user load of the traffic model, the aggregated average Iub link throughput is larger and moreover less bursty which also needs less extra bandwidth, and as a result a higher link utilization.

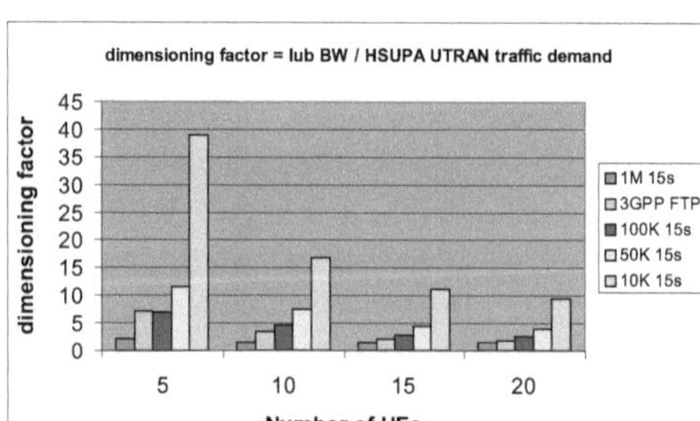

Figure A.18: *Dimensioning factor*

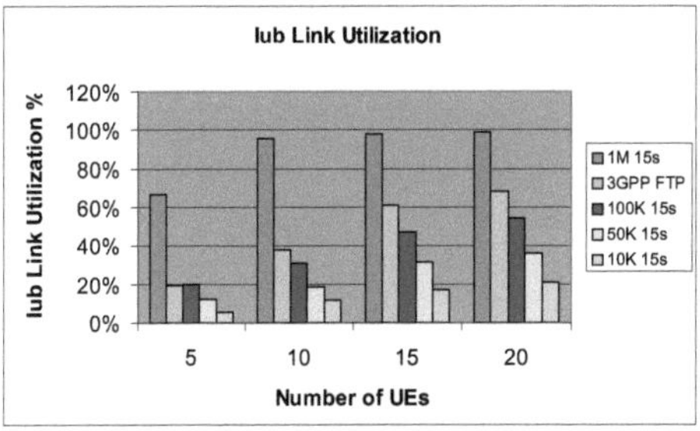

Figure A.19: *Iub link utilization*

Dimensioning over Different QoS Targets

This part discusses the Iub dimensioning over user QoS requirements. Figure A.20 presents the dimensioning factor as a function of the desired user application throughput as the QoS target. It is represented by the normalized average application throughput of all users in the cell, which is calculated as the achieved application throughput under certain limited Iub link bandwidth as a percentage of the maximum application throughput achieved with sufficient link bandwidth.

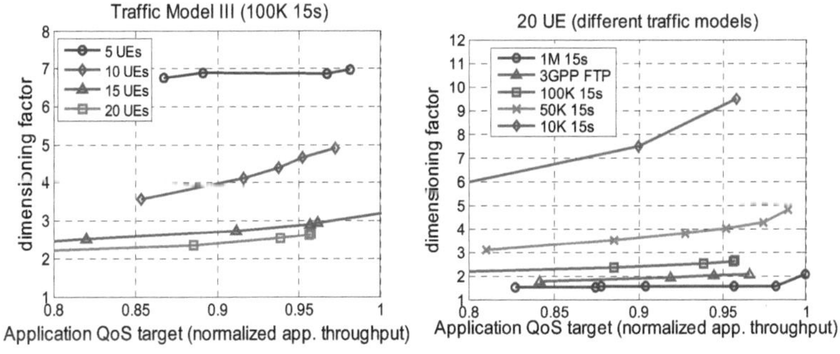

Figure A.20: *Dimensioning factor vs. QoS target*

The left diagram gives an example of different number of HSUPA UEs with the traffic model III, and the right one shows the results of 20 UEs with various traffic models. From both figures, it is seen that in general the required dimensioning factor increases with the increased QoS requirement. But the increment curve differs for different number of UEs as well as for different traffic models. By fixing the traffic model, more UEs results in a lower dimensioning factor range; by fixing the number of UEs, less traffic demand per UE leads to a higher dimensioning factor and the tendency of increment over the QoS increases as well.

From the above dimensioning results of HSUPA obtained from simulations, it can be concluded that the dimensioned Iub bandwidth is not a linear relation to the offered HSPA traffic demand, but strongly dependent on the aggregated Iub traffic characteristic, which is an outcome of the HSPA scheduling influenced by different number of HSPA users distributed in the cell and the traffic demand of each UE. For a general dimensioning approach, we suggest following steps:

(1) For the configured traffic model and number of UEs in the cell, calculate the total HSPA traffic demand in the cell, depicted as ρ;
(2) For the configured traffic model and number of UEs in the cell, estimate the bursty level of the aggregated traffic in terms of coefficient of variation of the cell throughput, depicted as β. It is dependent on the implementation of HSPA scheduler that determines each user's data rate;
(3) Determine the dimensioning factor η for different traffic bursty levels. The calculation of η is a function of the bursty level in terms of coefficient of variation of the cell throughput β, the transport network layer protocol overheads *TNL_OH*, the previously introduced Iub congestion factor φ and the QoS target Q_s.
(4) By knowing the dimensioning factor η, the dimensioned Iub bandwidth can be calculated as $Iub\_BW = \rho \cdot \eta$.

A.16 Use of Traffic Separation for Transport of HSPA and Rel99

So far, HSPA services have been already supported in the existing ATM-based UMTS networks to enhance data transmissions. Besides, the UMTS system still accommodates a significant amount of Rel99 traffic such as voice telephony. However, Rel99 and HSPA services have rather different QoS requirements: Rel99 mainly carries delay sensitive traffic like voice or streaming services; while HSPA traffic is primarily interactive and background traffic which is insensitive to the delay. This chapter presents traffic separation approach to transmit HSPA traffic in the existing ATM-based UMTS network, together with Rel99 traffic in the same radio access network. The traffic separation technique enables QoS differentiations of HSPA and Rel99 traffic, providing a differentiated QoS support for each type of traffic according to its individual QoS requirements while at the same time aims to achieve a maximum utilization of the transport resources in the radio access network. The potential benefit of applying traffic separation and its impact on the performance of the transport network as well as the end users are explored in this chapter. The quantitative evaluations are provided by means of simulations.

I. Problem without Traffic Separation

Figure A.21 illustrates the evolved UMTS system with integrated REL99 and HSPA services. It is seen that one UMTS cell supports (1) normal UMTS Rel99 users like traditional voice users; (2) HSDPA users who require HSDPA service for high-speed data transfer on the downlink, e.g. Internet access; (3) HSUPA users who only uses HSUPA service for uplink data transmissions, e.g. ftp upload; (4) or HSPA users who use HSUPA on the uplink and HSDPA on the downlink simultaneously. HSPA technology is integrated directly into the existing UMTS nodes, i.e. Node B and RNC, via software/hardware updates. Thus, the Iub interface between the RNC and Node B carries both HSPA and Rel99 traffic.

Rel99 and HSPA traffic have different delay requirements on the transport network. There is an extremely strict delay constraint on the Iub interface for DCH channels of Rel99, not only due to the delay requirements of the user traffic itself but also because of the requirements derived from supporting radio control functions such as outer-loop power control and soft handover. The excessively delayed Frame Protocol packets (their delay is larger than predefined delay boundaries) will be discarded at the Node B as they become too late to be sent over the air interface for the allocated time slot. However, HSPA traffic has significantly lower delay requirement on the Iub interface. Because for both HSDPA and HSUPA a fast scheduling is introduced at the Node B which reserves the time slot on the air interface replacing the scheduling at RNC in Rel99, and furthermore there is buffering in the Node B which supports fast HARQ. Thus, the delay requirements for HSPA are essentially only due to the service itself, which are mainly delay-tolerant best effort services that have loose constraints on the delay and delay variations. Thanks to the Rel99 traffic having a much more stringent delay requirement on the Iub interface, the Rel99 traffic is usually given a higher priority to transmit over the HSPA traffic.

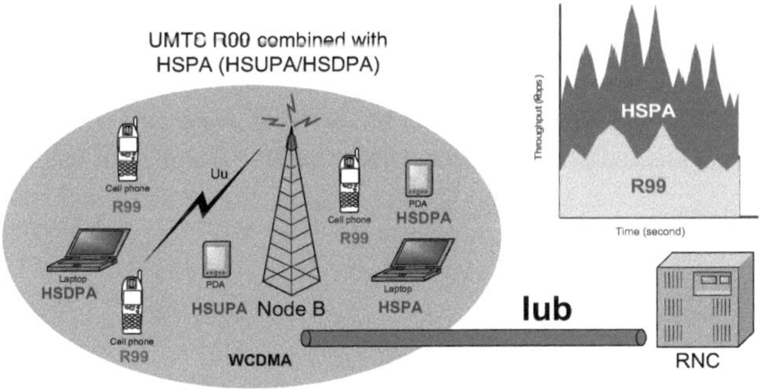

Figure A.21: *UMTS Network supporting Rel99 and HSPA traffic*

In the currently deployed UMTS system, the UTRAN transport network is ATM-based. In the case without using traffic separation at the Iub interface, the Rel99 traffic and HSPA traffic are carried within a single ATM CBR (Constant Bit Rate) VP (Virtual Path). In this case, it is assumed that a pipe with guaranteed bandwidth is established between the RNC and the Node B. So the intermediate ATM switches should not discard any ATM cells (neither of Rel99 nor of HSPA) as long as the RNC and the Node B comply with ATM service level agreement for CBR traffic. Thus, to use this high quality ATM service category CBR for the transport of HSPA and Rel99 traffic, the required bandwidth is overspecified and this causes unnecessary high costs in terms of leased ATM bandwidths.

In order to save the transport costs for the fixed lines, applying traffic separation techniques is a very popular and cost-efficient solution for the network operators or service providers. Instead of paying for overestimated CBR VPs to transmit all traffic types, cheap UBR (Unspecified Bit Rate) or UBR+ (UBR with a minimum guaranteed rate) VPs can be used separately to transport HSPA traffic whereas the CBR VPs are only used to transmit Rel99 traffic.

II. Concept of Traffic Separation

The basic idea of traffic separation technique is to apply separate ATM Virtual Paths (VPs) or Virtual Circuits (VCs) with different ATM QoS categories to transmit different traffic types. One example of using traffic separation to transmit Rel99, HSDPA and HSUPA traffic at the Iub interface is depicted in Figure A.22. In this example, each traffic type is carried by one individual ATM VP. Rel99 traffic is transported with ATM CBR (Constant Bit Rate) service category. It is defined as high priority traffic class, where bandwidth is reserved up to requested Peak Cell Rate (PCR) with guaranteed cell loss ratio and cell transfer delay. This also means high transport costs. While the transport of HSDPA and HSUPA traffic uses ATM traffic class UBR (Unspecified Bit Rate) or UBR+. UBR is a best effort service and is the lowest class of service in ATM. It is defined as low priority traffic class, which utilizes all bandwidth unused by the high priority traffic. Therefore it does not provide any guarantees for bandwidth, cell loss ratio and cell transfer delay. This traffic class has much lower transport costs. UBR+ is similar to UBR, but bandwidth is guaranteed up to a minimum rate MDCR (Minimum Desired Cell Rate). With UBR+, the HSPA traffic can be guaranteed up to MDCR.

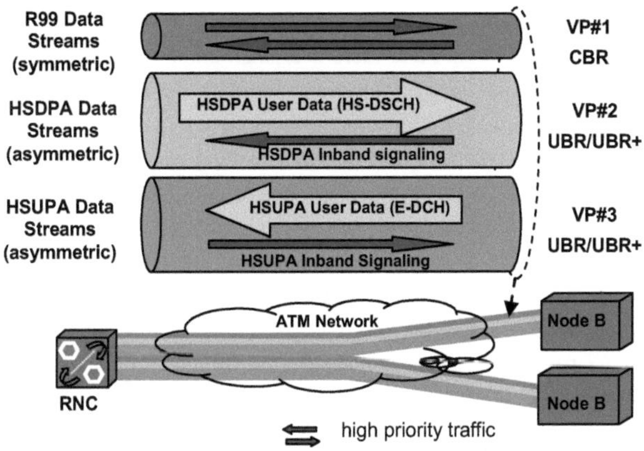

Figure A.22: *Concept of Traffic Separation*

With the use of traffic separation technique to differentiate the HSPA and Rel99 traffic over different paths with different priorities, the transport of Rel99 traffic is separated from the HSPA traffic and its stringent delay requirements can be guaranteed by using the CBR VP. On the other hand, using the UBR+ VPs for the transport of HSPA traffic can result in cell discards or long delays on the HSPA traffic if the MDCR is exceeded. But since the QoS requirements of the HSPA traffic is usually lower than the Rel99 traffic, some decreased performance of HSPA by using a separated UBR or UBR+ VP can be tolerated. In this way, compared to not using any traffic separation, the overall network cost is reduced at expenses of possible degraded HSPA performance.

III. Configuration of HSPA and Rel99 Traffic Separation

For setting up a traffic separation scenario, the following ATM parameters need to be configured:

- PCR (*Peak Cell Rate*) is the upper limit of the traffic that can be sent to the link.
- MDCR (*Minimum Desired Cell Rate*) defines a minimum guaranteed cell rate on UBR VC. It is optionally configured on either a VC or VP connection.

PCR is required to configure for both ATM CBR and UBR/UBR+ service categories. Maximum allowed bandwidth can be set different for the uplink and downlink by means of an asymmetric PCR configuration of VPs and VCs. MDCR is only configurable for UBR+ VP/VCs.

As the Rel99 traffic consists of a considerable amount of symmetric voice traffic, CBR traffic class is elected for providing high QoS for the real time services and also a symmetric PCR is configured for both directions. On the HSDPA and HSUPA path, it allows an asymmetric configuration of UBR/UBR+ VPs or VCs, e.g. asymmetric PCR or MDCR settings, to support the asymmetric traffic property of HSPA traffic, i.e. HSDPA user data is only transmitted on the downlink and there is a small amount of inband signaling traffic on the uplink, and HSUPA user data is only transmitted on the uplink with a small amount of inband signaling on the downlink.

To transport HSDPA, HSUPA and Rel99 traffic simultaneously in the UTRAN transport network, there are mainly four possible scenarios to be considered:

(1) 3 VPs: 1 CBR VP for Rel99, 1 UBR/UBR+ VP for HSDPA, 1 UBR/UBR+ VP for HSUPA;

(2) 2 VPs: 1 CBR VP for Rel99, 1 UBR/UBR+ VP for HSPA with separated VCs to transmit HSDPA and HSUPA;

(3) 2 VPs: 1 CBR VP for Rel99, 1 UBR/UBR+ VP for HSPA without separated VCs to transmit HSDPA and HSUPA;

(4) 1 VP: 1 Common CBR VP or VC to carry all traffic types.

Scenario 1 applies three VPs each transferring one traffic type. Scenario 2 and 3 uses two VPs: 1 VP is assigned for Rel99 and the other one for the HSPA traffic. For these two cases, the HSDPA data traffic will be mixed with HSUPA inband signaling traffic and the HSUPA data traffic will be mixed with HSDPA inband signaling traffic. The difference of scenario 2 and 3 is whether to use separate VCs for transmitting HSDPA and HSUPA traffic. With separated VCs, each UBR/UBR+ VC can be configured with different PCR or MDCR for HSDPA and HSUPA individually. Moreover, in order to protect the HSPA inband signaling traffic which has high priority, Cell Loss Priority bit (CLP) that is defined in the ATM cell header can be used to select which cell to discard in case of congestion: CLP=1: for low priority traffic, cell may be discarded by ATM network in case of congestion; CLP=0: for high priority traffic, cell should not be discarded by ATM network. So we can set different CLP value for the separated VCs to differentiate the inband signaling traffic and HSPA traffic so that the HSPA inband signaling traffic can be protected. In scenario 4, all Rel99, HSUPA and HSDPA traffic share one common CBR VP/VC, i.e. there is no traffic separation in this case. For scenario 1, 2 and 3, the transport of HSPA traffic can either be on a UBR or UBR+ VP. If UBR+ VP is used, there is a guaranteed minimum bandwidth for transmitting the HSPA traffic, with which a minimum QoS is assured for the requested HSPA services.

IV. Impact of Applying Traffic Separation

This section investigates the impact of using traffic separation for transmitting the HSPA and Rel99 traffic. The HSDPA and Rel99 traffic scenario is investigated as an example here for the investigations. In the following, the results of applying traffic separation to transport both HSDPA and Rel99 at the Iub interface is presented and compared to the scenario without traffic separation. The parameter settings for the traffic separation and its impact on the dimensioning will be also discussed. Additionally one example of the Iub dimensioning with traffic separation and without traffic separation is given and their transmission efficiency is compared.

Simulation Scenarios

The simulation model of HSPA and Rel99 are introduced in Chapter 5. The modeling of combining HSPA and Rel99 traffic is described in section 5.7. The simulation scenario consists of one Node B and one RNC. In the HSDPA model, a Round Robin air interface scheduler is used in the simulations. In addition, in order to protect the congestion on the Iub link, flow control and congestion control schemes are applied on the Iub. The HSDPA traffic is modeled with 20 Internet users browsing the web. The web traffic model is defined by ETSI standards (see Table 5.3). Each user requests multiple pages where the inactive time between pages follows the geometric distribution. The same traffic model is used for generating the Rel99 traffic where multiple packet-witched RABs are available for transmitting the data.

When no traffic separation is applied in the Iub interface, Rel99 and HSDPA traffic are sharing one common ATM CBR VP, where the AAL2 priority is applied which assigns higher priority to Rel99 traffic over the HSPA traffic. While in the case of using traffic separation, two ATM VPs are established: the transport of Rel99 traffic is over one ATM CBR VP and the transport of HSDPA traffic is on an ATM UBR+ VP. Here UBR+ VP is set to low priority.

The following metrics are used for performance evaluation:

- **Application Throughput**: the average throughput of transferring a web page at the application layer, excluding the reading time period. Throughput indicates the transaction speed, i.e. how long it takes to transfer a certain amount of data. It is directly related to the application delay and the volume of corresponding data transaction. The normalized application throughput is given in simulation results defined as the ratio of the application throughput under certain Iub link bandwidth to the maximum application throughput under an ideal Iub capacity.
- **Cell Discard Ratio**: in case of congestion of the Iub link, the ATM cells are discarded. The packet discard ratio is measured as the ratio of discarded ATM cells to the total ATM cells sent to the Iub link.
- **TCP Retransmission Counts**: the total number of TCP retransmissions.
- **Link Utilization**: the Iub link throughput over the given Iub link bandwidth. The link throughput includes transport network overheads as well as all TCP/RLC retransmissions.

Impact of Traffic Separation

In this section, the influence of traffic separation (TS) is investigated by comparing to the scenario without traffic separation technique in use in the transport network. In this example, there is in average 815.9kbps HSDPA traffic and 968.7kbps Rel99 packet-switched traffic on the Iub link. In both with and without traffic separation cases, the offered HSDPA and Rel99 traffic is fixed while the common Iub link rate is step by step increased. For the configuration with traffic separation, the PCR of CBR VP for transport of the Rel99 traffic is set to 1600kbps, whereas the MDCR of UBR+ VP for transmitting the HSDPA traffic is increased from 0kbps up to 1400kbps which results in the increase of the total Iub link bandwidth.

Figure A.23 compares the performance difference of using and not using traffic separation. It shows that with the usage of traffic separation technique, the end user application throughput is improved while the cell losses and resultant TCP retransmissions are reduced significantly. The major reason is that traffic separation provides a minimum bandwidth guarantee for HSDPA traffic, thus the HSDPA traffic will get less influence from the Rel99 traffic. Though the link utilization is similar in both scenarios, there is more link load contributed by RLC and TCP retransmissions in the case of no traffic separation.

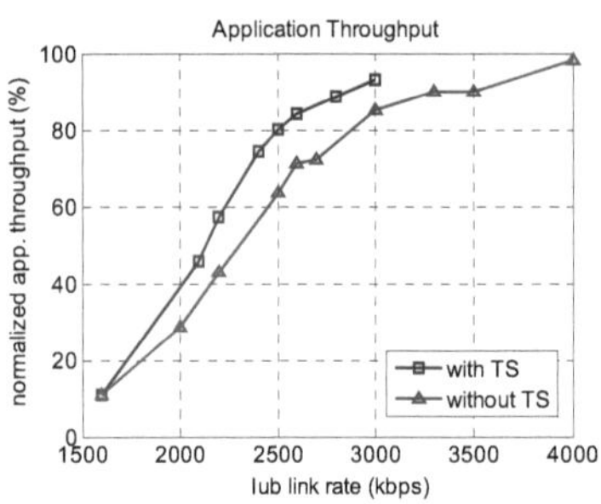

(a) Normalized application throughput

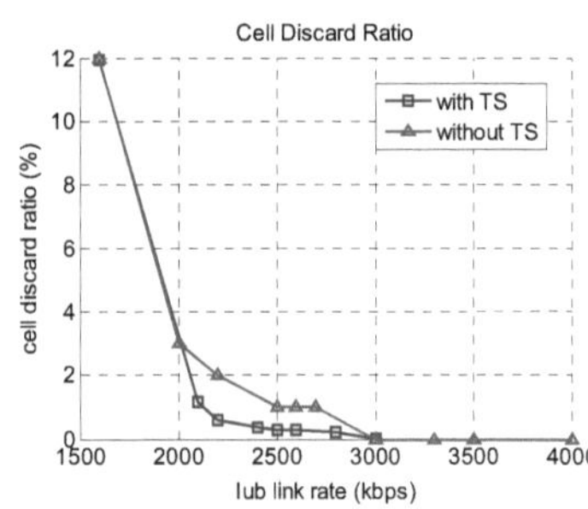

(b) Cell discard ratio

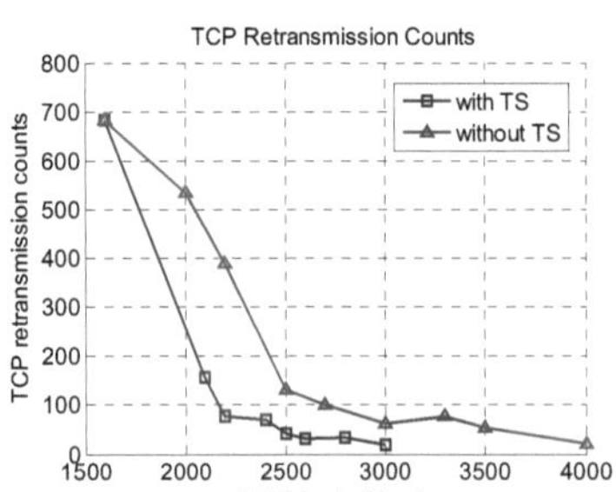

(c) TCP retransmission counts

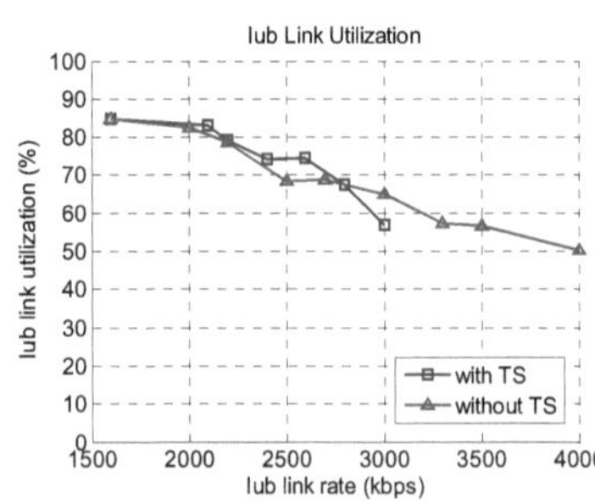

(d) Iub link utilization

Figure A.23: *Performance comparisons: with TS and without TS*

From these results, we can conclude that to achieve the same application throughput or cell discard ratio target, using traffic separation needs less bandwidth on the Iub link, which means a more efficient utilization of the transport resources. For example, to achieve 90% normalized application throughput, applying traffic separation requires 2800kbps while no traffic separation requires 3300kbps on the Iub link. The obtained bandwidth saving is 15%. To guarantee less than 1% cell discard ratio, using traffic separation requires minimum 2100kbps bandwidth while no traffic separation requires minimum 2500kbps on the Iub link. The obtained bandwidth saving is 16%.

Impact of MDCR Settings for UBR+ VP/VC

This part discuses the influence of MDCR settings of ATM UBR+ VP/VC on the overall performance, based on the results of the traffic separation scenario in the above example shown in Figure A.23. As the PCR of CBR VP for transport of the Rel99 traffic is fixed to 1600kbps, the MDCR of UBR+ VP (MDCR = the total Iub link rate – allocated bandwidth on Rel99 path) for transmitting the HSDPA traffic varies from 0kbps up to 1400kbps. It can be observed from Figure A.23 that with the increased MDCR rates the end user application performance is improved considerably: the normalized application throughput is increased from 11% to 95%. Because with a higher MDCR rate, there is more bandwidth reserved for HSDPA traffic, and therefore the performance is better. Besides the improvement of application performance with a higher MDCR setting, the network performance is also enhanced. It is observed that RLC delays, cell discard ratio, number of TCP retransmissions are all decreased when MDCR increases. But on the other hand, the link utilization drops down due to a higher Iub link bandwidth caused by larger MDCR rates is configured to transfer the same offered traffic. Therefore, MDCR should be chosen as a compromise of the system performance and the Iub link utilization. That means, MDCR rate should be set properly to achieve the maximum link utilization while stratifying the QoS target.

Moreover it is observed that the application performance is much more sensitive to the MDCR setting than transport network performances. When MDCR is larger than 500kbps (i.e. Iub link rate = 2100kbps), the transport network performance such as cell discard ratio, TCP retransmissions, has been improved drastically. And afterwards, with further increased MDCR rate, the pace of the improvement is reduced and becomes more stable. But the application throughput is still quite low with 500kbps MDCR rate: only 46% of normalized application throughput is achieved. In order to achieve more than 90% of the application throughput, the MDCR need to be set higher than 1200kbps. So it is basically a choice of network operation to decide the MDCR rate based on its predefined QoS target. If the transport network performance is more important, then a smaller MDCR is adequate. If the end user application performance is the main target of dimensioning, the MDCR rate needs to be configured to a relative higher value.

Dimensioning with vs. without Traffic Separation

This section presents the results of dimensioning of the Iub link, which transmits the HSDPA and Rel99 traffic either with or without Traffic Separation (TS) technique. In the following example, Rel99 traffic contains 50% web traffic (web traffic model is defined in Table 1) and 50% voice traffic with AMR codec. The voice model is given in Table 5.2. HSDPA consists of purely web traffic. In the following results, we fix the Rel99 traffic load and gradually increase the offered HSDPA traffic to the Iub link, and investigate the bandwidth demand for transferring the combined HSDPA and Rel99 traffic satisfying the predefined QoS targets of both traffic types. In this example, the QoS target for Rel99 traffic is 1% packet discard ratio and for HSDPA 95% normalized application throughput.

Figure A.24(a) shows the required Iub link bandwidth over different offered UTRAN traffic loads in kbps. The offered UTRAN traffic is the total sum of traffic entering UTRAN network including HSDPA and Rel99. It shows that with the increased traffic demand, the required Iub bandwidth to achieve the predefined QoS targets is increasing. It can be also obviously seen that, the required Iub bandwidth for the traffic separation scenario is much lower than that for the case without traffic separation. Therefore it is concluded that applying the traffic separation technique brings a significant bandwidth saving for the Iub dimensioning, which reduces the transport cost.

The required capacity can be also expressed in terms of "Over-provisioning factor", β, which is same as "overdimensioning factor" or "normalized capacity" defined in section 4.3. This parameter indicates in addition to the mean traffic load on the Iub link how much extra bandwidth is needed in order to fulfill the QoS requirements. Figure A.24(b) shows the obtained over-provisioning factor in percentage of the mean Iub traffic. It can be seen that the degree of over-provisioning decreases for higher traffic load on the Iub link in both with TS and without TS scenario. That means, with a larger traffic load a higher multiplexing gain is achieved which results in decreased over-provisioning factor. Furthermore, with traffic separation technique less extra bandwidth is required for transmitting the same amount of the traffic on the Iub link. And moreover, at the lower traffic load range, the over-provisioning factor of without traffic separation is much higher than that of the traffic separation scenario, and with the increase of the aggregated traffic load their gap is slowly reduced. This implies the traffic separation is able to achieve more bandwidth savings (compared to without traffic separation) at a lower mean Iub traffic load, where the room for the potential multiplexing gain is more.

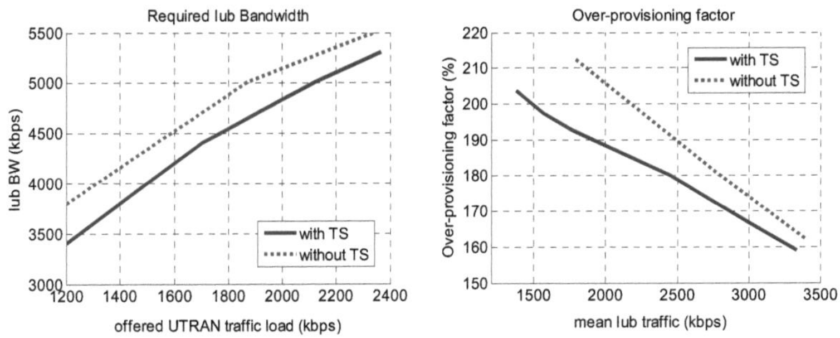

(a) Required Iub Bandwidth (b) over-provisioning factor

Figure A.24: *Dimensioning for different UTRAN load*

V. Comparing Different Configurations for Traffic Separation

This section presents the simulation results of transmitting all HSDPA, HSUPA and Rel99 traffic simultaneously in the UTRAN transport network. Two different traffic separation solutions are compared: (1) 2 VPs among which one CBR VP for Rel99 and one UBR+ VP for both HSDPA and HSUPA without using separated VCs; (2) 3 VPs where each VP carries a separate traffic type: one CBR VP for Rel99, one UBR+ VP for HSDPA and one UBR+ VP for HSUPA. The goal of the comparisons is to find out the performance differences of the two traffic separation configurations, and discuss a more suitable configuration to apply in the UTRAN ATM transport network, which can achieve high bandwidth utilization while guaranteeing the QoS of each traffic type.

Simulation Scenarios

The scenario consists of one Node B and one RNC. In the simulated scenario, there are six HSPA users in the cell: 6 HSDPA in downlink and 6 HSUPA in uplink. For both HSUPA and HSDPA, an ftp traffic model is used which generates heavy traffic on both uplink and downlink. The ftp traffic model parameters are defined in Table A.6. With this ftp traffic model, all users are downloading or uploading very large files continuously. That means, all users have always data to be transmitted at any time and demanding the network resources constantly. Such a traffic model can be considered as a worst case traffic scenario from the network point of view. In addition there is a 2 Mbps Rel99 traffic transmitted on the Iub as well. In this setup, the Iub is heavily loaded, which is considered to be in a worst case scenario. In such overloaded traffic scenario, the investigated gain of applying traffic separation will be more significant. The simulations scenarios are classified into two categories depending on the configurations of the transport network (2 VPs or 3 VPs). The transport network is configured with a line rate of 4 Mbps. The configurations of each VP are given in Table A.7. In both scenarios, the Rel99 traffic is fully utilizing its allocated 2 Mbps capacity.

File size	Constant Distribution Mean file size = 5 Mbyte
Inter-arrival time	~ 0.0 seconds which means immediately after the first file downloading of the second file is started

Table A.6: *ftp Traffic Model for HSPA*

2 VPs	1 CBR VP for Rel99: PCR = 2 Mbps 1 UBR+ VP for HSDPA/HSUPA with their inband signaling traffic, PCR = 4 Mbps, MDCR = 0.4 Mbps
3 VPs	1 CBR VP for Rel99: PCR = 2 Mbps 1 UBR+ VP for HSDPA and HSUPA inband signaling traffic, PCR = 4 Mbps, MDCR = 0.4 Mbps 1 UBR+ VP for HSUPA and HSDPA inband signaling traffic, PCR = 4 Mbps, MDCR = 0.4 Mbps

Table A.7: *Simulation Scenarios*

All simulation scenarios run for 1000 seconds. Following the main parameter configurations for the simulations are given:

- Radio configuration
  - TTI = 2 ms HSDPA, 10 ms HSUPA
  - Noise Rise = 6dB (for HSUPA)
  - Others-to-own interference factor = 0.6 (for HSUPA)
- Number of HARQ processes per user flow: 4
- RLC protocol: Operate in RLC AM mode
- TCP protocol: TCP New Reno version

Performance Analysis

In the following the comparison results are presented. Special attentions are paid to two main different performance aspects: transport network performance and the end user performance. In the transport network, the network-specific QoS measures such as packet transport delay and packet losses need to be controlled to meet the agreed quality of a network, which is targeted to low delay and low loss. On the other hand, the end user performances like application throughput also need to be guaranteed to satisfy the user's particular QoS requirements. In the following presented results, the total discarded ATM cells and the link throughput represent the transport network performance. The user application throughput and the experienced TCP retransmissions indicate the end user performances.

Figure A.25 shows the total discarded ATM cells on the downlink (left diagram) and the uplink (right diagram) individually. The ATM cell discard is due to the ATM buffer overflow, which is caused by the link overload. In Figure A.25 the ATM cell discards only happen to HSDPA traffic on the downlink and HSUPA traffic on the uplink. The main reason of cell losses of HSPA traffic is because that with both 2 and 3 VPs configurations, the HSPA traffic is assigned to cheap UBR+ VP. In the heavy loaded situations, the intermediate ATM switches may drop HSPA cells if the MDCR rate has been exceeded by the amount of the injected HSPA traffic. In this case, the traffic which is above the MDCR has no bandwidth guarantee at all and therefore can be discarded in favor of Rel99 traffic. While on the Rel99 path, there is no cell loss on both directions as the Rel99 traffic is using a separate CBR VP which has a guaranteed bandwidth between the Node B and RNC.

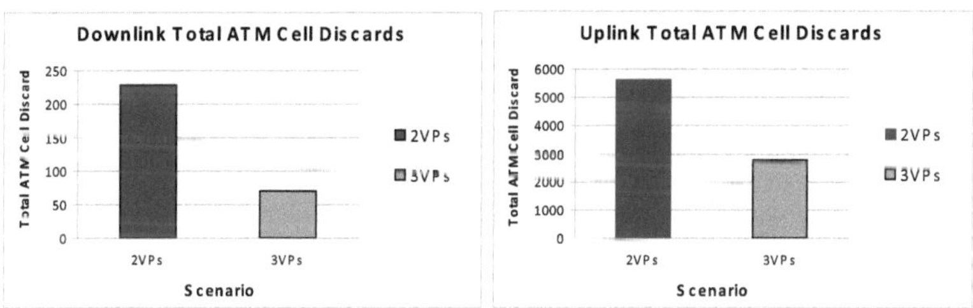

Figure A.25: *Total ATM cell discards on downlink and uplink*

When comparing the cell losses of using 2 and 3 VPs solution, it can be found that the 2 VPs scenario has more cell losses than the 3 VPs scenario on both uplink and downlink directions. This is because, in the configured 2 VPs scenario the HSDPA and HSUPA traffic are sharing one UBR+ VP but without separated VCs, i.e. HSDPA data and HSUPA signaling traffic are mixed on the downlink while HSUPA user data and HSDPA signaling traffic are combined on the uplink. Thus, there is no way to distinguish the signaling traffic from the user data traffic and therefore the signaling traffic can be dropped irrespective of its importance. Usually the signaling traffic consists of control messages sent by HSPA flow control or congestion control function, which are used to adapt the user data rate according to the available bandwidth in order to protect the Iub interface from heavy congestions. If the signaling traffic is dropped, the system will not react properly on the congestion situations and therefore the transport path keeps overloaded. But with 3 VPs solution, the signaling traffic is protected from the user data traffic as different VPs are used for transmitting HSDPA and HSUPA traffic separately. For each individual UBR+ VP to transmit either HSDPA or HSUPA traffic, the configured MDCR will guarantee the bandwidth for the signaling traffic. Therefore, the signaling traffic is with a high probability not dropped in this case and hence the system can respond more correctly to the congestion situations. As a consequence, the resultant losses within the transport network in the 3 VPs scenario are less due to a better controlled transport network.

The cell discard on the ATM layer can further have influence on the TCP transport layer of the users. When the discarded cells are not able to be recovered by the RLC retransmissions, they will cause TCP retransmissions. It is seen in Figure A.26 that resultant TCP retransmissions on HSDPA and HSUPA traffic is higher in the 2 VPs scenario than the 3 VPs scenario due to a higher cell losses. The TCP retransmissions will reduce the corresponding TCP congestion window and as a consequence decrease the user throughput.

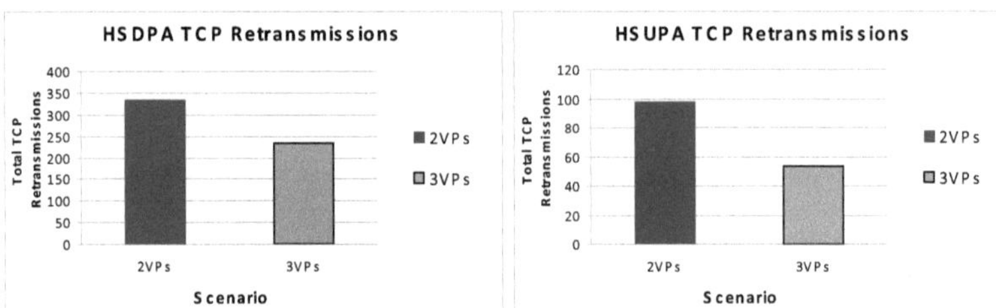

Figure A.26: *HSDPA and HSUPA TCP retransmissions*

Figure A.27 presents the HSDPA per user application throughput (left figure) and the average application throughput of all HSDPA users (right figure). It can be obviously seen that the achieved average HSDPA user application throughput is slightly lower in the 2 VPs scenario. This is due to more TCP retransmissions, as explained above. Moreover, since there are more cell losses in the transport network as shown in Figure A.25, the Iub congestion control function of HSDPA will be triggered more often to reduce the data rate of the users on the downlink direction. Similar is for the HSUPA traffic.

Figure A.28 compares the resultant ATM link throughput on the downlink (left diagram) and uplink (right diagram) individually. It shows that the link throughput obtained in the 3 VPs scenario is a little higher than the 2 VPs one, but the gap is not significant.

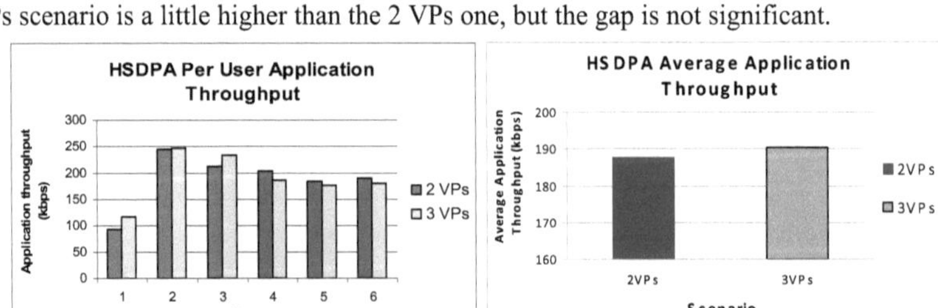

Figure A.27: *HSDPA application throughput*

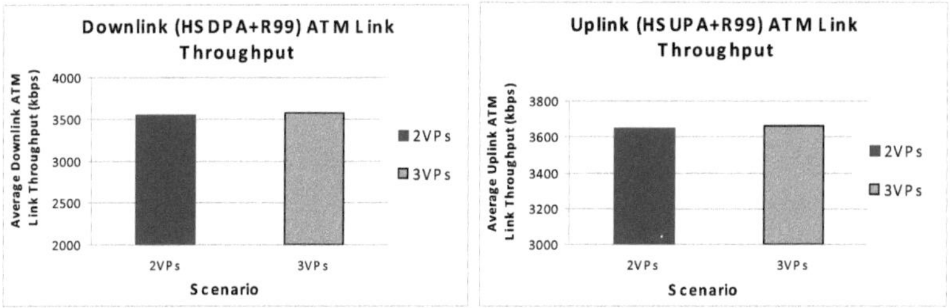

Figure A.28: *Downlink and uplink ATM link throughput*

As a summary, through comparing the network and user performances of using 2 and 3 VPs as the transport solutions, it can be concluded the obtained user and transport network performances of using 3 VPs is slightly better than the 2 VPs due to a better protection for the signaling traffic at the expenses of one additional VP. From the performance perspective, the 3 VP setup is the best transport solution as it provides a clear traffic separation and QoS differentiation. However, it also requires additional cost for buying a separate VP. By taking considerations of both QoS and network costs, both 2 and 3 VPs solutions can be considered for the network operators or service providers to design a UMTS access network to transport both Rel99 and HSPA traffic.

A.17 Dimensioning for Mixed Traffic Cases (ATM-Based Iub)

The following results in Figure A.29 validate the proposed dimensioning approach for the mixed traffic scenario in the ATM-based Iub. The analytical dimensioning model is presented in section 6.3.7.5 and its corresponding dimensioning approach is given in section 6.3.9. In the given example, there is 40% voice traffic and 60% web traffic. The traffic model of the voice traffic is according to Table 5.2. And web users request a constant 50 kbyte web page with RAB 128 kbps. For dimensioning process, the desired QoS target for the elastic web traffic is the end-to-end delay factor of 1.23. It can seen in Figure A.29 that the analytical calculated bandwidth based on M/G/R-PS model is able to provide accurate estimations on the bandwidth requirements (shown in the left diagram) with a relative error less than 10% (shown in the right diagram).

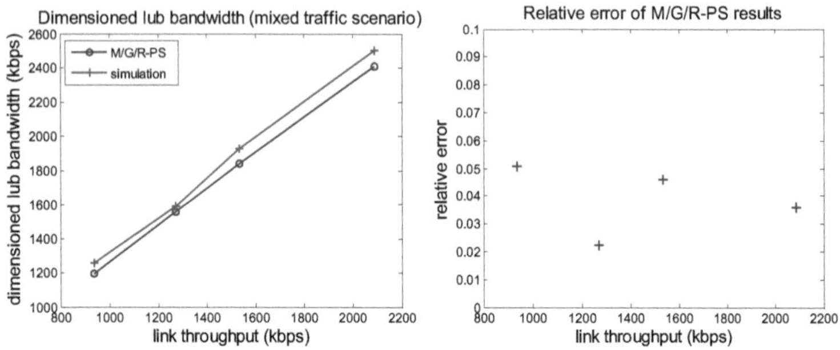

Figure A.29: *Validating the dimensioning approach for the mixed traffic scenario for web traffic: RAB 128 kbps, no CAC, page size =50 kbyte, delay factor f = 1.23 for voice traffic: AMR 12.2 kbps codec*

A.18 Validating Dimensioning Approaches for Multi-Iub IP RAN

I. Validation for Approach (1)

The approach (1) is normally for the cases of a desired end-to-end delay factor close to 1. For validation, in the following examples the target end-to-end delay factor f_{target} varies between 1.1 to 1.3. Thus, the expected end-to-end delay $T_t = T_{\min} \cdot f_{target}$. The offered traffic amount on the last mile is 1060kbps, and thus the total carried traffic at the backbone link is 1060kbps * 4 (number of Node Bs). The validation of the applicability of the approach (1) for dimensioning the last mile link and the backbone link is performed with the following steps:

- Step 1: given the desired end-to-end delay factor f_{target}, assign the target delay factor for the last mile link and backbone link individually according to the approach (1), i.e. $f_{ac\_i} = f_{bb} = f_{target}$;
- Step 2: for each link, applying the proposed extended M/G/R-PS model (to resolve the equation (6.26) and (6.27) numerically) to calculate the required link bandwidth for the assigned target delay factor;
- Step 3: apply the calculated bandwidths into the simulation;
- Step 4: after running the simulation, measure the average application delays from the simulation and compare it with the expected end-to-end delays.

The graph in Figure A.30 (a) shows the average end-to-end delays T_t for different desired end-to-end delay factors as the QoS criteria. Every point of the curve represents the delay measured in the simulation for the corresponding target end-to-end delay factor. In addition to the simulated mean values, the expected end-to-end delays (target delays) are given in the same graph for comparison. Figure A.30 (b) evaluates the differences of the simulated delays and the expected end-to-end delay. It is seen from both graphs that with the bandwidths calculated from the proposed approach (1) for the last mile link and the backbone link (given in Table A.8) in the simulation, the obtained average end-to-end delays from the simulations are close to the target end-to-end delays, i.e. the desired user application QoS is met well. Though it can be observed that the simulated delays are a little higher than the defined delay targets. This gap is due to that the proposed approach (1) neglects the correlations between concatenated links, as it dimensions each link independently according to the end-to-end delay factor assuming that this link is the bottleneck while neglecting the impact of the other link. Nevertheless, the presented results demonstrate that the proposed approach (1) can work sufficiently well for assigning appropriate bandwidths for the individual links in the network for satisfying small end-to-end delay factors (which are close to 1), despite the independence assumption. And secondly, the simulation results also verify that the extended M/G/R-PS model can predict the achievable application QoS accurately and thus can be used for dimensioning links and networks for elastic traffic for a guaranteed user application QoS.

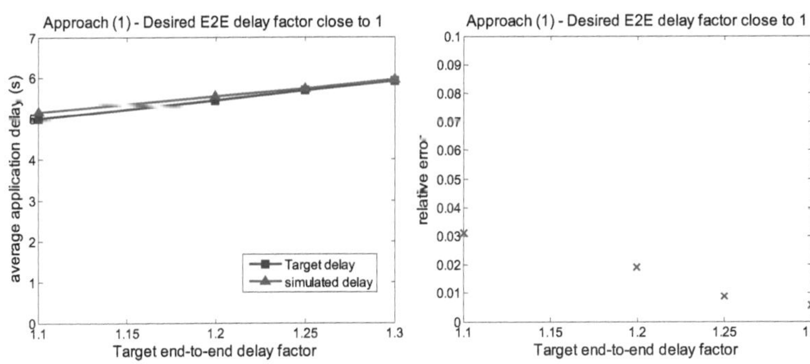

(a) End-to-end delays (b) relative error

Figure A.30: *Validating Approach (1):E2E delays over different target delay factors*

Target end-to-end delay factor f_{target}	Last mile link bandwidth (kbps)	Backbone link bandwidth (kbps)	Overbooking factor for backbone link
1.1	1458.8	4817.7	1.21
1.2	1297.6	4416.8	1.18
1.25	1279.3	4289.6	1.19
1.3	1232.6	4229.9	1.17

Table A.8: *Dimensioning results for different target delay factors with approach (1)*

In Table A.8, given the last mile link bandwidth and the backbone link bandwidth calculated from the approach (1), the overbooking factors of the backbone link are calculated with equation (8.1). It is observed that the obtained overbooking factors with the suggested dimensioning approach are larger than 1. That means the proposed dimensioning approach (1) is capable of attaining overbooking on the backbone link.

II. Validation for Approach (2)

The following simulation investigates the accuracy of the approach (2). The approach (2) is used for the case when the desired end-to-end delay factor is much larger than 1 corresponding to a relatively low application QoS requirement. In the presented simulation example, the target end-to-end delay factor f_{target} is set to 2. Thus, the expected end-to-end transfer delay for a page size of 50 kbyte $T_t = T_{\min} \cdot f_{target} = 9.1$ seconds, which is the target delay. The validation of the applicability of the approach (2) for dimensioning follows the same procedure as for validating the approach (1), only that the assignment of the target delay factors for the individual links is according to the approach (2): the target delay factor for the last mile link is set to f_{target} and the one for the backbone link is set to 1.1 in this example.

Figure A.31(a) shows the average end-to-end delays T_t under different offered traffic load per last mile link. Every point of the curve represents an average end-to-end delay for a given offered traffic load. In the graph, the simulated values are compared to the target delay. Figure A.31(b) evaluates the deviation of the simulated delays to the expected end-to-end delay. As can be seen clearly from the graph (a), the average end-to-end delays obtained from the simulations fit quite well with the target end-to-end delays. The observed gap of the simulated delays with the expected end-to-end transfer delay is fairly small. This proves that the proposed approach (2) can provide sufficiently good QoS guarantee for large delay factors for the elastic traffic flows.

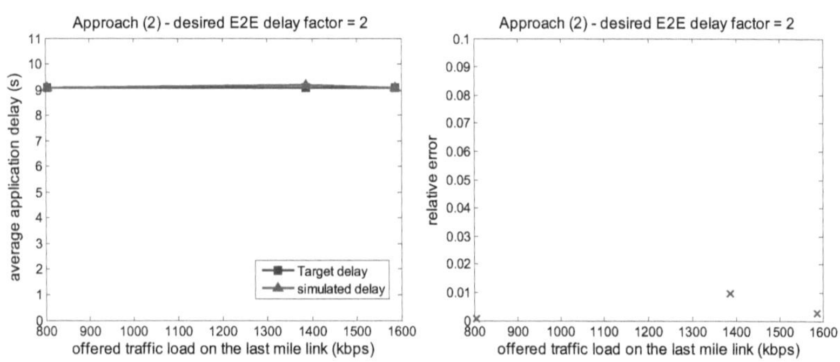

(a) End-to-end delays (b) relative error

Figure A.31: *Validating Approach (2): E2E delays over different offered last mile loads*

The calculated bandwidths of the last mile link for different offered traffic load is given in Table A.9, as well as the corresponding achieved overbooking factor on the backbone link.

From the outcome given in Table A.9, it can be seen that the approach (2) indeed cannot achieve much overbooking on the backbone link, since the goal of this approach is to limit the last mile link bandwidth as much as possible while allocating a relative sufficient bandwidth for the backbone link.

Offered traffic load at the last mile link (kbps)	Last mile link bandwidth (kbps)	Overbooking factor for backbone link
804.1	915	1
1385.3	1480	1
1586.8	1700	1

Table A.9: *Dimensioning results for different offered traffic loads with approach (2)*

A.19 Investigating Overbooking in the ATM-Based UTRAN

The relations of overbooking with different traffic types, last mile utilizations, last mile link capacities and number of connected Node Bs are discussed in the following. The investigations and qualitative evaluations are based on extensive simulations.

Simulation Scenarios

In the following investigations, the simulation scenarios are categorized upon the following aspects:

(1) Various traffic profiles: voice traffic only, packet-switched data traffic only (web traffic), mixed voice and packet-switched traffic with examples of packet-switched dominant and voice dominant traffic scenario respectively;
(2) Different last mile utilizations, i.e. varied offered loads per Node B on the last mile, given a certain last mile link rate;
(3) Different last mile link capacities or link rates;
(4) Different number of Node Bs.

For the packet switched data services, web traffic is simulated using the ETSI traffic model (see Table 5.3) and for voice services a traffic model specified in Table 5.2 are used. The criteria of choosing the optimum overbooking factor in the following investigations is to maximally utilize the backbone bandwidth while satisfying the QoS of UTRAN transport network as well as the end users. In the following investigations, the QoS requirements in UTRAN is that the acceptable FP PDU delay value considered for voice service is 10ms for 99% of voice packet transmissions and 30ms for data services for 99% of packet-switched packet transmissions. Those packets whose experienced FP PDU delays violate the given values would be discarded. That means, we allow maximum 1% packet loss ratio for both voice and packet-switched services. For the end users, the required average transfer delay for transferring average 25Kbyte page should be within a maximum defined value, e.g. 2.5 seconds in the presented examples, and the voice packet delay should be lower than 100ms [3GP06c].

Impact of overbooking

Figure A.32 shows the impact of setting of overbooking factors. The given scenario consists of 4 Node Bs where the last mile link is set to 1 E1 (2 Mbps). Each Node B transmits a mixture of packet-switched web and circuit-switched voice traffic with 85% web traffic, and the total offered load per Node B is around 40% of the last mile link.

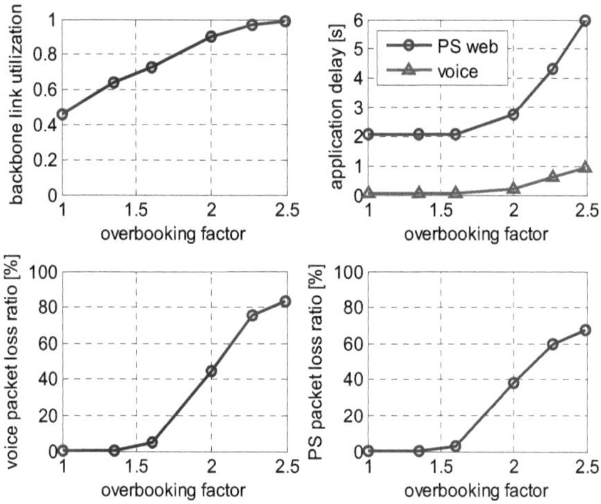

Figure A.32: *Impact of overbooking*

It is seen from Figure A.32 that when *OB_factor* is increased from 1 to 2.5, the backbone link utilization is improved by more than 50%. But the UTRAN network performance in terms of the packet loss ratio of both voice and packet-switched packets are gravely depredated with higher *OB_factor*, especially when *OB_factor* is larger than 1.6. When *OB_factor* is set to 2.5, the packet loss ratio for voice reaches above 80% and for packet-switched about 70%. To satisfy 1% packet loss ratio as the QoS target for both services, *OB_factor* need to be configured below 1.4 in this example. Furthermore, the end user performance is also seriously influenced by the overbooking. It can be seen that there is a rapid increase of the application delays for both packet-switched web and circuit-switched voice services when the *OB_factor* is above 2. This investigation demonstrates the importance of selecting optimum *OB_factor* to provide a good balance between the costs, i.e. to utilize the bandwidth resource as much as possible, and satisfy the predefined QoS objectives. If several QoS requirements need to be fulfilled, then the minimum *OB_factor* should be chosen which satisfies all QoS constrains.

Overbooking under different last mile utilizations

This part evaluates the impact of different last mile utilizations on the setting of the optimum *OB_factor*. In the investigated scenarios, there are 4 Node Bs and the last mile link is set to 1 E1 (1.6 Mbps for user plane). Here investigates four different traffic profiles: pure packet-switched web traffic, pure circuit-switched voice, traffic mix with packet-switched traffic dominant (85% web) and voice dominant (60% voice) individually. The last mile utilization varies between 10% and 60%.

In Figure A.33, the left diagram shows the selected optimum overbooking factors over different last mile utilizations and the right one shows the resultant backbone link utilizations. It shows that with the increased last mile utilization in terms of higher offered load per Node B, the selected optimum *OB_factor* declines for all traffic profiles. This is because when the offered traffic on the last mile is larger, the room for statistical multiplexing in the aggregated traffic on the backbone link is less and therefore the potential overbooking capability is lower. Moreover, by comparing different traffic profiles we can observe that the selected optimum *OB_factor* for packet-switched traffic only and packet-switched traffic dominant scenario are similar, and when there is more voice traffic the selected optimum *OB_factor* gets increased and for voice only scenario it becomes the highest. The is because that for the defined 1% packet loss ratio, the voice traffic requires less bandwidth due to its less bursty nature than the web traffic. So for this network QoS target the voice traffic can be set to a higher *OB_factor* than the web traffic. Consequently, the larger the configured *OB_factor*, the higher is the resultant backbone link utilization.

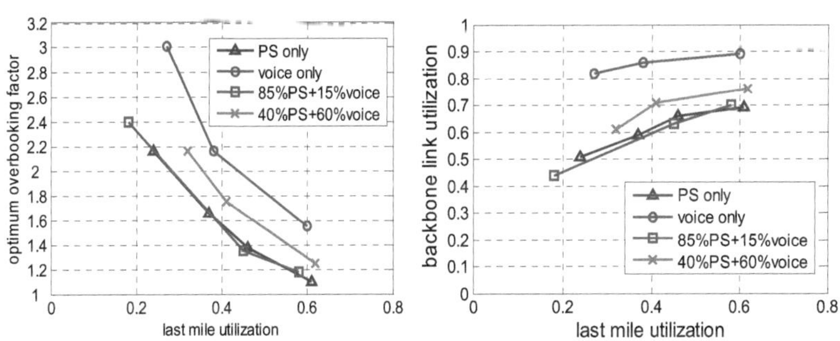

Figure A.33: *Optimum OB_factor (left), link utilization (right) vs. last mile utilizations*

Overbooking under different last mile link capacities

Figure A.34 and Figure A.35 present the impact of the last mile link capacity on the selection of the optimum *OB_factor*. Figure A.34 compares the optimum *OB_factors* over different last mile utilizations under two different last mile link rates, 1.6 Mbps and 2.5 Mbps (for user plane) with 4 Node Bs connected to the backbone link. It is seen that for the packet-switched and packet-switched traffic dominant scenarios, with 2.5 Mbps last mile link the obtained optimum *OB_factor* is considerably higher than with 1.6 Mbps, but there is no significant gain for the voice only and voice dominant scenarios. For the bursty packet-switched data traffic, the self-similar traffic property allows a high probability of utilizing the spare resources of other users that are not transmitting. Using a higher last mile link resulting in more simultaneous connections can achieve more statistical multiplexing gain, and hence a higher overbooking is allowed. But the voice traffic is Poisson like traffic, i.e. the voice traffic will be smoothed under a high level aggregation, so with enough number of voice users the variations of the aggregated traffic remain relatively stable, therefore the last mile link capacity does not play an important role on overbooking in this case, so the statistical multiplexing achieved by 1.6 Mbps is similar to that achieved by 2.5 Mbps link.

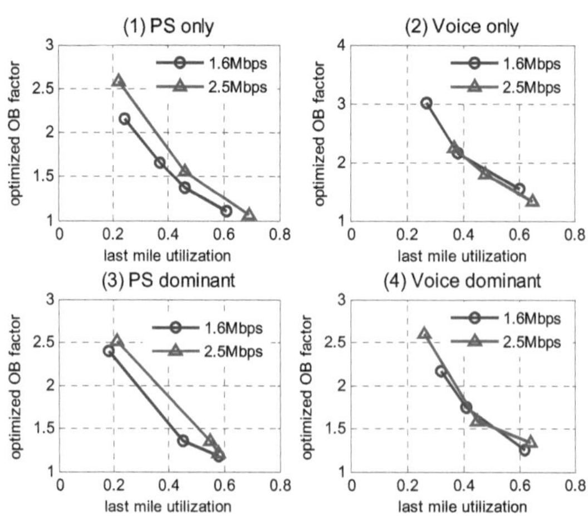

Figure A.34: *Optimum OB_factor vs. last mile utilization (various last mile links)*

Figure A.35 shows the overbooking as a function of the last mile link capacity for 2 Node B and 4 Node Bs, given a packet-switched traffic dominant scenario where the offered per Node B traffic consists of 87% web traffic, occupying in average 30% of the last mile link bandwidth. The left chart is the selected optimum *OB_factor* and the right one is the corresponding backbone link utilization. It is seen from Figure A.35 that when the last mile link rates increases from 1.6 Mbps to 3 Mbps, the allowed maximum *OB_factor* gets increased as well as the backbone link utilization for both 2 and 4 Node Bs. And having 4 Node Bs can obtain higher overbooking than having 2 Node Bs. The reason of improved overbooking with a higher last mile link bandwidth has been explained, i.e. more statistical multiplexing gain is achieved with a higher last mile link bandwidth for the bursty data traffic.

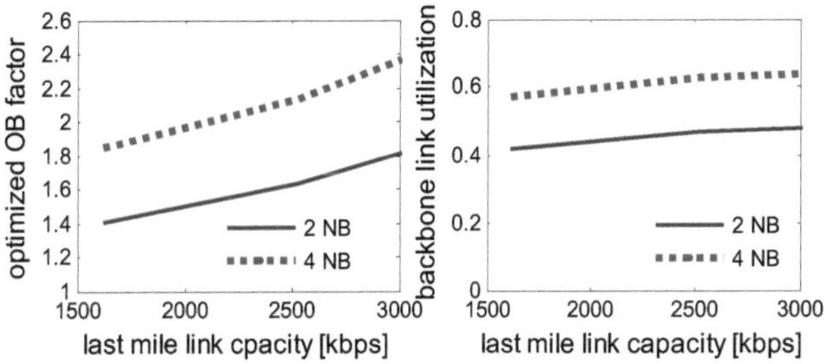

Figure A.35: *Optimum OB_factor vs. last mile link capacity (packet-switched dominant)*

Overbooking under different number of Node Bs

This part evaluates the impact of different number of Node Bs on the setting of the optimum *OB_factor*. The investigated scenario is a packet-switched dominant traffic with 85% web traffic generated by each Node B. The last mile link is set to 1 E1 for each Node B. In Figure A.36 the left diagram shows the chosen optimum overbooking factors over different last mile utilizations, for 2, 4, 16 and 50 Node Bs respectively. The right one presents the relations of optimum overbooking factors over different number of Node Bs for various last mile utilizations. It can be observed from both figures that more Node Bs connected to the backbone link results in a better overbooking. The left figure demonstrates that the optimum overbooking factor also strongly depends on the last mile utilization and the relation of optimum *OB_factor* with the last mile utilizations differs for different number of Node Bs: (1) under lower last mile utilization, the optimum *OB_factor* is more sensitive to the configured number of Node Bs, i.e. the improvement of overbooking is more significant with the increased number of Node Bs; (2) under higher last mile utilization, the optimum *OB_factor* converges to 1 regardless of number of Node Bs. When the last mile utilization reaches 60%, there is almost no overbooking possible in this example. This is because that with an increased offered load per Node B, the statistical multiplexing gets more difficult and it slowly reaches a maximum value. The right figure also shows the same conclusion, it can be obviously seen that the lower the last mile utilization, the higher is the optimized *OB_factor* value. The overbooking is improved with higher number of Node Bs, and when the number of Node Bs is greater than 16 the increment of the optimum *OB_factor* gets much more slowly converging to a maximum value, which should be lower than 1/last mile utilization.

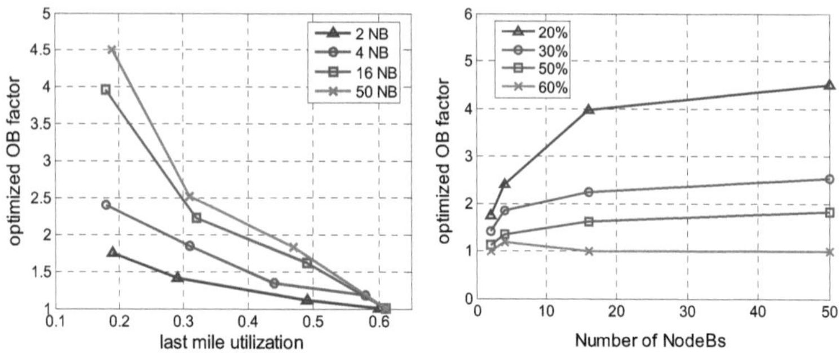

Figure A.36: *Impact of number of Node Bs*

A.20 Statistical Evaluation: Confidence Interval

The *Confidence interval* (CI) is a statistical range with a specified probability that a given parameter lies within the range. In statistics, confidence intervals are the most prevalent form of interval estimation. Confidence intervals contain two parts:

- An interval within which the population parameter is estimated to fall (estimate ± margin of error).
- A confidence level which states the probability that the method used to calculate the interval will contain the population parameter.

There are two methods to evaluate CI: *normal distribution* is used for large sample size reps; and *Student's t-distribution* (or simply the *t*-distribution) is used to small sample size (e.g. the number of samples are less than 30).

The following gives some examples of evaluating CIs for the statistical results measured from simulations in this thesis. From the observations of the estimated CIs, the reliability and stability of the simulation results can be determined.

In Figure A.37 a web scenario is investigated in the ATM-based Iub. In this example, the confidence intervals for the average application delay are estimated. The graph shows the average application delays and their 95% CI over different loads. For each load in the graph, 8 simulation runs were performed and each run used a different random seed. For each simulation run *i*, the average application delay is measured as T_i. In Figure A.37, every point of the curve represents the mean delay value $\overline{T}$ over 8 average application delays measured from the 8 different seeds simulations. It is calculated with equation (A-8) where $n = 8$. Since the number of samples is less than 30, the *Student distribution* is used in this case.

$$\overline{T} = \frac{1}{n}\sum_{i=1}^{n} T_i \qquad \text{(A-8)}$$

The confidence interval for a confidence level (1-α) is calculated below:

$$\left[\overline{T} - t_{\left(\alpha/2,\ n-1\right)} \frac{s}{\sqrt{n}},\ \overline{T} + t_{\left(\alpha/2,\ n-1\right)} \frac{s}{\sqrt{n}}\right] \tag{A-9}$$

$t_{\left(\alpha/2,\ n-1\right)}$ is the upper critical value of the t-distribution with (n-1) degrees of freedom for the desired confidence level (1-α). Its value can be look up from the t-distribution table. s is the sample standard deviation computed with an unbiased estimator as follows:

$$s = \sqrt{\frac{1}{n-1} \sum_{i=1}^{n} (T_i - \overline{T})^2} \tag{A-10}$$

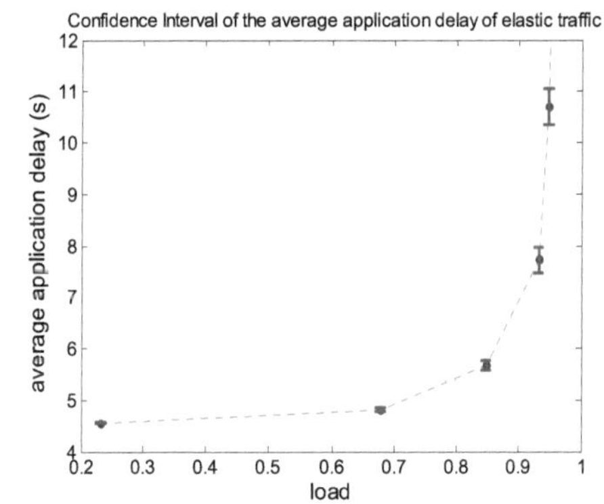

Figure A.37: *Average application delay with 95% CI: web scenario, RAB 128 kbps, no CAC, constant file size of 50 kbyte, Iub link = 1 E1 line (ATM)*

The calculated values of 95% CI for the average application delays and their corresponding margin of error are given in Table A.10. It can be observed from Figure A.37 and Table A.10 that the confidence intervals get larger with a higher load (link utilization). However, when the load reaches 95% of the link capacity the relative margin of errors remain below 5%. It demonstrates that the measured average application delay statistical results from simulations are reliable and stable giving a sufficient confidence. For the same scenario, the confidence intervals for the packet loss ratios in the transport network were also investigated in Figure A.38. The graph shows the packet loss ratios and their 95% CI over different loads. Same as above, every point of the curve represents the mean packet loss ratio of 8 different seeds simulations. The calculations of the confidence intervals are same as above.

Load (normalized)	mean application delay (s)	95% Confidence Interval for the mean application delay (s)	margin of error (s)	relative margin of error
0.23	4.5517	[4.5409, 4.5625]	0.0108	0.0024
0.68	4.8119	[4.7739, 4.8499]	0.038	0.0079
0.85	5.6766	[5.5759, 5.7773]	0.1007	0.0177
0.93	7.722	[7.474, 7.97]	0.248	0.0321
0.95	10.7047	[10.3602, 11.0492]	0.3445	0.0322

Table A.10: 95% *CI for application delays of a 50kbyte web page (RAB 128kbps)*

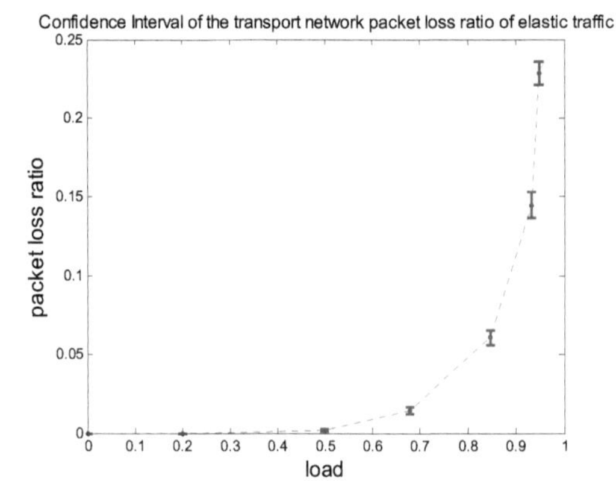

Figure A.38: *Packet loss ratio with 95% CI: web scenario, RAB 128 kbps, no CAC, constant file size of 50 kbyte, Iub link = 1 E1 line (ATM)*

Similarly, the 95% confidence intervals were also investigated for the CAC connection reject ratio (see Figure A.39) for the circuit-switched voice traffic. For each point shown in the graph, 8 simulation runs were performed and each run with a different random seed. The graph also shows that the confidence intervals get wider in higher loads.

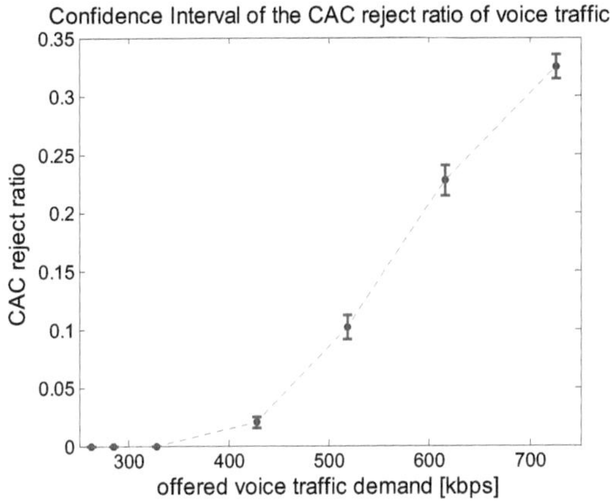

Figure A.39: *CAC reject ratio with 95% CI: voice scenario, Iub link =1 Mbps (ATM)*

In all investigated example scenarios, the obtained confidence intervals for different QoS measures have demonstrated that the simulation results are reliable and confident (with relative margin of error <10%).

A.21 Statistical Evaluation: Limited Relative Error

The *Limited Relative Error* (LRE) is another important statistical evaluation method to evaluate the statistical significance of correlated random sequences. Moreover, with the LRE method the simulation run time can be controlled by comparing the estimated error from the simulation results with a predefined maximum error. In this thesis, the LRE method is used to investigate whether the configured simulation run time is long enough to provide reliable and stable statistical evaluations of the simulation results. The LRE method is one of the class in the *Communication Network Class Library* (CNCL) developed at the Chair of Communication Networks, RWTH Aachen.

In this thesis, the *Discrete Limited Relative Error* (DLRE) is used, which is the LRE optimized for discrete, correlated sequences. The DLRE is used in this thesis to estimate the relative error $d(x)$ for the queuing delay distribution function $F(x)$, in order to investigate if the simulation run time is sufficient for providing reliable queuing delay distribution function. The queuing delay distribution is important for estimating the packet loss ratio. Figure A.40 gives an example of investigating the queuing delay distribution function of the voice traffic in the AAL2 RT buffer. In this example, there is only voice traffic transmitting on a 1.7 Mbps ATM Iub link, with 72.8% link utilization. The simulation run time is 3600 seconds. The maximum allowed relative error is set to be 1%. It can be observed from Figure A.40 that the estimated relative error $d(x)$ for the queuing delay distribution function $F(x)$ has rather low relative errors (lower than 0.3%) for different queuing delay ranges. This example demonstrates that for 3600 seconds simulation run time, the obtained queuing delay distribution function is reliable with relative errors far below the maximum allowed relative error of 1%.

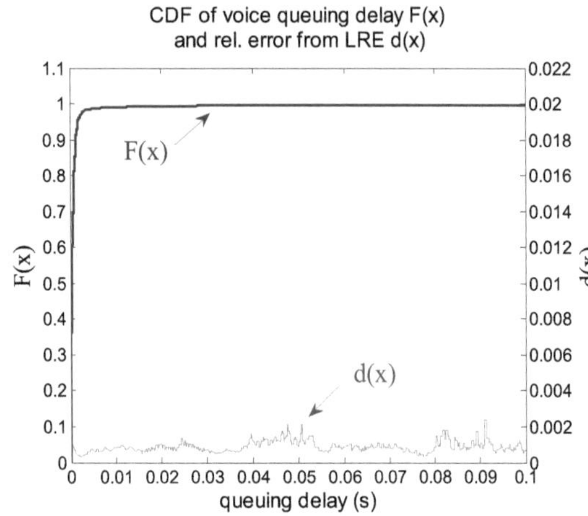

Figure A.40: *LRE distribution function F(x) and relative error d(x) for queuing delays of circuit-switched voice: voice scenario, Iub link =1.7 Mbps (ATM)*

A.22 Comparing ATM and IP-Based UTRAN

The following compares the performance and bandwidth requirement of the ATM and IP based UTRAN. Figure A.41 compares the calculated bandwidth efficiency of the ATM-based and the IP-based UTRAN of without using multiplexing (IP), using *Composite IP* (CIP) and *Lightweight IP Encapsulation* (LIPE) multiplexing schemes. The comparison is given for each RAB type. The LIPE and CIP are the proposed multiplexing schemes by 3GPP [3GP04a] for implementing the IP-based UTRAN (see section 2.3.4.1), which are optional. The detailed introduction of the CIP and LIPE schemes are given in Appendix A.11. If not using multiplexing, each radio frame (i.e. FP PDU) is encapsulated into one UDP/IP packet. With CIP and LIPE schemes, it allows multiplexing multiple radio frames (FP PDU) into one UDP/IP packet, and thus reduces the relative UDP/IP transport overheads.

It can be seen that for low RAB type such as AMR voice and 64 kbps, the ATM-based UTRAN achieves higher efficiency than the IP-based UTRAN of without using multiplexing scheme. This is because that the radio frame sizes of low RAB types are relatively small, and thus the added UDP/IP/Ethernet overhead of the IP transport network is larger compared to the ATM transport. On the other hand, the bandwidth efficiency of the IP-based UTRAN increases for higher RAB rate and even exceeds the efficiency of the ATM-based UTRAN when the RAB rates are 128 kbps or above. This is because that with higher RAB rate, the radio frame sizes are larger and thus the overhead of UDP/IP/Ethernet is relatively smaller than the one in the ATM-based UTRAN. When multiplexing scheme is applied in the IP-based UTRAN, with either CIP or LIPE scheme, it can be seen that the bandwidth efficiency is significantly better compared with not using the multiplexing scheme in the IP-based UTRAN, and also better than ATM-based UTRAN. This is because a multiplexing gain is achieved with combining several radio frames into one IP packet, which achieves improved bandwidth efficiency. Besides, the overhead of LIPE is one byte smaller than that of CIP, and therefore the bandwidth efficiency of LIPE is slightly better than CIP.

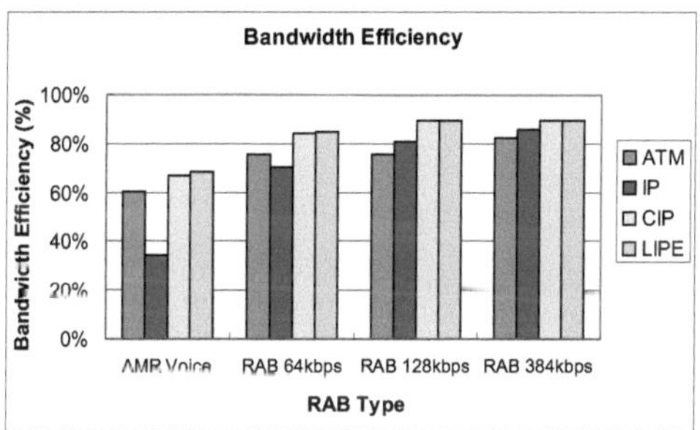

Figure A.41: *Calculated bandwidth efficiency - ATM vs. IP-based UTRAN*

The following present several examples of the simulation results. These results have been presented in [LC+07]. In the IP-based UTRAN scenario, 10BaseT Ethernet cable is used at the physical layer, while the allocated Iub link bandwidth is 2Mbps

Voice Scenario

For voice scenario, AMR codec is used for generating voice frame. The voice traffic model is defined in Table 5.2. Multiple users are configured in the simulations. In the shown example, each user contributes for around 20 kbps link load, and then the total traffic is scaled up with the increased number of users. In IP multiplexing scenario, only LIPE is investigated below, we set maximum 35 FP PDUs encapsulated into one LIPE or CIP container for voice traffic. In addition, we set a multiplexing timer of 5 ms and 10 ms as an example.

Figure A.42 presents the measured bandwidth efficiency of different transport layer schemes, as a function of the number of simultaneously active voice users. It shows that the IP-based UTRAN with LIPE multiplexing scheme with both timers achieve a little better efficiency than the one of ATM-based UTRAN model, and the IP-based UTRAN model without multiplexing schemes has the lowest efficiency due to relatively larger overhead. This conclusion is the same as the calculation results shown in Figure A.41. And with the increased number of voice users, the bandwidth efficiency is also increased due to more number of FP PDUs arrive before the MUX timer expires.

Figure A.43 gives the measured FP PDU delay of different transport layer schemes, as a function of the number of active voice users. It is seen in this figure that both ATM and IP-based UTRAN without using multiplexing scheme have rather low FP PDU delay, but with LIPE multiplexing scheme the FP PDU delay is higher and in the range of multiplexing timer. For all cases, the FP PDU delays are increased with more number of active voice users. When comparing 5ms and 10ms timer results, it is found that 5ms timer reaches the similar efficiency as 10ms timer as shown in Figure A.42, but on the other hand it gets much lower FP PDU delays. So, a smaller MUX timer is preferred for the voice traffic.

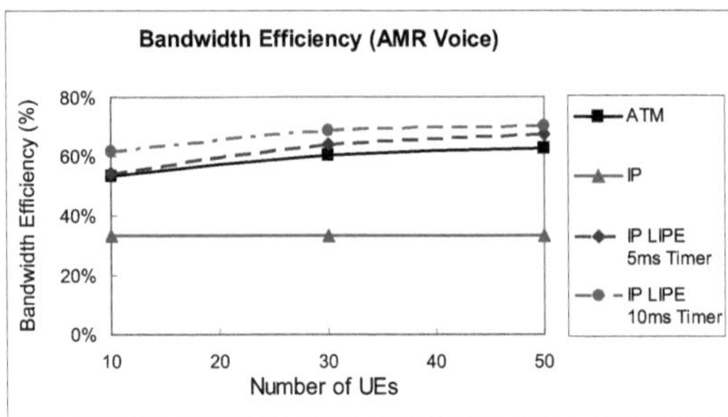

Figure A.42: *Simulated bandwidth efficiency - ATM vs. IP-based UTRAN (voice traffic)*

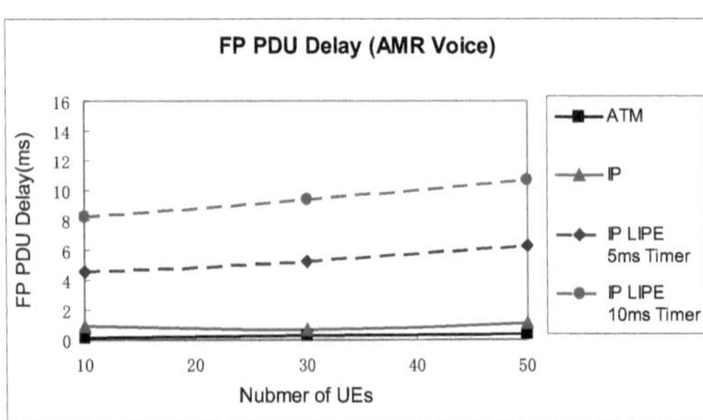

Figure A.43: *Simulated FP PDU delay - ATM vs. IP-based UTRAN (voice traffic)*

Web Traffic Scenario

For the given web traffic example, RAB 64 kbps is configured for transferring web pages. Multiple numbers of users are configured in the simulations. In this example, each user contributes for around 27kbps link load, and then the total traffic is scaled up with the increased number of users. In IP multiplexing scenario, only LIPE is investigated below, we set for the packet switched data maximum 8 numbers of FP PDUs encapsulated into one LIPE or CIP container. In addition, we set a multiplexing timer of 5ms and 10ms as an example. Figure A.44 compares the measured bandwidth efficiency of different transport layer schemes, as a function of the number of simultaneously active users. It shows that IP-based UTRAN with LIPE multiplexing scheme greatly increases the bandwidth efficiency when comparing to the ATM-based UTRAN. Moreover, a longer multiplexing timer results in a higher efficiency. When the load is increased in terms of more active users in the system, the efficiency becomes even better as the average number of FP PDUs increases which arrive before the MUX timer expires. Without IP multiplexing scheme, the bandwidth efficiency is the lowest in this case due to larger UDP/IP overhead compared to relatively smaller FP PDU frame.

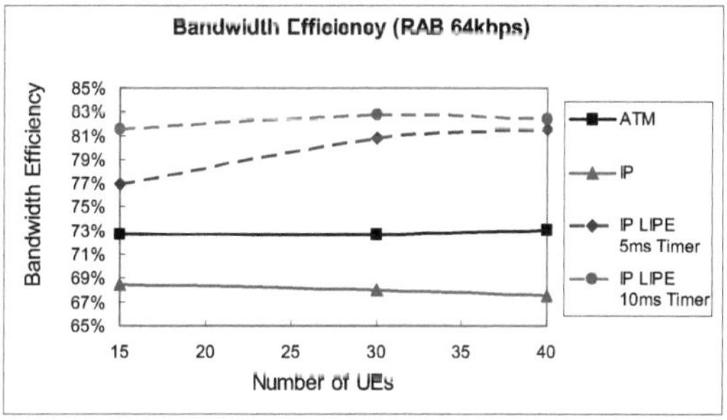

Figure A.44: *Simulated bandwidth efficiency - ATM vs. IP-based UTRAN (web traffic)*

Figure A.45 shows the measured FP PDU delay of different transport layer schemes, as a function of the number of active users. It shows that in this case the IP-based UTRAN with LIPE multiplexing scheme gets a higher FP PDU delays than the ATM-based UTRAN due to a higher multiplexing timer value (ATM-based UTRAN uses 1ms timer at the AAL2 layer) . However, IP-based UTRAN without multiplexing achieves a slightly lower FP PDU delay than the one in the ATM-based UTRAN when the load is lower than 30 numbers of UEs. But when the load is increased, the FP PDU delay is increased significantly. This is due to more resource is occupied for transferring the overhead information and no multiplexing gain is achieved in this case. We can also observe that with longer multiplexing timer in the IP-based UTRAN, the bandwidth efficiency is higher, but the FP PDU delay gets worse. Therefore, there is a tradeoff between the packet delay and the efficiency.

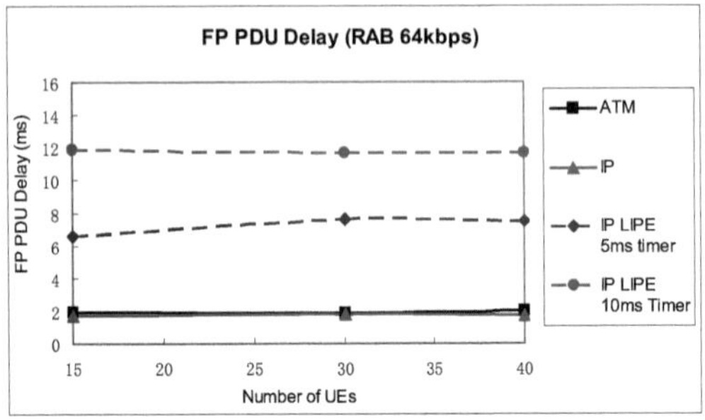

Figure A.45: *Simulated FP PDU delay - ATM vs. IP-based UTRAN (web traffic)*

Based on the above investigations, it is found that there is also a tradeoff between the efficiency and the delay. Therefore, optimization of the multiplexing parameters is important for the dimensioning.

Bibliography

[3GP99a] 3GPP, Technical Specification Group Services and System Aspects. *Mandatory Speech Codec Speech Processing Functions AMR Speech Codec; General Description*. 3GPP TS 26.071 version 3.0.1, August 1999.

[3GP99b] 3GPP, Technical Specification Group RAN. *Radio Interface Protocol Architecture*. 3GPP TS 25.301 version 3.3.0, December 1999.

[3GP99c] 3GPP, Technical Specification Group RAN. *Physical Channels and Mapping of Transport Channels onto Physical Channels (FDD)*. 3GPP TS 25.211 version 3.1.0, December 1999.

[3GP01a] 3GPP, Technical Specification Group RAN. *High Speed Downlink Packet Access (HSDPA); Overall UTRAN description*. 3GPP TR 25.855 version 5.0.0, October 2001.

[3GP01b] 3GPP, Technical Specification Group RAN. *Architecture for an All IP Network*. 3GPP TR 23.922 version 4.0.0, March 2001.

[3GP01c] 3GPP, Technical Specification Group RAN. *Delay Budget within the Access Stratum*. 3GPP TR 25.853 version 4.0.0, April 2001.

[3GP02b] 3GPP, Technical Specification Group RAN. *UTRAN Iur Interface User Plane Protocols for Common Transport Channel Data Streams*. 3GPP TS 25.425 version 3.7.0, April 2002.

[3GP02c] 3GPP, Technical Specification Group RAN. *UTRAN overall Description*. 3GPP TS 25.401 version 3.10.0, June 2002.

[3GP02d] 3GPP, Technical Specification Group RAN. *Packet Data Convergence Protocol (PDCP) specification*. 3GPP TS 25.323 version 3.10.0, Sep. 2002.

[3GP02e] 3GPP, Technical Specification Group RAN. *Synchronisation in UTRAN Stage 2 Release 1999)*. 3GPP TS 25.402 version 3.10.0, June. 2002.

[3GP04a] 3GPP, Technical Specification Group RAN. *IP Transport in UTRAN*. 3GPP TR 25.933 version 5.4.0, January 2004.

[3GP04b] 3GPP, Technical Specification Group RAN. *Radio Link Control (RLC) Protocol Specification*. 3GPP TS 25.322 version 3.18.0, June 2004.

[3GP04c] 3GPP, Technical Specification Group RAN. *Medium Access Control (MAC) Protocol Specification.* 3GPP TS 25.321 version 3.17.0, June 2004.

[3GP04d] 3GPP, Technical Specification Group RAN. *Radio Resource Control (RRC); Protocol Specification.* 3GPP TS 25.331 version 3.21.0, Dec. 2004.

[3GP04e] 3GPP, Technical Specification Group RAN. *Common Test Environments for User Equipment (UE); Conformance Testing.* 3GPP TS 34.108 version 3.16.0, June 2004.

[3GP04f] 3GPP, Technical Specification Group RAN. *Feasibility Study for Enhanced Uplink for UTRA FDD.* 3GPP TR 25.896 version 6.0.0, March 2004.

[3GP05a] 3GPP, Technical Specification Group RAN. *All-IP network (AIPN) Feasibility Study.* 3GPP TR 22.978 version 7.1.0, June 2005.

[3GP05b] 3GPP, Technical Specification Group RAN. *UTRAN High Speed Downlink Packet Access (HSDPA).* 3GPP TR 25.950 version 4.0.1, July 2005.

[3GP06a] 3GPP, Technical Specification Group RAN. *FDD Enhanced Uplink; Overall Description; Stage 2.* 3GPP TS 25.309 version 6.6.0, Apr. 2006.

[3GP06b] 3GPP, Technical Specification Group RAN. *Requirements for Evolved UTRA (E-UTRA) and Evolved UTRAN (E-UTRAN).* 3GPP TR 25.913 version 7.3.0, March 2006.

[3GP06c] 3GPP, Technical Specification Group RAN. *Quality of Service (QoS) Concept and Architecture.* 3GPP TS 23.107 version 6.4.0, March 2006.

[3GP06d] 3GPP, Technical Specification Group RAN. *Iub/Iur Congestion Control.* 3GPP TR 25.902 version 6.1.0, October 2006.

[3GP08] 3GPP, Technical Specification Group RAN. *3GPP System Architecture Evolution (SAE): Report on Technical Options and Conclusions.* 3GPP TR 23.882 version 1.15.1, March 2008.

[AL02a] F. Agharebparast and V.C.M. Leung. *QoS Support in the UMTS/GPRS Backbone Network Using DiffServ.* IEEE Global Telecommunications Conference, Vol.2, pp.1440-1444, Nov. 2002.

[AL02b] F. Agharebparast and V.C.M. Leung. *A Framework for QoS Support in the UMTS/GPRS Backbone Network Using DiffServ.* Computing and Informatics, Vol. 21, No. 2, 2002.

[ATM96] ATM Forum White Papers. *ATM Service Categories : The Benefits to the User.* August 1996.

[BBC+98] S. Blake, D. Black, M. Carlson, E. Davies, Z. Wang, and W. Weiss. *An Architecture for Differentiated Services.* IETF RFC 2475, December 1998.

[BBJ99] S. Borst, O. Boxma, and P. Jelenkovic. *Generalized Processor Sharing with Long-Tailed Traffic Sources.* In Proc. of the International Teletraffic Congress (ITC 16), Edinburgh, Scotland, 1999.

[BCS94] R. Braden, D. Clark, and S. Shenker. *Integrated Services in the Internet Architecture: An Overview.* IETF RFC 1633, June 1994.

[BHK+01] J. V. L. Beckers, I. Hendrawan, R. E. Kooij, and R. D. van der Mei. *Generalized Processor Sharing Performance Models for Internet Access Lines.* In Proc. Of the 9th IFIP Conference on Performance Modeling and Evaluation of ATM & IP Networks, Budapest, Hungary, 2001.

[Bra69] P. T. Brady. *A Model for Generating On-Off Patterns in Two-Way Conversations.* Bell Systems Technical Journal, pp. 2445-2472, 1969.

[Bre02] L. Breuer. *An EM algorithm for Batch Markovian Arrival Processes and its Comparison to a Simpler Estimation Procedure.* Annals of Operations Research 112, pp. 123-138, 2002.

[BZB+97] R. Braden, L. Zhang, S. Berson, S. Herzog, and S. Jamin. *Resource Reservation Protocol (RSVP), version 1 Functional Specification.* IETF RFC 2205, September 1997.

[Cha00] J. Charzinski. *HTTP/TCP Connection and Flow Characteristics.* Performance Evaluation Vol. 42 No. 2-3, pp. 149-162, September 2000.

[CI01] C. Chen, R. Izmailov. *The Notion of Overbooking and Its Application to IP/MPLS Traffic Engineering.* IETF draft, 2001.

[CM98a] C. M. Pazos and M. Gerla. *Improving Internet Traffic Transport over ABR Backbones Through Bandwidth Overbooking.* In Proc. of the IEEE Globecom '98, Sidney, Australia, 1998.

[CM98b] C. M. Pazos and M. Gerla. *Improving IP over ATM Efficiency Using Bandwidth Overbooking.* In Proc. of LANMAN'98. Alberta, Canada, 1998.

[CNRS98] E. Crawley, R. Nair, B. Rajagopalan, and H. Sandick. *A Framework for QoS-Routing in the Internet.* IETF RFC 2386, August 1998.

[Coh79] J.W. Cohen. *The Multiple Phase Service Network with Generalized Processor Sharing.* Acta Informatica, 12:245-284, 1979.

[DGA01] S. Dixit, Y. Guo, and Z. Antoniou. *Resource Management and Quality of Service in Third-Generation Wireless Networks.* IEEE Communication Magazine, pp. 125-133, Feb. 2001.

[DKS89] A. Demers, S. Keshav, and S. Shenker. *Analysis and Simulation of a Fair Queueing Algorithm.* In Proc. of the ACM SIGCOMM, pp. 3-12, 1989.

[DPSB07] E. Dahlman, S. Parkvall, J. Sköld and P. Beming. *3G Evolution: HSPA and LTE for Mobile Broadband.* Elsevier, 2007.

[EHA48] E. Brockmeyer, H.L. Halstrøm, and A. Jensen. *The Life and Works of A.K.Erlang.* Transactions of the Danish Academy of Technical Sciences. No. 2, pp. 277, Copenhagen 1948.

[ETSI98] ETSI. Universal Mobile Telecommunications System (UMTS): *Selection Procedures for the Choice of Radio Transmission Technologies of the UMTS (UMTS 30.03 version 3.2.0).* TR 101 112 version 3.2.0, April 1998.

[ETSI01] ETSI. *Universal Mobile Telecommunications System (UMTS): Physical Layer Aspects of UTRA High Speed Downlink Packet.* 3GPP TR 25.848 version 3.2.0, March 2001.

[Fan02] Z. Fan. *Dimensioning Bandwidth for Elastic Traffic.* Networking 2002, pp. 826-837, 2002.

[FGM+99] R. Fielding, J. Gettys, J. Mogul, H. Frystyk, L. Masinter, P. Leach, and T. Berners-Lee. *Hypertext Transfer Protocol - HTTP/1.1.* IETF RFC 2616, June 1999.

[FH99] S. Floyd and T. Henderson. *The NewReno Modification to TCP's Fast Recovery Algorithm.* IETF RFC 2582, April 1999.

[FJ93] S. Floyd and V. Jacobson. *Random Early Detection Gateway for Congestion Avoidance.* IEEE/ACM Trans. Networking, Vol. 1, pp. 397-413, Aug.1993.

[FK78] R. F. Farmer and I. Kaufmann. *On the Numerical Evaluation of Some Basic Traffic Formulae.* Networks, Vol.8, pp. 153-186, 1978.

[Flo01] S. Floyd. *A Report on Recent Developments in TCP Congestion Control.* IEEE Communications Magazine, 39(4):84-90, April 2001.

[FMH92] W. Fischer and K. Meier-Hellstern. *The Markov-modulated Poisson Process (MMPP) Cookbook.* Performance Evaluation 18, pp. 149-171, 1992.

[FPR00] T. Ferrari, G. Pau, and C. Raffaelli. *Priority Queueing Applied to Expedited Forwarding: A Measurement-Based Analysis.* First COST 263 International Workshop on Quality of Future Internet Services, pp. 167-181, 2000.

[GAC+02] A. B. García, M. Alvarez-Campana, E. Vázquez, J. Berrocal. *Quality of Service Support in the UMTS Terrestrial Radio Access Network.* In Proc. of the 9th HP Openview University Association Workshop (HP-OVUA), 2002.

[GZFO07] H. Galeana-Zapién, R. Ferrús, J. Olmos. *Transport Capacity Estimations for Over-provisioned UTRAN IP-based Networks*. IEEE WCNC, 2007.

[HDB+04] F. Huang, C. Deccio, R. Ball, M. Clement, Q. Snell. *A Piecewise Linear Approach to Overbooking*. High Performance Switching and Routing (HPSR) 2004 Workshop, pp. 326-330, 2004.

[HT04] H. Holma, A. Toskala. *WCDMA for UMTS*. John Wiley & Sons, Inc., Chichester, UK, 3rd edition, 2004.

[HT06] H. Holma, A. Toskala. *HSDPA/HSUPA for UMTS: High Speed Radio Access for Mobile Communications*. John Wiley & Sons, Chichester, UK, 2006.

[HZ05] P. G. Harrison and Y. Zhang. *Delay Analysis of Priority Queues with Modulated Traffic*. In Proc. of the 13th IEEE International Symposium on Modeling, Analysis and Simulation of Computer and Telecommunication Systems, pp. 280-287, 2005.

[ITU93] ITU-T, Geneva, Switzerland. *ITU-T Recommendation E.800: Terms and Definitions Related to Quality of Service and Network Performance Including Dependability*. August 1993.

[ITU99] ITU-T Recommendation Q.2630.1. *AAL type 2 Signaling Protocol (Capability Set 1)*. 1999.

[ITU00] ITU-T Recommendation Q.2630.2. *AAL Type 2 Signaling Protocol (Capability Set 2)*. December 2000.

[Ive04] V. B. Iversen. *Teletraffic Engineering and Network Planning*. COM Course 34340, Technical University of Denmark, September 2004.

[Jac88] V. Jacobson. *Congestion Avoidance and Control*. ACM Computer Communication Review; Proc. of the Sigcomm '88 symposium in Stanford, CA, Vol. 18, No. 4, pp. 314-329, 1988.

[KAL+01] H. Kaaranen, A. Ahtiainen, L. Laitinen, S. Naghian, V. Niemi. *UMTS Networks - Architecture, Mobility and Services*. John Wiley & Sons, 2001.

[KKSC02] S. H. Kang, Y. H. Kim, D. K. Sung, and B. D. Choi. *An Application of Markovian Arrival Process (MAP) to Modeling Superposed ATM cell Streams*. IEEE Trans. Commun., Vol. 50, No. 4, pp. 633-642, April 2002.

[Kle75] L. Kleinrock. *Queueing Systems, Volume I: Theory*. Wiley Interscience, New York, 1975.

[Kle76] L. Kleinrock. *Queueing Systems Volume II: Computer Applications*. Wiley-Interscience, New York, 1976.

[KLL01] A. Klemm, C. Lindemann and M. Lohmann. *Traffic Modeling and Characterization for UMTS Networks.* In Proc. of the Globecom, Internet Performance Symposium, San Antonio, TX, pp. 1741-1746, 2001.

[KLL03] A. Klemm, C. Lindemann, and M. Lohmann. *Modeling IP Traffic Using the Batch Markovian Arrival Process.* Performance Evaluation, Vol. 54, No. 2, pp. 149-173, 2003.

[KSC98] S. H. Kang, D. K. Sung, and B. D. Choi. *An Empirical Real-Time Approximation of Waiting Time Distribution in MMPP(2)/D/1.* IEEE Communications Letters, Vol. 2, No. 1, pp. 17-19, January 1998.

[LBG+08] X. Li, W. Bigos, C. Görg, A. Timm-Giel and A. Klug. *Dimensioning of the IP-based UMTS Radio Access Network with DiffServ QoS Support.* In Proc. of the 19th ITC Specialist Seminar on Network Usage and Traffic (ITC SS 19), October, 2008.

[LC+07] X. Li, W. Cheng, A. Timm-Giel, C. Görg, R. Schelb and B. Kracker. *Modeling IP-based UTRAN for UMTS in OPNET.* In Opnetwork 2007, Washington, August 2007.

[LF03] H.-L. Lu and I. Faynberg. *An Architectural Framework for Support of Quality of Service in Packet Networks.* IEEE Communications Magazine, Vol. 41, No. 6, pp. 98-105, June 2003.

[Lin99] K. Lindberger. *Balancing Quality of Service, Pricing and Utilisation in Multiservice Networks with Stream and Elastic Traffic.* In Proc. of the International Teletraffic Congress (ITC 16), Edinburgh, Scotland, 1999.

[LK05] H.-J. Lee, J.-H. Kim. *Analysis of Bandwidth Gain Over Various Timer$_$CU of AAL2 for Voice Traffic Multiplexing.* Vehicular Technology, IEEE Transactions on Vol. 54, No. 4, pp. 1438-1446, 2005.

[LM97] T.V. Lakshman and U. Madhow. *The Performance of TCP/IP for Networks with High Bandwidth-Delay Products and Random Loss.* IEEE/ACM Transactions on Networking, Vol. 5, No. 3, pp. 336-350, June 1997.

[LL01] C. Lindemann, and M. Lohmann. *Numerical Robust Parameter Estimation for the Batch Markovian Arrival Process Using Randomization.* 2001.

[LL03] C. Lindemann and M. Lohmann. *IP2BMAP Software Package.* http://www.ip2bmap.de/, release year 2003, last visited in January 2009.

[LLGT06] X. Li, S. Li, C. Görg and A. Timm-Giel. *Traffic Modeling and Characterization for UTRAN.* In Proc. of the 4th International Conference on Wired/Wireless Internet Communications, Bern Switzerland, 2006, and Lecture Notes in Computer Science, Springer, 2006.

[LLSTG06] X. Li, S. Li, R. Schelb, A. Timm-Giel and C. Görg. *Delay in UMTS Radio Access Networks: Analytical Study and Validation.* In Proc. of the Australian Telecommunication Networks and Applications Conference (ATNAC), Melbourne, Australia, 2006.

[LSGT05] X. Li, R. Schelb, C. Görg and A. Timm-Giel. *Dimensioning of UTRAN Iub Links for Elastic Internet Traffic.* In Proc. of the 19th International Teletraffic Congress, Beijing, 2005.

[LSGT06] X. Li, R. Schelb, C. Görg and A. Timm-Giel. *Dimensioning of UTRAN Iub Links for Elastic Internet Traffic with Multiple Radio Bearers.* In Proc. of the 13th GI/ITG Conference Measuring, Modeling and Evaluation of Computer and Communication Systems, Nürnberg, March 2006.

[LSGT08] X. Li, R. Schelb, C. Görg and A. Timm-Giel. *UMTS HSPA and R99 Traffic Separation.* In Proc. of the 10th IFIP International Conference on Mobile and Wireless Communications Networks (MWCN), Toulouse, France, 2008.

[Luc91] D. M. Lucantoni. *New Results on the Single Server Queue with a Batch Markovian Arrival Process.* Commun. Statist.-Stochastic Models, Vol. 7, No. 1, pp. 1-46, 1991.

[Luc93] D. M. Lucantoni. *The BMAP/G/1 Queue: A tutorial.* In Models and Techniques for Performance Evaluation of Computer and Communications Systems, Joint Tutorial Papers of Performance '93 and Sigmetrics '93, pp. 330-358, 1993.

[LWS+07] X. Li, L. Wang, R. Schelb, T. Winter, A. Timm-Giel and C. Görg. *Optimization of Bit Rate Adaptation in UMTS Radio Access Network.* In Proc. of the IEEE 65th Vehicular Technology Conference VTC2007-Spring, Dublin, Ireland, April 2007.

[LZK+08] X. Li, Y.Zeng, B. Kracker, R.Schelb, C.Görg and A. Timm-Giel. *Carrier Ethernet for Transport in UMTS Radio Access Network: Ethernet Backhaul Evolution.* In Proc. of the IEEE 67th Vehicular Technology Conference VTC2008-Spring, Singapore, May 2008.

[LZW+08] X. Li, Y. Zaki, T. Weerawardane, A. Timm-Giel, C. Görg. *HSUPA Backhaul Bandwidth Dimensioning.* In Proc. of the 19th IEEE International Symposium on Personal, Indoor and Mobile Radio Communications (PIMRC), September 15-18, in Cannes France, 2008.

[LZW+09] X. Li, Y. Zaki, T. Weerawardane, A. Timm-Giel, C. Görg and G. C. Malafronte. *Use of Traffic Separation Techniques for the Transport of HSPA and R99 Traffic in the Radio Access Network with Differentiated Quality of Service.* International Journal of Business Data Communications and Networking (IJBDCN), Vol. 5, No. 2, pp. 84-100, 2009.

[Man03] J. Manner. *Provision of Quality of Service in IP-based Mobile Access Networks.* PhD thesis (Report number A-2003-8), University of Helsinki, Department of Computer Science, November 2003.

[Mcl97] M. McLoughlin, et al. *A Management Briefing on Adapting Voice For ATM networks an AAL2 Tutorial.* General DataComm, 1997.

[Mei87] K. S. Meier. *A Fitting Algorithm for Markov-Modulated Poisson Processes having two Arrival Rates.* European J. Oper. Res. 29, pp. 370-377, 1987.

[Mes06] C. O. Mesa. *WCDMA - Enhanced Uplink Performance Evaluation.* MSc thesis University of Twente, the Netherlands, 2006.

[Mot07] Motorola. Long *Term Evolution (LTE): Overview of LTE Air-Interface* Technical White Paper. 2007. http://www.motorola.com/mot/doc/6/6993_MotDoc.pdf, last visited in 2009.

[MR99] L. Massoulie and J. Roberts. *Arguments in Favour of Admission Control for TCP flows.* In Proc. of the ITC 16, pp. 1-12, Edinburgh, UK, 1999.

[MWIF01] Mobile Wireless Internet Forum (MWIF). *IP in the RAN as a Transport Option in 3rd Generation Mobile Systems.* Technical Report MTR-006, Rel. V2.0.0, June 18, 2001.

[NBBB98] K. Nichols, S. Blake, F. Baker, D. Black. *Definition of the Differentiated Services Field (DS Field) in the IPv4 and IPv6 Headers.* RFC 2474, December 1998.

[NBM99] R. Nunez, H. van den Berg and M. Mandjes. *Performance Evaluation of Strategies for Integration of Elastic and Stream Traffic.* PNA-R9903, 1999.

[NRM+05] Sz. Nádas, S. Rácz, Sz. Malomsoky, S. Molnár. *Connection Admission Control in the UTRAN Transport Network.* Telecommunication Systems, Vol. 28, No. 1, pp. 9-29, 2005.

[NSV99] A. Nogueira, P. Salvador, R. Valadas. *Fitting Algorithms for MMPP ATM Traffic Models.* In Proc. of the Broadband Access Conference, Cracow, Poland, pp. 167-174, October 1999.

[OAS90] T. Okuda, H. Akimaru, and M. Sakai. *A Simplified Performance Evaluation for Packetized Voice Systems.* Trans. IEICE, Vol. E73, No. 6, 1990.

[OKM96] T. J. Ott, J. Kemperman, and M. Mathis. *The Stationary Behavior of Ideal TCP Congestion Avoidance.* 1996. http://citeseer.ist.psu.edu/old/ott96stationary.html, last visited in February 2009.

[Opnet03] OPNET Modeler version 10.0. OPNET technologies Inc., *OPNET Modeler Accelerating Networks R&D.* http://www.opnet.com, release year 2003.

[Ott84] T. J. Ott. *The Sojourn-time Distribution in the M/G/1 Queue with Processor Sharing.* Journal of Applied Probability 21, pp. 360-378, 1984.

[Pos81] J. Postel. *Transmission Control Protocol.* IETF RFC 793, September 1981.

[[Qvd+99a] R. Núñez Queija, J.L. van den Berg, and M.R.H. Mandjes. *Performance Evaluation of Strategies for Integration of Elastic and Stream Traffic.* In Proc. of the International Teletraffic Congress (ITC 16), Scotland, 1999.

[Qvd+99b] R. Núñez Queija, J.L. van den Berg, and M.R.H. Mandjes. *Performance Evaluation of Strategies for Integration of Elastic and Stream Traffic.* Research Report - Probability, Networks and Algorithms PNA-R9903, Centrum voor Wiskunde en Informatica, February 1999.

[RBPP00] A. Riedl, T. Bauschert, M. Perske, A. Probst. *Investigation of the M/G/R Processor Sharing Model for Dimensioning of IP Access Networks with Elastic Traffic.* In First Polish-German Teletraffic Symposium PGTS 2000.

[RFC04] IETF RFC 3916: *Requirements for Pseudo-Wire Emulation Edge-to-Edge (PWE3).* September 2004.

[RFC05] IETF RFC 3985: Pseudo *Wire Emulation Edge-to-Edge (PWE3) Architecture.* March 2005.

[RFC06] IETF RFC 4717: *Encapsulation Methods for Transport of Asynchronous Transfer Mode (ATM) over MPLS Networks.* December 2006.

[Rie04] A. Riedl. *Routing Optimization and Capacity Assignment in Multi-Service IP Networks.* PhD thesis at Technische Universität München, 2004.

[Ris02] A. Riska. *Aggregate Matrix-analytic Techniques and their Applications.* PhD thesis, the College of William and Mary, 2002.

[RM98] J. Roberts and L. Massoulié. *Bandwidth Sharing and Admission Control for Elastic Traffic.* In ITC Specialist Seminar, Yokohama, Japan, October 1998.

[RM+04] R. Ball, M. Clement, F. Huang, Q. Snell, C. Deccio. *Aggressive Telecommunications Overbooking Ratios.* IEEE International Performance, Computing, and Communications Conference, pp. 31-38, 2004.

[Rob97] J. Roberts. *Realizing Quality of Service Guarantees in Multiservice Networks.* In Proceedings of the IFIP Conference PMCCN'97, Tsukuba Science City, Japan, November 1997.

[RPBP00] A. Riedl, M. Perske, T. Bauschert, and A. Probst. *Dimensioning of IP Access Networks with Elastic Traffic.* In Networks 2000, Toronto, Canada, 2000.

[RVC01] E. Rosen, A. Viswanathan, and R. Callon. *Multiprotocol Label Switching Architecture.* IETF RFC 3031, January 2001.

[Ryd96] T. Ryden. *An EM Algorithm for Parameter Estimation in Markov Modulated Poisson Processes.* Computational Statistics and Data Analysis, Vol. 21, No. 4, pp. 431-447, 1996.

[Sch03] A. Schieder. *Echtzeitdienste in Paketvermittelnden Mobilfunknetzen.* PhD thesis, at the department of Communication Networks (ComNets), RWTH Aachen University, 2003.

[Sie06] Siemens Communication – Fixed Networks Access. *Converging an 'ALL IP' Mobile Transport on Carrier Ethernet Networks.* White paper, 2006.

[Ste94] W. R. Stevens. *TCP/IP Illustrated, Volume 1: The Protocols.* Addison-Wesley, Reading, MA, 1994.

[TNC+01] F. Tobagi, W. Noureddine, B. Chen, A. Markopoulou, C. Fraleigh, M. Karam, JM Pulido, J. Kimura. *Service Differentiation in the Internet to Support Multimedia Traffic.* In Springer Verlag LNCS, Vol. 2170, September 2001.

[TS96] J. S. Turner. *Maintaining High Throughput During Overload in ATM Switches.* In Proc. of the IEEE INFOCOM 1996, San Francisco, California, pp. 287-295, April 1996.

[TWP+02] U.Türke, T. Winter, R. Perera, E. Lamers, E. Meijerink, E. R.Fledderus, et al. *Comparison of Different Simulation Approaches for Cell Performance Evaluation.* http://momentum.zib.de/paper/momentum-d22.pdf, last visited in January 2009.

[Tür07] U. Türke. *Efficient Methods for WCDMA Radio Network Planning and Optimization.* PhD Thesis at University of Bremen. Vieweg+Teubner, 1st Edition, September 2007.

[UMTS06] UMTS Forum. *3G/UMTS Evolution: Towards a New Generation of Broadband Mobile Services.* White Paper, 2006.

[VRF99] A. G. Valko, A. Racz, and G. Fodor. *Voice QoS in Third Generation Mobile Systems.* IEEE Journal on Selected Areas in Communications, Vol.17, No. 1, pp. 109-123, January 1999.

[WPT98] W. Willinger, V. Paxson, and M.S. Taqqu. *Self-similarity and Heavy Tails: Structural Modeling of Network Traffic.* In A Practical Guide to Heavy Tails: Statistical Techniques and Applications, eds Robert J. Adler et al., Birkhäuser, Boston, MA, pp. 27-53, 1998.

[WTG+06] T. L. Weerawardane, A. Timm-Giel, C. Görg and T. Reim. *Impact of the Transport Network Layer Flow Control for HSDPA Performance.* In Proc. of the IEE conference 2006, Sri Lanka, 2006.

[WL+06] T. L. Weerawardane, X. Li, A. Timm-Giel and C. Görg. *Modeling and Simulation of UMTS HSDPA in OPNET.* In OPNETWORK 2006, Washington DC, USA, September 2006.

[WSA03] B.Walke, P.Seidenberg and M.P.Althoff. *UMTS: The Fundamentals.* John Wiley & Sons, Inc., Chichester, UK, 2003.

[WTG+08] T. Weerawardane, A. Timm-Giel, G. Malafronte, D. Gianluca, S. Hauth and C. Görg. *Preventive and Reactive based TNL Congestion Control Impact on the HSDPA Performance.* In IEEE 67th VTC Spring, 2008.

[Wol82] R. W. Wolff. *Poisson Arrivals See Time Averages.* Operations Research, Vol. 30, pp. 223-231, 1982.

MIX
Papier aus verantwortungsvollen Quellen
Paper from responsible sources
FSC® C105338

If you have any concerns about our products,
you can contact us on
ProductSafety@springernature.com

In case Publisher is established outside the EU,
the EU authorized representative is:
Springer Nature Customer Service Center GmbH
Europaplatz 3, 69115 Heidelberg, Germany

Printed by Libri Plureos GmbH
in Hamburg, Germany